Longitudinal Research
with Latent Variables

Kees van Montfort
Johan H.L. Oud
Albert Satorra

Editors

Longitudinal Research with Latent Variables

Springer

Editors

Professor Dr. Kees van Montfort
Vrije Universiteit Amsterdam
Department of Econometrics and Operations Research
De Boelelaan 1105
1081 HV Amsterdam
Netherlands
kvmontfort@feweb.vu.nl

Professor Dr. Albert Satorra
Universitat Pompeu Fabra
Department of Economics and Business
Ramon Trias Fargas 25-27
08005 Barcelona
Spain
albert.satorra@upf.edu

Professor Dr. Johan H.L. Oud
Radboud University Nijmegen
Behavioural Science Institute
Montessorilaan 3
6525 HR Nijmegen
Netherlands
j.oud@pwo.ru.nl

ISBN 978-3-642-11759-6 e-ISBN 978-3-642-11760-2
DOI 10.1007/978-3-642-11760-2
Springer Heidelberg Dordrecht London New York

Library of Congress Control Number: 2010926488

Cover design: WMXDesign GmbH, Heidelberg, Germany

Printed on acid-free paper

Springer is part of Springer Science+Business Media (www.springer.com)

Preface

Since Charles Spearman published his seminal paper on factor analysis in 1904 and Karl Jöreskog replaced the observed variables in an econometric structural equation model by latent factors in 1970, causal modelling by means of latent variables has become the standard in the social and behavioural sciences. Indeed, the central variables that social and behavioural theories deal with, can hardly ever be identified as observed variables. Statistical modelling has to take account of measurement errors and invalidities in the observed variables and so address the underlying latent variables.

Moreover, during the past decades it has been widely agreed on that serious causal modelling should be based on longitudinal data. It is especially in the field of longitudinal research and analysis, including panel research, that progress has been made in recent years. Many comprehensive panel data sets as, for example, on human development and voting behaviour have become available for analysis. The number of publications based on longitudinal data has increased immensely. Papers with causal claims based on cross-sectional data only experience rejection just for that reason.

The chapters in this book combine longitudinal research and latent variable research. They all explain how longitudinal studies with objectives formulated in terms of latent variables should be performed. The emphasis is on exposing how the methods are applied. Because currently longitudinal research with latent variables follows different approaches with different histories, different types of research questions, and different computer programs to perform the analysis, the book is divided into nine, rather self sufficient chapters. The chapters give an up to date overview of the current state of the approach. Each chapter is written by one or more experts in the approach. In addition to some background information about the specific approach (short history and main publications), the chapter describes the type of research questions the approach is able to answer and the kind of data to be collected, gives the statistical and mathematical explanation of the models used in the analysis of the data, discusses the input and output of the programs used in the analysis, and provides one or more examples with typical data sets enabling the reader to apply the programs themselves. Data sets and computer code for the analysis with various software programs are a very important component of the book and partly made available at the book website `http://www.econ.upf.edu/~satorra/longitudinallatent/readme.html` .

The chapters present an up to date overview of the current state of the approach in such detail that readers get the means for application in their own research. The emphasis is not on new results. The main purpose is to give a state of the art explanation of longitudinal research methodology with latent variables and to show how this methodology is implemented in practice with current state of art software and real data sets. Each of the chapters is supposed to be rather complete for the specific approach and the chapters together are meant to cover the field exhaustively.

The book "Longitudinal Research with Latent Variables" addresses the great majority of researchers in the behavioural and related sciences, in academic as well as non-academic environments. This includes readers who are involved in research in psychology, sociology, education, economics, management, and medical sciences. It is meant as a reference work for all those actually doing longitudinal research. The book also addresses methodologists and statisticians, who are professionally dealing with longitudinal research, to provide standards for state of the art practices. It specially offers PhD students in the fields indicated the means to carry out longitudinal research with latent variables.

Kees van Montfort, Han Oud, and Albert Satorra

Contents

List of Contributors

Daniel J. Blake
Department of Political Science, 2140 Derby Hall, 154 N. Oval Mall, Ohio State University, Columbus, OH 43210-1373, USA
E-mail: blake.165@polisci.osu.edu

Kenneth A. Bollen
Odum Institute for Research in Social Science, CB 3355 Manning Hall, University of North Carolina, Chapel Hill, NC 27599-3355, USA
E-mail: bollen@unc.edu

Janet M. Box-Steffensmeier
Department of Political Science, 2140 Derby Hall, 154 N. Oval Mall, Ohio State University, Columbus, OH 43210-1373, USA
E-mail: steffensmeier.2@polisci.osu.edu

Jacques J. F. Commandeur
Department of Econometrics, VU University Amsterdam, De Boelelaan 1105, 1081 HV Amsterdam, The Netherlands
E-mail: jcommandeur@feweb.vu.nl
and
Dutch National Road and Safety Research Institute (SWOV), Duindoorn 32, 2262 AR Leidschendam, The Netherlands
E-mail: jacques.commandeur@swov.nl

Marc J. M. H. Delsing
Praktikon, Radboud University Nijmegen, Postbus 9104, 6500 HE Nijmegen, The Netherlands
E-mail: m.delsing@acsw.ru.nl

Kevin J. Grimm
Department of Psychology, University of California, Davis, One Shields Avenue, Davis, CA 95616, USA
E-mail: kjgrimm@ucdavis.edu

Jacques A. Hagenaars
Department of Methodology and Statistics, Tilburg University, Warandelaan 2, 5037 AB Tilburg, The Netherlands
E-mail: jacques.a.hagenaars@uvt.nl

Siem Jan Koopman
Department of Econometrics, VU University Amsterdam, De Boelelaan 1105, 1082 HV Amsterdam, The Netherlands
E-mail: s.j.koopman@feweb.vu.nl

Nicholas T. Longford
SNTL, Barcelona, Spain
E-mail: NTL@sntl.co.uk
and
Department d'Economia i Empresa, Universitat Pompeu Fabra, Ramon Trias Farga 25-27, 08005 Barcelona, Spain
E-mail: nick.longford@upf.edu

John J. McArdle
Department of Psychology, University of Southern California, Los Angeles, CA 90089, USA
E-mail: jmcardle@usc.edu

Geert Molenberghs
Interuniversity Institute for Biostatistics and Bioinformatics (I-BioStat), Universiteit Hasselt, Agoralaan, B-3590 Diepenbeek, Belgium
E-mail: geert.molenberghs@uhasselt.be
and
Interuniversity Institute for Biostatistics and Bioinformatics (I-BioStat), Katholieke Universiteit Leuven, Kapucijnenvoer 35, B-3000 Leuven, Belgium
E-mail: geert.molenberghs@med.kuleuven.be

Johan H. L. Oud
Behavioural Science Insititute, Radboud University Nijmegen, Postbus 9104, 6525 HR Nijmegen, The Netherlands
E-mail: j.oud@pwo.ru.nl

Dimitris Rizopoulos
Department of Biostatistics, Erasmus University Medical Center, NL-3000 CA Rotterdam, The Netherlands
E-mail: d.rizopoulos@erasmusmc.nl

Albert Satorra
Department of Economics and Business, Universitat Pompeu Fabra, Ramon Trias Fargas 25-27, 08005 Barcelona, Spain
E-mail: albert.satorra@upf.edu

Kees van Montfort
Department of Econometrics and Operations Research, VU University Amsterdam, De Boelelaan 1105, 1081 HV Amsterdam, The Netherlands
E-mail: kvmontfort@feweb.vu.nl
and
Nyenrode Business Universiteit, Straatweg 25, 3621 BG Breukelen, The Netherlands
E-mail: k.van.montfort@nyenrode.nl

Geert Verbeke
Interuniversity Institute for Biostatistics and Bioinformatics (I-BioStat), Katholieke Universiteit Leuven, Kapucijnenvoer 35, B-3000 Leuven, Belgium
E-mail: geert.verbeke@med.kuleuven.be
and
Interuniversity Institute for Biostatistics and Bioinformatics (I-BioStat), Universiteit Hasselt, Agoralaan, B-3590 Diepenbeek, Belgium
E-mail: geert.verbeke@uhasselt.be

Jeroen K. Vermunt
Department of Methodology and Statistics, Tilburg University, Warandelaan 2, 5037 AB Tilburg, The Netherlands
E-mail: j.k.vermunt@uvt.nl

Byungwon Woo
Department of Political Science, 2140 Derby Hall, 154 N. Oval Mall, Ohio State University, Columbus, OH 43210-1373, USA
E-mail: woo.54@polisci.osu.edu

Catherine Zimmer
Odum Institute for Research in Social Science, CB 3355 Manning Hall, University of North Carolina, Chapel Hill, NC 27599-3355, USA
E-mail: cathy_zimmer@unc.edu

Chapter 1
Loglinear Latent Variable Models for Longitudinal Categorical Data

Jacques A. Hagenaars

Abstract Errors and unreliability in categorical data in the form of independent or systematic misclassifications may have serious consequences for the substantive conclusions. This is especially true in the analysis of longitudinal data where very misleading conclusions about the underlying processes of change may be drawn that are completely the result of even very small amounts of misclassifications. Latent class models offer unique possibilities to correct for all kinds of misclassifications. In this chapter, latent class analysis will be used to show the possible distorting influences of misclassifications in longitudinal research and how to correct for them. Both simple and more complicated analyses will be dealt with, discussing both systematic and independent misclassifications.

1.1 Introduction

Longitudinal data come in many varieties and many different approaches exist for their analysis. In this chapter, analysis techniques will be proposed for strictly longitudinal data, that is, data that result from repeated observations over time of the same subjects. Moreover, the emphasis will be on data from panel surveys in which the subjects' scores are only known for particular discrete points in time, essentially, the time points t_i at which the panel waves were conducted. In agreement with this, time will be treated as discrete (Coleman, 1964). This is not to say that it is assumed that the changes occur always and only at discrete points in time. But with discrete time observations, what happens in-between the time points t_i and t_{i+1} is in principle unknown. Therefore, the application of continuous time models to such data essentially amounts to the imputation of the missing in-between values by assuming the validity of some continuous, usually smooth process of change. If this process is

Jacques A. Hagenaars
Department of Methodology and Statistics, Tilburg University, The Netherlands
e-mail: jacques.a.hagenaars@uvt.nl

K. van Montfort et al., *Longitudinal Research with Latent Variables*,
DOI 10.1007/978-3-642-11760-2_1, © Springer-Verlag Berlin Heidelberg 2010

misspecified, the resulting estimated change parameters can be seriously distorted (Carlsson, 1972; Kohfeld, & Salert, 1982; Schoenberg, 1977). On the other hand, a lot of, perhaps most characteristics undergo continuous changes (though often irregular) and not only changes at particular discrete time points and certainly not only at those points in time that happen to coincide with the moments of observation. Therefore, in the discrete time models discussed here, a pragmatic position is assumed in which the consequences of the possibly continuous, but mostly irregular processes of change will be studied at particular discrete moments in time without making restrictive assumptions about the nature of the process itself (Hagenaars, 1990, p. 16; continuous time models are explicitly dealt with in other chapters of this volume).

The proposed models are not only discrete time, but also discrete state models: the characteristics involved will be treated as truly categorical (Coleman, 1964). The categorical data may pertain to truly discrete variables, such as political party preference, religious denomination, or number of children, but also to categorizations of possibly continuous variables but measured by means of just a few categories like "1. yes, 2. no" or "1. completely agree, ..., 5. completely disagree" or "1. always, 2. sometimes, 3. never". These categorizations of possibly continuous variables can be analyzed in two ways: either one treats them as truly categorical, discrete variables (the Yulean approach after George Udny Yule) or one explicitly interprets the categories as realizations of some underlying continuous variable (the Pearsonian approach, after Karl Pearson). The Yulean approach will be adopted here. The main reason not to use the Pearsonian approach is that this approach always involves largely untestable assumptions about the data, such as a priori assumptions about the underlying (normal) distribution of the not observed continuous variable or assumptions about the nature of the process that transforms the scores on the continuum into the observed categorical response. In the Yulean approach, no such assumptions are necessary as it treats the observed scores as given, as the categories they are. Note that this does not imply that the discrete states are always or necessarily treated as categories of a nominal level variable, as is often thought. Categorical variables can be treated as nominal level, but also where appropriate as ordinal, interval or ratio level variables. Although most of the examples below involve nominal level variables, this is only done to keep the expositions simple. In the last section this issue will be taken up again.

Three general ways of analyzing categorical panel data are often distinguished (Bergsma, Croon, & Hagenaars, 2009; Molenberghs & Verbeke, 2005; Vermunt & Hagenaars, 2004). First, there is the marginal modeling approach in which the net changes are studied. Essentially, the research question is "how different are the time one from the time two scores"? The dependencies among the observations arising from repeatedly interviewing the same persons are of no substantive interest and are just treated as a (statistical) nuisance. Second, there are the subject specific or random effect models, in which the dependencies among the observations are taken care of by means of introducing random components at the subject level. Finally, there is the conditional or transition approach in which the dependencies over time are the express purpose of the investigation. In this chapter, only conditional models will be extensively dealt with, leaving the brief discussion of the other two approaches to the final section.

Causal models, or rather: Structural Equation Models (SEMs) – the latter term being preferred here to avoid the often too sloppy use of the causal terminology – are the most general and flexible forms of conditional analysis (and include multiple regression or multiple logit models as special cases). As, moreover, loglinear modeling may be considered as the most general and flexible method for the analysis of categorical data, loglinear SEMs for categorical panel data will be the focus of this chapter, and then, especially, loglinear SEMs with (categorical) latent variables. The possibility to include latent variables is especially important in longitudinal analyses as the patterns of observed change do not only result from the true changes, but also from random and systematic measurement error and misclassifications. Latent variable SEMs offer the possibility to discover the true patterns of change and to evaluate the distorting influences of many kinds of errors and biases on the patterns of change.

The basic latent class model and its extensions form a first step towards the building of a loglinear SEM with one or more latent variables. This basic latent class model will be discussed in the next section. It will be shown how even small amounts of random measurement errors in the form of independent misclassifications may lead to very misleading conclusions about the nature of the changes over time, when these conclusions are only and directly based on the observed data. First in Section 1.2, the changes in one simple dichotomous characteristic will be discussed. After that, slightly more complicated latent class models will be presented to study the changes over time involving two or more indicators.

Because SEMs for categorical data in the Yulean tradition may not be familiar to many readers, the basic principles will be outlined in Section 1.3, first loglinear SEMs without and thereafter SEMs with latent variables. Substantive applications of loglinear SEMs with latent variables will be presented in Section 1.4. It will be shown how true changes can be discovered by explicitly modeling the nature of independent and systematic misclassifications together with models for the nature of the true changes and their causes and consequences. Finally, some discussion points, extensions, and new developments will be presented in the last Section 1.5.

1.2 Latent Class Models: Separating Unreliability and True Change

Imagine a two-wave panel study into the drinking behavior of young people. A sample of 1100 respondents was interviewed first at age eleven and again one year later at age twelve. Now assume that the use of alcohol remained completely the same for each individual during the whole period of investigation and that 200 of the respondents used alcohol at least once a month ("regular users") and 900 less than once a month ("nonusers"). This true state of affairs is represented by the dichotomous, not directly observed variable X with the latent categories (or latent classes): 1. True "regular users" and 2. True "nonusers". The scores on X are not directly observed, only the respondents' manifest answers are. When the respondents are asked about their drinking behavior, the reliability of their answers is assumed to be very high, although not perfect: the probability that a respondent answers in agreement with the true position, that is, with the latent class the respondent belongs to equals .90. This response probability of .90 is assumed to be true both for the true "regular users" and for the true "nonusers" and both for the time one and the time two observations. Moreover, independent misclassifications are assumed. i.e., given the true scores, the misclassification of one respondent is independent of the misclassification of another respondent and also, for each and every respondent, the misclassifications in the first and the second panel wave are independent of each other. The observed drinking behavior in the first wave will be denoted as (dichotomous) variable A and in the second wave as (dichotomous) variable B.

Given that all of the above is true, how will the observed data look like? First, a look at the observed marginal distributions. The total number of respondents that will be registered as regular user in the first wave will consists of the respondents who are truly regular users (X = 1) and are expected to answer accordingly (.90 × 200 = 180) plus those who are expected to give the "wrong answer" compared to their true position of nonuser (X = 2) (.10 × 900 = 90). Therefore, in total, there will be (180 + 90 =) 270 regular users (A = 1) and 830 nonusers (A = 2). Because of the absence of any latent change and the constant response probability of .90 of answering correctly plus the independence of the misclassifications, the same marginal distribution will be obtained in the second wave for variable B.

Because of the assumed pattern of misclassifications, the observed distribution of A (or B) will be less peaked than the true distribution (of X). If the response probability of .90 had been .50 (the maximum unreliability), the marginal distribution of A or B would have been the uniform distribution. Here, according to the assumed state of affairs, the observed percentage of nonusers will be 75.5% (= 100 × 830 / 1100) at each point in time, while the true percentage of nonusers equals 81.8% (= 100 × 900 / 1100). For the distribution of a dichotomous variable, less peakedness implies a larger variance. This is analogous to the consequences of unreliability in continuous measurements. In the classical error theory for continuous variables, random measurement error adds to the true variance, as the observed variance is the sum of the true variance plus the error variance (Allen & Yen, 1979; Lord & Novick, 1968). Although with categorical data, the observed variance is no longer a simple sum of the true and error variance, a similar increase in variance can be seen. Using the scores 1 and 2 for the categories, the variance of latent variable X equals .149, while the variance of the less peaked observed variable A (and B) equals .185.

However no such straightforward analogue exists between the categorical and the classical error theory concerning the mean or expected value. "Less peakedness" also implies that the mean of observed variable A (or B) will be different from the mean of the latent variable X (here: observed mean: 1.755 *vs.* latent mean: 1.818). According to classical error theory for continuous data, the observed mean score will be an unbiased estimate of the true mean. But this is generally not true with categorical data, not even with independent misclassifications. A frequently (but not necessarily) occurring pattern is that the percentages belonging to the smaller latent categories, the minorities (here: the true regular users) will be overestimated in the observed data and underestimated for the larger latent categories, the majorities (here: the true nonusers).

So, the latent and the observed distributions will generally differ from each other. On the other hand, given the assumptions of the stability of the true scores and the identical probabilities of misclassifications over time, the two observed marginal distributions of A and B will be the same and it will be correctly concluded on the basis of the observed data that there was no *net change* from age eleven to age twelve. However, due to the misclassifications, there will be observed *gross change*, despite the complete stability of the true scores X. The calculation of the expected observed cell entries of the 2 × 2 observed turnover table AB under the proposed model is straightforward. For example, the number of respondents that will belong to cell (AB = 11) of this table, that is, the observed number of stable regular users will be: .90 × 200 × .90 = 162 from the truly regular users (X = 1) plus .10 × 900 × .10 = 9 from the truly nonusers (X = 2), summing up to 171 respondents. The complete observed table, expected under the model, then looks as in Table 1.1.

Table 1.1 Observed transition table alcohol use (simulated data; see text)

A-t_1 / B-t_2	*Regular user* 1	*Nonuser* 2	*Total*
1	171 (63.3%)	99 (11.9%)	270 (24.5%)
2	99 (36.7%)	731 (88.1%)	830 (75.5%)
Total	270 (100%)	830 (100%)	1100 (100%)

A researcher relying only on the observed cell entries in Table 1.1 would (wrongly) conclude that 18% (= 100 × (99 + 99) / 1100) of the respondents changed their alcohol use between the ages eleven and twelve. Especially the regular users are seemingly prone to change: more than one third (36.7%) of those that were observed as regular users the first time changed to nonuse the second time, while the corresponding transition probability for the nonusers is just above ten percent (11.9%). This difference in relative stability is statistically significant as can be seen by changing the categories of variable B in Table 1.1 into 1. "Same answer as A" and 2. "Different answer from A", by rearranging the cell entries accordingly and testing for independence in the rearranged table. The test results are (for the maximum-likelihood χ^2:) $G^2 = 75.50$, df = 1, p = .00 (and for the Pearson χ^2: $X^2 = 84.74$). Researchers will be and have been tempted to find substantive explanations for such changes and differences in transition probabilities. But these changes and differences are very misleading here. In reality, there were no true changes at all and, as far as unreliability and misclassifications might be interpreted as "random change", both the true regular users and true nonusers had the same misclassification probability and, in that sense, the same amount of random changes. A consequence of the misclassification pattern assumed above is that the larger the misclassification probability is, the larger the amount of observed change. And especially, the patterns of observed change will be such that the smaller, the minority categories will always seem to change comparatively more than the larger, the majority ones. This latter phenomenon is yet another appearance of the "regression to the mode". This "regression to the mode" is closely related to the notorious "regression to the mean", which is well known for its misleading consequences in the analysis of continuous data, but is certainly as misleading for categorical characteristics. The "regression to the mode" phenomenon has led many researchers astray in simple situations as the one discussed here, but also in more complicated situations, for larger tables, for comparisons of changes in several subgroups and in related characteristics, and for quasi-experimental designs (Hagenaars, 1990, 2005).

So far, it was assumed that the true state of the world and the parameters that govern this world were known and from this knowledge, the observed data could be derived. In practice, researchers have only the observed data at their disposal and have to work the other way around: for a particular observed table (AB) and assuming a particular model, the parameters, essentially the entries in the complete table ABX have to be estimated. Most, if not all, models for separating true changes in categorical data from changes due to misclassifications can be formulated as some variant of the general latent class (Clogg, 1995; Goodman, 1974a, 1974b; Haberman, 1979; Lazarsfeld & Henry, 1968; Wiggins, 1973; Hagenaars & McCutcheon, 2002). The latent class model implied by the simple model above can be depicted as in Figure 1.1. In this figure, X represents the dichotomous, stable characteristic "true alcohol use" and A and B refer to the observed alcohol use at age eleven and age twelve, respectively. Crucial is further that, as explained in the model above, there is no direct influence, no arrow from A to B. Variables A and B are only correlated with each other because they are both influenced by X.

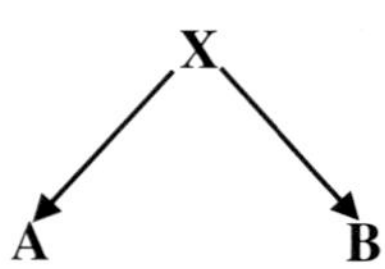

Fig. 1.1 Basic latent class model.

The model in Figure 1.1 can be parameterized in terms of (conditional response) probabilities as follows, where π_{ijt}^{ABX} refers to the probability that a respondent belongs to cell (i, j, t) of table ABX, π_t^X indicates the probability of being in latent class t of X, and $\pi_{it}^{A|X}$ the conditional response probability of being in category of i of A, given that respondent belongs to X = t (and the other parameters have obvious, analogous meanings)

$$\pi_{ijt}^{ABX} = \pi_t^X \pi_{it}^{A|X} \pi_{jt}^{B|X} \tag{1.1}$$

In agreement with Figure 1.1, equation (1.1) embodies the basic latent class assumption of local independence: within the latent classes, A and B are statistically independent from each other. All A and B have in common is their being an indicator of X. When X is controlled, held constant, the relationship between A and B disappears. How this basic latent class assumption is represented in equation (1.1) can be easily seen if one realizes that from elementary rules of probability, it follows that by definition $\pi_{ijt}^{ABX} = \pi_t^X\, \pi_{ijt}^{AB|X}$. However, the joint conditional response probability $\pi_{ijt}^{AB|X}$ is written in equation (1.1) as the product $(\pi_{it}^{A|X} \pi_{jt}^{B|X})$ which is valid if and only if A and B are conditionally independent of each other, given X.

Because X is a latent, not directly observed variable, table ABX is not an observed table. The link between the observed frequencies and the parameters at the right hand side of equation (1.1) follows from summing the complete table ABX over the latent variables to obtain the observed table AB ($\sum_t \pi_{ijt}^{ABX} = \pi_{ij+}^{ABX} = \pi_{ij}^{AB}$). To be able to identify the parameters, first standard identifiability restrictions have to be imposed. In equation (1.1), it has to be taken into account that the parameters are probabilities and have to sum to one wherever appropriate, e.g., $\sum_t \pi_t^X = \sum_i \pi_{it}^{A|X} = 1$. However, these minimal identifying restrictions are not sufficient here to be able to estimate the parameters from the observed Table 1.1. There are still more unknown parameters to estimate than known observed cell probabilities. Table 1.1 contains three independent cell entries, as the cell probabilities have to sum to 1. But after the identifying restrictions have been imposed, the LCA model above still has five independent parameters left to estimate (e.g., the set π_1^X, $\pi_{11}^{A|X}$, $\pi_{12}^{A|X}$, $\pi_{11}^{B|X}$, and $\pi_{12}^{B|X}$). In our little alcohol use example above, extra "restrictions" have been imposed by making particular response probabilities equal to each other: $\pi_{11}^{A|X} = \pi_{22}^{A|X} = \pi_{11}^{B|X} = \pi_{22}^{B|X}$. If Table 1.1 had been a normally observed table, all these restrictions together would yield one degree of freedom that may be used to test the model. In this simulated case, for the observed frequencies in Table 1.1, a perfectly fitting model is found with the parameter "estimates" given above.

General overviews of identifiability of latent class models and of procedures to obtain the maximum likelihood parameter estimates (EM-algorithm, scoring, and Newton-Raphson methods) and to test the models can be found in the general latent class literature given above. User friendly and very flexible programs are available to check the identifiability of the models and to estimate its parameters and test the goodness of fit, e.g., Vermunt's LEM, Vermunt and Magidson's Latent Gold and Muthén's Mplus (Muthén & Muthén, 2006; Vermunt, 1997b; Vermunt & Magidson, 2005).

Essentially the same estimation and testing procedures (and programs) can be employed for another useful and flexible parameterization of the same latent class model in Figure 1.1, but now in terms of a loglinear model (in multiplicative or additive form):

$$\pi_{ijt}^{ABX} = \eta\, \tau_i^A\, \tau_j^B\, \tau_t^X\, \tau_{it}^{AX}\, \tau_{jt}^{BX} \tag{1.2a}$$

$$\ln \pi_{ijt}^{ABX} = \vartheta + \lambda_i^A + \lambda_j^B + \lambda_t^X + \lambda_{it}^{AX} + \lambda_{jt}^{BX} \tag{1.2b}$$

In equation (1.2), the constant parameter η (or θ) has to do with the average cell probability and is directly related to the sample size. The one-variable parameter τ_i^A (or λ_i^A) reflects the average distribution of variable A within the categories of the other variables in the equation (here: B and X) and the other one-variable parameters in the equation have obvious, analogous interpretations. The two-variable parameter τ_{it}^{AX} (or λ_{it}^{AX}) indicates the direct association between A and X and is a function of the partial odds ratio(s) between A and X. More about the precise meaning of the parameters of the loglinear models can be found in one of the many elementary or intermediate introductions into the field, e.g., Knoke & Burke (1980) and Hagenaars (1990). The basic local independence assumption of the latent class model is now reflected in equation (1.2) by the fact that there is no direct association between A and B, that is, there is no parameter τ_{ij}^{AB} (or λ_{ij}^{AB}) in equation (1.2).

The necessary general identifying restrictions on the loglinear parameters take the form of the usual dummy or effect coding restrictions. In this chapter, effect coding will be used as default: the product of each τ-parameter over any of its subscripts equals one (or: the sum of each λ-parameter over any of its subscripts equals zero).

Equations (1.1) and (1.2) are equivalent representation of the same latent class model: they imply the same (conditional) independence restrictions on the data, and as such yield the same expected frequencies. The parameters in equation (1.1) can be expressed in terms of the loglinear parameters and vice versa. In this connection, it is interesting to note that the conditional response probability, e.g., $\pi_{it}^{A|X}$ turns out to be a function of both the loglinear two-variable parameter τ_{it}^{AX} and the one-variable parameter τ_i^A, i.e., a function of the association between A and X, but also of the level or popularity of A = *i*. Not surprisingly then, the extra restrictions made in the above simulated example ($\pi_{11}^{A|X} = \pi_{22}^{A|X} = \pi_{11}^{B|X} = \pi_{22}^{B|X}$) are equivalent to the loglinear restrictions: $\tau_{it}^{AX} = \tau_{it}^{BX}$ and $\tau_i^A = \tau_i^B = 1$. More and more precise information on the relationships between the parameterizations in equations (1.1) and (1.2) is provided by Hagenaars (1990) and Heinen (1996).

After this extremely simple example, a more elaborate, real world example is due to illustrate further the flexibility and usefulness of latent class models for the analysis of change. In the discussion of this and other examples, the relevant models will often be indicated in the short hand notation that is usual for hierarchical loglinear models, namely by means of the highest order interactions in the model. The model in Figure 1.1 is then denoted as model {AX,BX} implying the presence of the terms τ_{it}^{AX} and τ_{it}^{BX} in the model equation, plus all lower order parameters that can be formed by the superscripts of each particular higher order term (τ_i^A, τ_j^B, and

τ_t^X) (and always including the overall effect parameter η in the models to reflect the sample size).

The real world example concerns data from a Dutch election study in which during the six months before the elections the same group of respondents was interviewed once every month. The data are about the respondents' Political Party Preference and their Candidate Preference for Prime Minister and are taken from the waves conducted three (t_1) and two months (t_2) before the elections. The categories used for both characteristics are 1. Christian-Democrat, 2. Left Wing, 3. Other (mainly Right Wing). The data, presented in Table 1.2, have been analyzed before from similar and different angles (Hagenaars, 1990; Bergsma, Croon, & Hagenaars, 2009). For the convenience of the reader the data and necessary LEM program files have been made available on the book website `http://www.econ.upf.edu/~satorra/longitudinallatent/readme.html`.

Table 1.2 Party and candidate preference (source Hagenaars, 1990)

	C	1	1	1	2	2	2	3	3	3	*Total*
	D	1	2	3	1	2	3	1	2	3	
A	B										
1	1	84	9	23	6	13	7	24	8	68	242
1	2	0	1	0	0	8	1	2	2	3	17
1	3	3	1	2	0	2	3	2	3	9	25
2	1	1	1	0	1	2	2	1	0	1	9
2	2	2	4	0	1	293	6	1	22	21	350
2	3	1	0	0	1	8	7	0	0	9	26
3	1	6	1	1	4	5	0	9	1	16	43
3	2	0	1	1	0	31	0	2	9	7	51
3	3	14	1	15	3	48	23	12	21	200	337
Total		111	19	42	16	410	49	53	66	334	1100

A - Party Preference t_1
B - Party Preference t_2
C - Candidate Preference t_1
D - Candidate Preference t_2

1. Christian- Democrat
2. Left Wing
3. Other

In principle, the data in Table 1.2 provide a wealth of information, e.g., on the association between Party and Candidate Preference, on the changes in this association, on the net changes and on the gross changes in the Preferences etc. However, the observed changes and associations between the two characteristics can be very misleading due to measurement errors. Possible misclassifications must be taken into account and latent class models are useful for this purpose. In all latent class models in this section, it will be assumed that the misclassifications are independent of each other (below this assumptions will be relaxed).

Two questions will be the focus of this little investigation. The first question concerns the nature of the true changes: are the true Preferences stable over time or do true changes take place? The second question concerns the nature of the observed variables as indicators: are they measuring one and the same concept (Political Orientation) or do they refer to two different concepts (Party Preference and Candidate Preference)?

The most restrictive answer to these two questions is that both Party and Candidate Preference are just indicators of one and the same underlying concept Political Orientation and that, moreover, this underlying variable is stable over time, i.e., stable during the period of investigation. These two elements of this composite hypothesis can be tested simultaneously by estimating latent class model {AX,BX,CX,DX}, depicted in Figure 1.2a. In this model, latent variable X is supposed to represent the underlying, stable Political Orientation with three (latent) categories in agreement with the categories of the observed variables and A through D are its indicators. However, if model {AX,BX,CX,DX} is tested against the data in Table 1.2, it turns out that it does not fit the data at all: (ML-)G^2 = 362.71, df = 54, p = .00 (Pearson-X^2 = 491.23).

One way to arrive at a better fitting model might be to enlarge the number of latent classes and treat X as a latent variable with four or five categories. Although this is not done here, as it would not be very logical given the data and the research problem, latent class models with latent variables that have a number of categories different from their indicator(s) may have very nice interpretations, also in the analysis of longitudinal data. Hagenaars (1990) discusses this issues further and provides several examples.

Given the test outcomes, obviously something is wrong with model {AX,BX,CX,DX}. Taking the three latent classes and the basic local independence assumptions for granted, it must be either with the assumption of perfect stability of Political Orientation or with the assumption about the observed variables as indicators of one and the same concept.

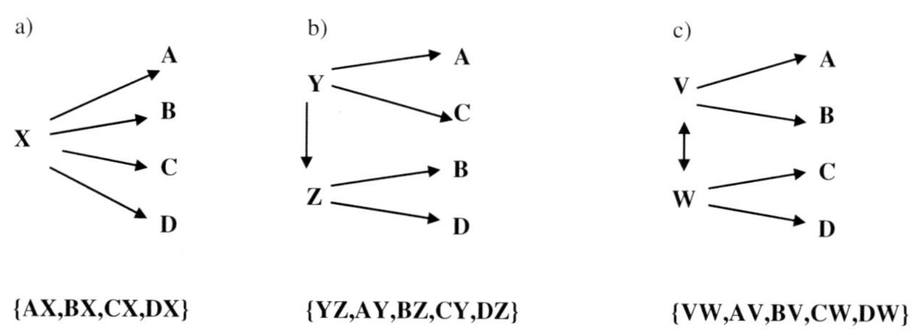

Fig. 1.2 Latent class models for political preference items.

The most unlikely part of the composite hypothesis was thought to be the stability assumption. After all, an election campaign was going on that might have caused people to change their preferences. More confidence was there in the validity of the other part, because it was believed that in the Netherlands preferences for parties and candidates are both just expressions of one's political orientation. These considerations lead to a two latent variable model with two underlying latent variables, where trichotomous latent variable Y is now meant to represent the respondents' true Political Orientation at time 1 and trichotomous latent variable Z their true Political Orientation at time 2. So people can change their true political orientation over time. The changes can be found in latent table YZ with entries π_{rs}^{VW}. Because the observed party and candidate preference are still considered to be indicators of one and the same concept Political Orientation, it is assumed that latent variable Y influences directly (and only) the scores on the indicators A and C and latent variable Z directly (and only) the scores on indicators B and D. Together with the usual local independence assumptions, all this leads to latent class model {YZ,AY,BZ,CY,DZ}. This model is depicted in Figure 1.2b where an arrow is inserted from Y to Z because, given the temporal order, Y is supposed to influence Z (rather than vice versa). However, contrary to the expectations, model {YZ,AY,BZ,CY,DZ} does not fit the data in Table 1.2 (G^2 = 351.71, df = 48, p = .00 (Pearson X^2 = 501.02)) and in terms of the sizes of the chi-square statistics, about as bad as the previous one latent variable model {AX,BX,CX,DX}.

Therefore, the problematic part of the one latent variable model in Figure 1.2a might not have been the latent stability of the true scores, but the idea that the observed candidate preference and party preference are just indicators of the same underlying concept. Perhaps Party Preference and Candidate Preference must be regarded as two distinct concepts with indicators (A,B) and (C,D) respectively. To investigate this possibility, latent class model {VW,AV,BV,CW,DW} (Figure 1.2c) is estimated for the data in Table 1.2. In model {VW,AV,BV,CW,DW}, the trichotomous latent variable V represents the underlying variable Party Preference which is now supposed to be stable for the two waves and the trichotomous latent variable W refers to the underlying stable variable Candidate Preference. The relationship between the two latent variables V and W is represented in Figure 1.2c by a double-headed arrow indicating an "unanalyzed correlation" between the (exogenous) latent variables, as it is not certain in which direction the causal order might go.

The equation for model{VW,AV,BV,CW,DW} is presented in full in equation (1.3), in terms of probabilities (1.3a) and in terms of loglinear parameters (1.3b), because its outcomes will be discussed further in this chapter. Although a bit more complicated, the particular form of equation (1.3) and the relationship between equation (1.3a) and (1.3b) follows from the above and from the same logic on which equations (1.1) and (1.2) were based (see also Hagenaars, 1990).

$$\pi_{rsijkl}^{VWABCD} = \pi_{rs}^{VW}\pi_{ir}^{A|V}\pi_{jr}^{B|V}\pi_{ks}^{C|W}\pi_{ls}^{D|W} \tag{1.3a}$$

$$\begin{aligned}\ln \pi_{rsijkl}^{VWABCD} &= \vartheta + \lambda_r^V + \lambda_s^W + \lambda_i^A + \lambda_j^B + \lambda_k^C + \lambda_l^D \\ &\quad + \lambda_{rs}^{VW} + \lambda_{ir}^{AV} + \lambda_{jr}^{BV} + \lambda_{ks}^{CW} + \lambda_{ls}^{DW}\end{aligned} \tag{1.3b}$$

Model {VW,AV,BV,CW,DW} fits the data in Table 1.2 much better than the previous two models, although still not good enough: G^2 = 84,74, df = 48, p = .00 (Pearson X^2 = 94.33). However, given its much better fit and the fact that the Bayesian information criterion BIC clearly showed that it had to be preferred to the saturated model and to the models in Figure 1.2a and 1.2b, some of its main implications and outcomes will be briefly discussed. (In Section 1.4, a slight but important modification of model {VW,AV,BV,CW,DW} will be discussed. that does fit the data excellently).

Table 1.3 Outcomes for model c) in Figure 1.2

a

$\hat{\pi}_{ir}^{A\|V}$ V / A	1	2	3
1	.866	.027	.041
2	.023	.871	.054
3	.111	.102	.905
	1.000	1.000	1.000

b

$\hat{\lambda}_{ir}^{AV}$ V / A	1	2	3
1	2.169	-1.321	-.849
2	-1.499	2.108	-.609
3	-.670	-.787	1.457

As always with latent variable models, the meanings of the latent variables have to be inferred from the relationships between the latent variables and the observed ones. The relationship between the latent Party Preference (V) and the manifest Party Preference at t_1 (A) is presented in terms of the estimated conditional response probabilities in Table 1.3a and in terms of the estimated two-variable loglinear parameters in Table 1.3b. The conditional response probabilities have the big advantage of easy interpretation, but at the same time, as discussed above, they are not only a function of the relationship between the latent and manifest variable but also depend on the difficulties of the items and the popularity of the manifest categories. For example, $\pi_{ir}^{A|V}$ (as well as the difference $(\pi_{iv}^{A|V} - \pi_{i(v+1)}^{A|V}))$ is not only a function of the (log)odds ratios in table AV, i.e., loglinear parameters λ_{ir}^{AV}, but also of the loglinear one-variable parameter λ_i^A that has nothing to do with the relationship between latent and manifest variables (Hagenaars, 1990; Heinen, 1996). But no matter how we measure the association between V and A, it is clear that V can be interpreted in terms of true Party Preference with the three categories Christian-Democratic, Left Wing and Other, certainly because the relation between V and B (not given here) shows the same pattern. Analogous conclusions can be drawn about the status of variable W as the true Candidate Preference, as the relationship between W and C or D follows largely the same pattern as found in Table 1.3.

Once the intended meanings of the latent variables have been confirmed, it makes sense to have a closer look at the outcomes. First, the relationships between the latent and the manifest variables show an interesting pattern over time. The "consistency" between V and B is stronger than between V and A, and stronger for W-D than for W-C. The conditional response probabilities as well as the loglinear two-variable parameters indicating a "correct" answer that is in agreement with the latent class a respondent belongs to, all increase from time 1 to time 2 (e.g., $\hat{\pi}_{11}^{A|V} = .866$, $\hat{\pi}_{11}^{B|V} = .941$; $\hat{\lambda}_{11}^{AV} = 2.169$, $\hat{\lambda}_{11}^{BV} = 2.595$). The relationship between a latent variable and its indicator, measured in terms of odds ratios or percentage differences turns out to be stronger at time two than at time one. As the election campaign evolves, the respondents make up their minds more firmly and less misclassification occur, in the sense that the influence of the true position on the expressed opinions becomes stronger and all kinds of "random events" have less influence. Whether this increase in "reliabilities" is statistically significant or not can be tested (not done here) by comparing model {VW,AV,BV,CW,DW} with the fit of the same model, but now with the extra restrictions that the corresponding conditional response probabilities or the relevant two-variable loglinear parameters remain the same over time. Note that restrictions on the conditional response probabilities (e.g., $\pi_{ir}^{A|V} = \pi_{ir}^{B|V}$), are generally not identical, as also indicated before, to restrictions on the two-variable loglinear parameters alone (e.g., $\lambda_{ir}^{AV} = \lambda_{ir}^{BV}$ and not $\lambda_i^A = \lambda_i^B$), yielding different models with different degrees of freedom and fit.

Further, regarding the relationships between the latent variable W (Candidate Preference) and its indicators C or D, it is seen that the probability (or odds) to choose a candidate in agreement with the latent position rather than a candidate of a different political color is highest for the Left Wing (e.g., $\hat{\pi}_{22}^{C|W} = .908$) and smallest for the Christian-Democratic candidate (e.g., $\hat{\pi}_{11}^{C|W} = .685$). This nicely reflects the fact that the Left Wing candidate was the Prime Minister at that time and an established leader of the Left, while the Christian-Democratic candidate was a newcomer. In such a situation, one can expect there to be more room for random fluctuations for expressing the preference for Christian-Democratic than for the Left Wing candidate.

What are the consequences of these misclassifications? First, regarding the net and gross changes, when model {VW,AV,BV,CW,DW} is accepted, it is also accepted that there is no net or gross change at all in the underlying characteristics Party and Candidate Preference, despite the many changes in the observed table. The observed changes completely result from the (independent) misclassifications. Further, in this example, the misclassifications do not result in large differences between the true underlying marginal distributions of Party and Candidate Preference compared to the observed ones. From the estimates of the probabilities π_{rs}^{VW} in equation (1.3a), the marginal distributions of the latent variables $\hat{\pi}_{r+}^{VW}$ and $\hat{\pi}_{+w}^{VW}$ can be computed, resulting in the following marginal probabilities for the categories 1. Christian-Democratic, 2. Left wing, 3. Other: .269, .373, .358 for

V-Party Preference and .197, .411, .392 for W-Candidate Preference. On the manifest level, the corresponding marginal distribution for A – Party Preference at time t_1 is: .258, .350, .392 and for C – Candidate Preference at time t_1: .156, .432, .412. Contrary to the simple example above, it is not true now that the (latent) largest categories are under- and the smallest categories overrepresented in the observed data. One reason is that there are no such really small minorities: the latent categories are much more evenly distributed. But at least as importantly, with three categories and changing response probabilities over time, the pattern of misclassifications is much more complex and leads to less obvious, although easily tractable results.

What, however, is very different at the latent and the manifest level (next to the complete absence of change at the latent level) is the association between Party and Candidate Preference. Very much akin to the phenomenon of attenuation of correlations due to unreliability in classical error theory, the consistency of the preferences in choosing parties and candidates is much larger at the latent than at the manifest level. For example, the "consistent" two-variable parameters in latent table VW are $\hat{\lambda}_{11}^{VW} = 1.821, \hat{\lambda}_{22}^{VW} = 2.204, \hat{\lambda}_{33}^{VW} = .799$, while at the manifest level the corresponding parameter estimates in table BD for the relation B-D (stronger than A-C) are much weaker $\hat{\lambda}_{11}^{BD} = 1.223, \hat{\lambda}_{22}^{BD} = 1.644, \hat{\lambda}_{33}^{BD} = .674$. In terms of odds ratios, comparing the odds ratios in manifest table BD table and latent table VW for the first two categories of the variables (Christian-Democratic *vs.* Left Wing), it is found that at the observed level, the odds in observed table BD of preferring the Christian-Democratic candidate above the Left Wing candidate are 157.6 times larger for those who prefer the Christian-Democratic Party than for those who support the Left Wing Party; at the latent level, on the other hand, in latent table VW, the similar odds ratio is almost nine times stronger, *viz.*: 1409.4.

Many more details might be provided about this real world example, but the above may be sufficient to show that latent class models are excellently suited to investigate the true relationships among categorical variables and the changes over time when measurement errors occur (which is almost always the case). However, it must be kept in mind that so far independent misclassifications have been assumed and that this is not always true in practice. Many forms of more systematic distortions of the data exist, certainly in panel data where systematic errors, correlated over time are often expected. To deal effectively with these kinds of misclassifications, usually a bit more complicated latent class models are required, often in the form of SEMs for categorical data. This will be the topic of the next section.

Another feature of the discussion of the latent class models so far, is that terms like "latent", "true", and "underlying" have been used as interchangeable expressions. However, one must be careful here, because mixing these terms is not always justified, neither are the meanings of these terms themselves always clear. Interpretations of the latent categories in terms of true scores usually makes only sense if there is a theoretical justification of treating the observed variables as indicators of the theoretical (latent) variables and if there is a substantively and empirically justified one-to-one correspondence between the categories of the latent and the manifest variables. Latent class analysis has many different uses, e.g., to correct for unobserved heterogeneity or to find clusters or typologies, for which it is usually not meaningful to think in terms of true scores and errors (Hagenaars & McCutcheon, 2002).

Moreover, if it is appropriate to think in terms of true scores and measurement errors, it must be kept in mind that unreliability and misclassifications not only refer to real mistakes, erroneous answers by the respondents, interviewing mistakes, processing errors, etc., but also to "random" but true behavior and attitudes (Converse, 1964, 1980; Hagenaars, 1990; Kendall, 1954; Lazarsfeld, 1972; Lord & Novick, 1968; Saris & Sniderman, 2004; Sutcliffe, 1965a, 1965b). Thinking about misclassifications as strictly and purely measurement errors or mistakes may be justified if the latent score can be regarded as a platonic score, that somehow really exists: people have truly voted for a particular political party, they truly have a certain occupation, are truly married or not, etc. An observation that differs from this existing true position might be classified as a real mistake, as measurement error in the strict sense. But this idea of a platonic true score is hardly applicable in terms of the kinds of variables discussed above, i.e., with attitudes, beliefs, preferences, etc. People's true positions on those kinds of variables fluctuate over time and all of the time. People have different moods and show "true" random fluctuations. Latent variable models do not separate the two sources of "randomness", strict random measurement error and "true" random behavior, and in that sense, the latent position, corrected for independent misclassifications adjusts for both strict measurement errors and "true" random fluctuations. When using latent variable models, researchers actually indicate that they are not interested in those volatile random fluctuations but in its stable component. The latent score is best seen as an imaginary "experimental score", i.e., the expected value obtained over of a series of hypothetical independent experiments or measurements. Whether the distinction between the platonic and the experimental true score matters, depends on the purposes of the investigation. To give a simple example, if the number of unemployed people is estimated for particular points in time using panel data and repeated measurements of the respondents' labor status, latent class models are useful to explain an enduring, latent propensity to unemployment and changes therein, correcting for measurement errors and "random behavior" (see below and Bassi et al., 2000). If the purpose of the same investigation is to estimate the true, existing amount of people entitled at a certain moment to unemployment benefits, only strict measurement errors should be taken into account and at each point in time several indicators are needed to estimate the true employment position.

1.3 Concomitant Variable Latent Class Models and SEMs for Categorical Data

From the nineteen seventies on when Goodman and Haberman developed present day latent class modeling, models have been proposed by many different authors in which (categorical) variables have been added to latent class models. Those variables function as covariates in the latent class model, i.e., as independent variables with the latent variable(s) as the dependent ones (for an overview and applications of these "concomitant variable latent class models", see several chapters in Hagenaars & McCutcheon, 2002). For example, one might extend observed Table 1.2 to include G-Gender to investigate how the differences are between men and women regarding their true political attitudes. The best model so far for Table 1.2, i.e., model {VW,AV,BV,CW,DW} (see Figure 1.2c) might then be extended to model {GVW,AV,BV,CW,DW} to investigate by means of table GVW the interactions between Gender and the latent variables true Party Preference and true Candidate Preference.

As a next step one may want to include more (and perhaps intervening) covariates and variables such as Voting Behavior that must be regarded as consequences of the latent variables. In other words, a researcher often wants to set up a "causal" model, a Structural Equation Model (SEM) in which the latent variable(s) play a central role. Such models include the concomitant latent class model as a special "simple" case, but most SEMs can no longer be represented by one loglinear model or equation, but will consist of several equations.

Most researchers are familiar with Structural Equation Models (SEMs) for continuous data, using well-known programs such as LISREL, AMOS, or EQS. However, also SEMs for categorical data have been developed within the loglinear framework a long time ago. Goodman explained the principles of loglinear SEMs for observed data already in the nineteen seventies (Goodman, 1973a, 1973b); how to integrate latent class models and latent variables into Goodman's "modified path models" has been shown by Hagenaars and others (Hagenaars, Heinen, & Hamers 1980; Hagenaars, 1990, 1993, 1998, 2002; Hagenaars & McCutcheon, 2002; Vermunt, 1997a); the incorporation of the principles of graphical modeling has made the approach more general and flexible (Cox & Wermuth, 1996; Kiiveri & Speed, 1982; Lauritzen, 1996; Pearl, 2000; Whittaker, 1990). The reader should consult these references for many important particulars, because below only the most basic elements of the categorical SEM approach will be outlined. The focus will be on standard recursive models without "causal" loops, with a special emphasis on SEMs for categorical panel data containing misclassifications.

Starting point is a set of four categorical variables A through D that have a clear asymmetrical order, denoted by >. For the time being, all variables are treated as observed variables; later latent variables will be added. The variables to the left of symbol > are strictly "prior" to the variables to right of that symbol, in some meaningful (causal, temporal, predictive) sense:

$$A > B > C > D$$

For ease of exposition, sometimes a "causal" terminology will be used with notions such as influence, effect etc., but the reader must remember what has been said above when preferring the term "structural equation model" above "causal model": be careful with causal conclusions.

The first variable in the "causal" chain is variable A. Variable A is the exogenous variable, not "influenced" or "determined" by any of the variables later in the chain.

Variable A may be a joint exogenous variable and consist of several exogenous variables A_1, A_2, A_3, etc. The next variable in the chain, variable B is only determined by A but not by C or D; variable C is only influenced by A or B; finally, variable D by all previous variables.

Following the order of the variables, the joint probability π_{iikl}^{ABCD} of belonging to a particular cell (i, j, k, l) of table ABCD can be decomposed as follows:

$$\pi_{iikl}^{ABCD} = \pi_i^A \pi_{ji}^{B|A} \pi_{kij}^{C|AB} \pi_{lijk}^{D|ABC} \tag{1.4}$$

Equation (1.4) is a tautological equation in the sense that the decomposition of the joint probability at the left hand side into the product of (conditional) probabilities at the right hand side is by definition true, as follows from elementary rules of probability calculus (*cf.* $\pi_{ij}^{PQ} = \pi_i^P \pi_{ji}^{Q|P}$). Other tautological decompositions can be given, e.g., starting from D and working in an analogous manner backward to A (*cf.* $\pi_{ij}^{PQ} = \pi_i^P \pi_{ji}^{Q|P} = \pi_j^Q \pi_{ij}^{P|Q}$). However, the decomposition in equation (1.4) is unique because it is the only one that reflects the presumed "causal", asymmetrical order of the variables. It corresponds to the adage that "later" variables cannot influence 'prior' ones and that one should never control for variables that appear later in the (causal) chain. Therefore, the relationships among the exogenous A (A_1, A_2, A_3, ...) variables in the population must be observed in marginal table A with entries π_i^A, obtained by collapsing table ABCD over B, C, and D. The way B depends on A has to be found in marginal table AB with entries $\pi_{ji}^{B|A}$, ignoring the later variables C and D. The influence of A and B on C must be investigated in marginal table ABC with entries $\pi_{kij}^{C|AB}$, where the direct effect of A on C is determined by only controlling for B and the direct effect of B on C by only controlling for A, but not for the later variable D. Finally, the way D varies with A, B, and C has to be investigated in table ABCD with entries $\pi_{lijk}^{D|ABC}$.

The effect parameters for the relationships among the variables can be found by parameterizing each of the probabilities at the right hand side of equation (1.4) in terms of loglinear or logit models. To follow the exposition below, remember that loglinear and logit models are completely equivalent models, in the sense that a particular logit model is equivalent to a loglinear model that has the same effect parameters involving the dependent variable as the logit model, plus all parameters necessary for reproducing the observed joint probability distribution of the independent variables (for more details, see Knoke & Burke, 1980; Hagenaars, 1990; Agresti, 1990, among many others). Further keep in mind that, the loglinear parameters for the effect of, e.g., A on B (λ_{ij}^{AB}) can be determined for table AB using the joint probabilities π_{ij}^{AB} or, with identical outcomes, the conditional probabilities $\pi_{ji}^{B|A}$.

If no restrictions are assumed for the right hand side elements in equation (1.4), i.e., if the effects of the variables upon each other is in no way further restricted, a set of saturated submodels (or loglinear equations) has to be used, one saturated submodel for each right hand side element. In that case, the right hand side elements can be replaced by their observed counterparts and these can be used to estimate the relevant loglinear parameters for the effects of the independent variables on the dependent ones. If particular restrictions are imposed, appropriate nonsaturated models must be defined for these right hand side elements. For example, it might be assumed that C is only directly influenced by B, but not by A. In that case loglinear model {AB,BC} must be valid for marginal table ABC and the entries $\pi_{kij}^{C|AB}$. Model {AB,BC} will be applied to the observed table ABC with entries f_{ijk}^{ABC}. The resulting estimated expected frequencies $\hat{F}_{ijk}^{ABC}$ can then be used to obtain an estimate for $\pi_{kij}^{C|AB}$ and for λ_{jk}^{BC}, the effect of B on C. An additional restriction might be that all variables have a direct effect on D but only in the form of the direct two-variable effects, excluding all higher order (interaction) effects. This no-interaction restriction implies that the entries $\pi_{lijk}^{D|ABC}$ must correspond to a logit model with only direct effects on D and no three- or four-variable interactions, in other words to loglinear model {ABC,AD,BD,CD}. When such nonsaturated submodels are defined, equation (1.4) is no longer a tautological equation that is by definition true, but only valid when the implied restrictions are true.

Starting point for the estimation of all these submodels and for testing the loglinear SEM as a whole is the complete observed table ABCD with observed frequencies f_{ijkl}^{ABCD}. The appropriate loglinear submodels are then applied to each of the observed (marginal) tables corresponding with the right hand side elements in equation (1.4). Assuming that the two sets of restrictions discussed so far are the only restrictions, a saturated loglinear submodel {A} is applied to observed marginal table A with entries f_i^A; further, a saturated submodel {AB} is defined for f_{ij}^{AB}; but a nonsaturated submodel {AB,BC} for f_{ijk}^{ABC}; and a nonsaturated submodel {ABC,AD,BD,CD} for f_{ijkl}^{ABCD}. In this way, maximum likelihood estimates for the pertinent loglinear (effect) parameters are obtained for each submodel. The estimated expected frequencies $\hat{F}$ for each submodel can be used to test the validity of each submodel by means of the G^2 test statistic. The saturated submodels have of course zero degrees of freedom and fit the observed data perfectly. The hypothesis that all restrictions implied by all submodels are true, in other words, that the whole SEM is valid can be obtained by simply summing the G^2 statistics (not the Pearson-X^2 statistics – see Goodman, 1968, 1970), as well as the degrees of freedom of all submodels.

There is another way to obtain the overall test statistic G^2 for testing the model as a whole. The estimated expected frequencies $\hat{F}$ for the different submodels can be used to estimate each of the (conditional) probabilities at the right hand side of equation (1.4). By means of these estimated right hand side probabilities elements, the maximum likelihood estimates $\hat{\pi}_{iikl}^{ABCD}$ at the left hand side of equation (1.4) can be computed, under the condition that all submodels are simultaneously valid in the population. By multiplying by sample size N and comparing $N\hat{\pi}_{iikl}^{ABCD}$ with f_{ijkl}^{ABCD} by filling them in in the usual formula for the (maximum likelihood) chi square, the overall test statistic G^2 can be obtained where the number of degrees of freedom is equal to the number of independent restrictions implied by the whole SEM. This latter overall G^2 test statistic has for the simple recursive models discussed above the same value and degrees of freedom as the former one obtained by summing the G^2s for the different submodels. However, this latter way of testing the model directly as a whole rather than summing the G^2s for the submodels is more generally applicable: it can be used in situations in which simply summing the G^2s for the submodels does not work.

This is for example true when "graphical simplifications" are introduced into a basic equation such as equation (1.4). Graphs are essentially defined in terms of the (conditional) independence restrictions they imply for the data. In directed graphs, there are asymmetrical relationships among the variables, as above, and the variables that are not (conditionally) independent of each other (in the relevant marginal table) are connected by a directed line or arrow. In undirected graphs, there is no order among the variables and the direct relationships, obtained by controlling for all other variables in the graph, are indicated by a straight line. Viewing a structural equation model as a (directed) graph has several advantages. In this way, it is more clearly seen that the basic principles underlying SEMs for categorical data are essentially the same as for SEMs for continuous variables (Kiiveri & Speed, 1982; Pearl, 2000). Only their parameterizations differ: for continuous variables, linear regression equations are used; for categorical data, loglinear and logit equations (and more recent developments make it possible to mix continuous and categorical variables in many ways; see the last section). Further, the introduction of the (conditional) independence restrictions into the basic equation, such as equation (1.4), makes the representation of SEMs simpler and, more importantly in practice, the estimation procedures often much more efficient enabling the researcher to estimate models that otherwise could not be handled. A very simple example can be given by means of the above discussion of equation (1.4). In that discussion, it was assumed that, in marginal table ABC, A did not have a direct effect on C. Therefore, submodel {AB,BC} was defined for f_{ijk}^{ABC}. Another way of expressing the same hypothesis is to say that A and C are conditionally independent of each other, conditional on B. The conditional independence restriction implies here that $\pi_{kij}^{C|AB}$ does not vary with A (not over subscript *i*), but only with B (over subscript *j*), that is, $\pi_{kij}^{C|AB} = \pi_{kj}^{C|B}$. Therefore equation (1.4) can be replaced by

$$\pi_{iikl}^{ABCD} = \pi_i^A \pi_{ji}^{B|A} \pi_{kj}^{C|B} \pi_{lijk}^{D|ABC} \tag{1.5}$$

In the same way as it is true after imposing nonsaturated submodels for one or more of the right hand side elements of equation (1.4), equation (1.5) is no longer a tautological equation but only valid if indeed C is conditionally independent of A, given B. The overall test of the whole model can, after this "graphical simplification", no longer be obtained as the sum of the G^2-statistics for the several submodels in equation (1.5). For example, saturated submodel {BC} is defined for marginal table BC to obtain the parameter estimates for the effect of B on C. But this saturated submodel would contribute 0 to the overall test statistic, despite the implied hypothesis in equation (1.5) that A has no direct effect on C. On the other hand, the alternative testing procedure in which the left hand side entries π_{iikl}^{ABCD} are estimated by means of the right hand side estimates in equation (1.5) encounters no such problems.

Not all restricted loglinear models can be expressed in terms of "graphical simplification". The other hypothesis in the above example that there were no three- or higher order interaction terms for the effects on D and that model {ABC,AD,BD, CD} was valid for $\pi_{lijk}^{D|ABC}$ cannot be formulated in a strict graphical context, because the absence of particular higher order interaction terms cannot be expressed in terms of (conditional) independence relations. It is of course also possible to further parameterize the "simplified" right hand side elements and to define nonsaturated loglinear models for these simpler "tables". For example, if it was also assumed that A had no direct effect on D and that moreover, B and C did not interact in their effects on D, $\pi_{lijk}^{D|ABC}$ in equation (1.5) can be replaced by $\pi_{ljk}^{D|BC}$ and model {BC,BD,CD} defined for table BCD.

Altogether, the Goodman SEM approach combined with the principles of graphical modeling offers a very powerful tool for testing and estimating models that involve categorical data. Especially so, because the logic underlying this approach can easily be extended to models in which some of the variables are latent. The estimation and testing procedures become more complicated, but are, on the other hand, rather straightforward extensions of the basic latent class models discussed in the previous section (Hagenaars, 1990, 1998, 2002). User friendly and flexible programs are available to obtain the maximum likelihood estimates for categorical SEMs with latent variables. Vermunt's program LEM is probably the best in this respect and can handle almost all categorical SEMs with categorical latent variables; Magidson and Vermunt's Latent Gold can be used for many and Muthén's Mplus for some models (Vermunt, 1997b; Vermunt & Magidson, 2005; Muthén & Muthén, 2006).

Tests of latent variable SEMs cannot be carried out by summing the G^2s of the several submodels because some of the tables for these submodels involve latent variables and are not completely observed. But, again, the other way of obtaining the overall G^2 poses no problem.

How such SEM analyses are actually carried out will be illustrated here below by means of a purely imaginary example, but typical of many kinds of panel analyses. The SEM involved takes only independent misclassifications into account. In the next section, some real world examples will be given, showing that SEMs with latent variables are excellently suited for handling all kinds of systematic measurement errors.

Imagine the model in Figure 1.3 for a two-wave panel study into Political Preference.

Variables Y-Political Preference at time 1 and Z-Political Preference at time 2 are latent variables, not directly measured or observed. Their (imperfect) indicators are Party and Candidate Preference at time 1 (A,C) and at time 2 (B,D). The researcher is especially interested in the changes in Political Preference and in the effect of the observed background variables E-Education and O-Occupation on this (changing) preference. The research hypotheses are reflected in Figure 1.3. The arrows in Figure 1.3 denote as usual in such "causal diagrams" the direct two-variable effects of one variable upon another, controlling for the appropriate antecedent and intervening variables. Absence of an arrow implies and is implied by the absence of a direct

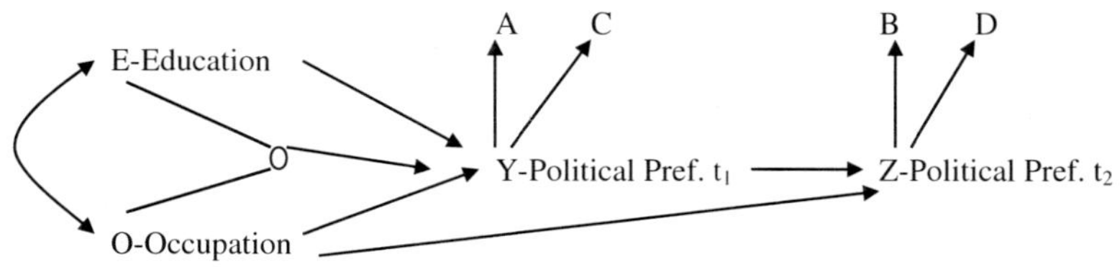

Fig. 1.3 Latent variable SEM for categorical data.

effect. Further, the curved double headed arrow between the two exogenous variables E and O represents a given, unanalyzed, undirected association. Higher order interactions (three- and higher order parameters) are indicated by a small circle or knot connecting the interacting variables. Absence of a knot implies and is implied by the absence of the corresponding interaction terms.

The order of the variables in agreement with Figure 1.3 is (E,O) > Y > A > C > Z > B > D, where the symbol (E,O) means that there are two exogenous variables E and O occupying the same position in the order. Further, the order between the two indicators A > C (A prior to C) is arbitrary here and can be reversed without consequences (see below), as is "B prior to D". Given this order, the tautological "starting" equation' is

$$\pi^{EOYACZBD}_{eoyaczbd} = \pi^{EO}_{eo}\pi^{Y|EO}_{yeo}\pi^{A|EOY}_{aeoy}\pi^{C|EOYA}_{ceoya}\pi^{Z|EOYAC}_{zeoyac}\pi^{B|EOYACZ}_{beoyacz}\pi^{D|EOYACZB}_{deoyaczb} \tag{1.6a}$$

$$\pi^{EOYACZBD}_{eoyaczbd} = (\pi^{EO}_{eo}\pi^{Y|EO}_{yeo}\pi^{Z|EOYAC}_{zeoyac})(\pi^{A|EOY}_{aeoy}\pi^{C|EOYA}_{ceoya}\pi^{B|EOYACZ}_{beoyacz}\pi^{D|EOYACZB}_{deoyaczb}) \tag{1.6b}$$

It might be insightful to rearrange the right hand side elements of equation (1.6a) into two parts, one being the structural part representing the "causal connections" among the variables and the other the measurement part representing the relations between the indicators and the other variables.

The SEM in Figure 1.3 implies a number of restrictions that can be formulated in terms of (conditional) independence relations. Latent variable Z is only directly influenced by Y and O and each of the indicators only directly by "their" latent variable. Assuming the validity of these restrictions, equation (1.6) can therefore be written, "graphically simplified", as

$$\pi^{EOYACZBD}_{eoyaczbd} = (\pi^{EO}_{eo}\pi^{Y|EO}_{yeo}\pi^{Z|OY}_{zoy})(\pi^{A|Y}_{ay}\pi^{C|Y}_{cy}\pi^{B|Z}_{bz}\pi^{D|Z}_{dz}) \tag{1.7}$$

To obtain the appropriate parameter estimates and test the model in equation (1.7), saturated loglinear submodels are defined for all but one of the right hand side elements of equation (1.7). For marginal table EO, the saturated model is submodel {EO}; for marginal table EOY, saturated submodel {EOY} and also saturated submodels {YA}, {YC}, {ZB}, and {ZD} for tables YA, YC, ZB, and ZD respectively. However, a nonsaturated submodel {OY,YZ,OZ} is required for table OYZ with entries $\pi_{zoy}^{Z|OY}$ because the model in Figure 1.3 contains no three-variable interaction term. (Note that this way of specifying the model in Figure 1.3 is completely equivalent to the following set of submodels and restrictions for the right hand side elements of equation (1.6): submodel {EO} for marginal table EO; submodel {EOY} for table EOY; submodel {EOY,YA} for table EOYA; submodel {EOYA,YC} for table EOYAC; submodel {EOYAC,OZ,YZ} for table EOYACZ; submodel {EOYACZ,ZB} for table EOYACZB; submodel {EOYACZB,ZD} for table EOYACZBD.)

All these saturated and nonsaturated submodels are estimated simultaneously by applying one of the algorithms (Newton-Raphson, Scoring, EM) for loglinear SEMs with latent variables using the observed table EOABCD with entries f_{eoabcd}^{EOABCD}. In this way, all right hand side elements in equation (1.7) are estimated, which can then be used to estimate the left hand side element $\pi_{eoyaczbd}^{EOYACZBD}$ under the assumption that the whole model is true. These estimated left hand side probabilities $\hat{\pi}_{eoyaczbd}^{EOYACZBD}$ are collapsed over the latent variables Y and Z to obtain the estimated probability distribution $\hat{\pi}_{eoacbd}^{EOACBD}$ for the observed variables under the postulated model. Finally, the estimated expected frequencies $\hat{F}_{eoacbd}^{EOACBD} = N\hat{\pi}_{eoacbd}^{EOACBD}$ can be compared to the observed frequencies f_{eoacbd}^{EOACBD} by means of G^2 to test the whole model with all its restrictions. The number of degrees of freedom is equal the total number of all independent restrictions in the model in Figure 1.3.

The choice between two hierarchically nested models can be made on the basis of a conditional test. One subtracts the G^2 for the less restricted model from the G^2 obtained for the more restricted model, as well as the their degrees of freedom to perform the conditional test that the restricted model is true in the population, given that the unrestricted is true. This conditional testing procedure can also be used to test just one particular restriction for one particular right hand side element. If the models to be compared are not nested, information based measured like BIC of AIC can be used, as well as any of the numerous descriptive fit measures.

1.4 SEMs for Dependent Misclassifications

In the fictitious example from the previous section and the model in Figure 1.3, the misclassifications in the indicators A through D were assumed to be independent of each other and of the exogenous variables E and O, given the scores on the latent variables. The relations of the indicators with the latent variables followed the classical latent class model. However, a large variety of more systematic patterns of misclassifications can be implemented in an easy and straightforward way following the basic principles outlined in the previous section. For example, continuing the fictitious example in Figure 1.3, it might be argued that the indicator Party Preference A is also directly influenced by Education and not only by the latent variable Y: higher educated people express different party preferences compared to lower educated people, regardless of their true Political Orientation Y. Right hand side element $\pi_{ay}^{A|Y}$ in equation (1.5) must then be replaced by $\pi_{aey}^{A|EY}$. If only a direct, main effect of E on A is assumed, logit model {EY,YA,EA} must be applied to marginal table EYA. If it is also expected that higher educated people give more reliable answers, i.e., that the influence of Y on A is larger for higher than for lower educated people, the saturated logit model {EYA} has to be estimated for table EYA. In an analogous manner, it would be possible and straightforward to introduce direct test-retest effects, such as A-B in Figure 1.3 and equation (1.5) (Hagenaars, 1988, 1990).

A real world application of a model with a test-retest or consistency effect can be given using the Dutch two-wave panel data presented in Table 1.2. In the previous analyses of the data in Table 1.2, two main questions were asked: Can Party and Candidate Preference be regarded as indicators of the same underlying concept Political Orientation or not, and: Are all manifest changes due to misclassifications or are they also coming from true changes at the latent level. The best latent class model so far for the data in Table 1.2 turned out to be model {VW,AV,BV,CW,DW} in which Party and Candidate Preference were regarded as two different, but stable underlying characteristics (Figure 1.2c). However, this comparatively best model did not really fit the data (G^2 = 84,74, df = 48, p = .00, Pearson X^2 = 94.33) (although it was accepted above for illustrative purposes). In the end, following the logic of the stated (composite) hypothesis for the data in Table 1.2, it seems necessary to conclude that Party Preference and Candidate Preference are two different concepts and that, moreover, these two underlying latent characteristics are not stable over time.

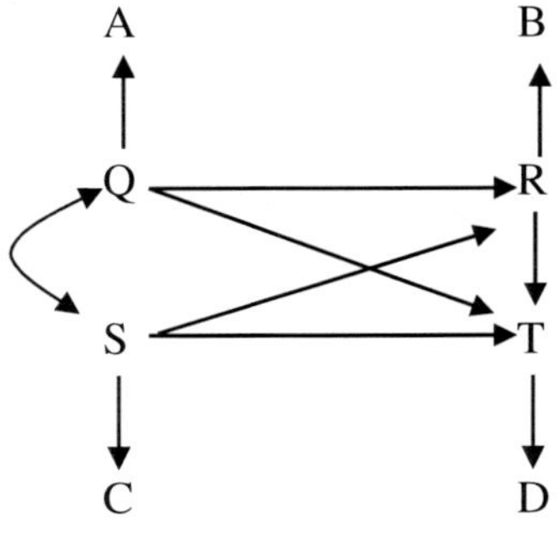

Fig. 1.4 Four latent variables SEM.

A possible model along these lines is presented in Figure 1.4. In this figure, variables Q and R are the trichotomous latent variables Party Preference in wave one, wave two, respectively. Variables S and T are the trichotomous latent variables Candidate Preference in the two waves. Manifest variables A through D are as before the four indicators, but now each one for a different latent variable. The arrows from

Q to R and from S to T indicate the true stability (or turnover) in Party Preference and Candidate Preference respectively. The cross-lagged direct effects represent the influence from wave one to wave two of Party on Candidate Preference (Q-T) and from Candidate on Party Preference (S-R). Further note that R influences T: It is assumed somewhat arbitrarily that the true Party Preference influences the true Candidate Preference in wave two, rather than vice versa. The appropriate equation after "graphical simplification" is

$$\pi_{qrstabcd}^{QRSTABCD} = (\pi_{qs}^{QS}\pi_{rqs}^{R|QS}\pi_{tqrs}^{T|QRS})(\pi_{aq}^{A|Q}\pi_{br}^{B|R}\pi_{cs}^{C|S}\pi_{dt}^{D|T}) \tag{1.8}$$

In the submodels for each of the right hand side elements in equation (1.8), all three-variable or higher order interactions terms have been left out in correspondence with the model in Figure 1.4. It turns out that this model fits the data in Table 1.2 well, with $G^2 = 27.96$ df = 24, p = .26 (Pearson $X^2 = 27.95$). This well fitting model is also an extreme model in the sense that there are as many latent as manifest variables. It may seem strange that this model is even identified. Identification is bought here by assuming that the three-variable and higher order interaction terms are absent from all submodels. It must be kept in mind that at least part of these restrictions on the higher order interactions are no longer empirically testable as they are needed for identification of the model and have to be assumed to be a priori true. Hagenaars provides an extensive discussion of the possibilities and limitations of this type of SEMs for the "cross-lagged panel correlation technique", applied to the data in Table 1.2 (Hagenaars, 1990, Chapter 5). The outcomes will not be further discussed here, as this extreme latent variable model is still an example of a model with independent misclassifications. Its main functions were to show that it is possible to relax simultaneously both key assumptions of true stability and of all indicators measuring one and the same concept by setting up the appropriate SEM, but especially to be able to contrast it with an alternative, more parsimonious way of improving model {VW,AV,BV,CW,DW}, *viz.* a model with dependent rather than independent misclassifications.

From inspection of the residuals of model {VW,AV,BV,CW,DW} for the data in Table 1.2 (Figure 1.2c) it was learned that especially the strength of the association (odds ratio) between Party and Candidate Preference in the first wave (in observed marginal table AC) was underestimated by the model (Hagenaars, 1990). For example, the odds ratio for the 2×2 part of marginal table AC involving the first two categories of A and of C is according to the observed data 105.22 while it is only 63.90 according to the estimated frequencies for model {VW,AV,BV,CW,DW}. It might be, especially in an early stage of the campaign, that the respondents have not yet developed a clear and consistent idea of their candidate preferences (see also the earlier discussions regarding the conditional response probabilities in model {VW,AV,BV,CW,DW}). In such a situation, perhaps less true in the second wave, later in the campaign, the respondent's answer to the (earlier) question on party preference might have influenced directly the answer to the (later) question on the candidate preference within the same interview at t_1. A model along these lines is depicted in Figure 1.5a.

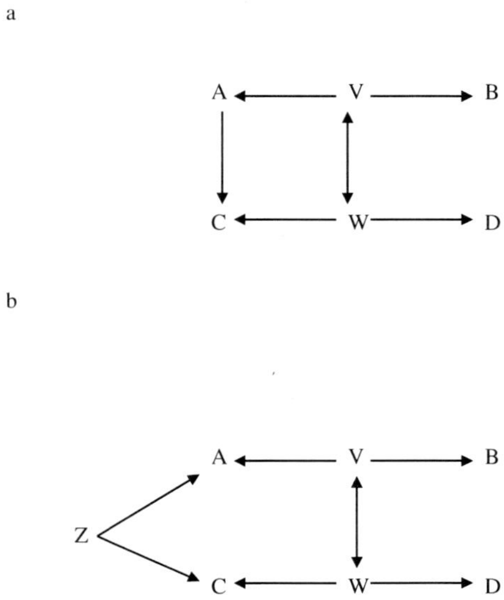

Fig. 1.5 Modeling consistency effect.

The direct effect A-C is a consistency effect, a kind of a test-retest effect but then within the same interview (Schuman & Presser, 1981). It is also a variant of "correlated errors". An alternative representation of correlated errors is a model with an extra third latent variable Z, uncorrelated with the other variables in the model, as depicted in model (1.5b); more will be said later about such a latent variable representation of correlated errors.

The basic equation for the model in Figure 1.5a is a simple extension of equation (1.3a), replacing $\pi_{ks}^{C|W}$ by $\pi_{kis}^{C|AW}$:

$$\pi_{rsijkl}^{VWABCD} = \pi_{rs}^{VW} \pi_{ir}^{A|V} \pi_{jr}^{B|V} \pi_{kis}^{C|AW} \pi_{ls}^{D|W} \tag{1.9}$$

However, the parameters of this model can no longer be obtained by means of one particular loglinear equation or model, not simply by means of loglinear model {VW,AV,BV,ACW,DW} or {VW,AV,BV,CW,AC,DW}, but a SEM, a set of loglinear models or equations is needed to represent the implications of the model in Figure 1.5a. This is different from all standard latent class models, previously discussed in Section 1.2. These standard latent class models could have been described and estimated according to the SEM principles outlined in Section 1.3, but an identical, simpler representation in terms of one particular loglinear model was possible. Whether or not a particular (recursive) SEM can be identically represented by means of just one loglinear model has to do with the collapsibility of loglinear models (Bishop, Fienberg, & Holland, 1975, p. 46) and with the question whether or not a directed acyclical graph (DAG) can be represented equivalently by means of a nondirected graph, where the nondirected graph can be represented by means of one particular loglinear model (Whittaker, 1990). The crucial point is whether or not the (conditional) independence implications of a directed graph are the same in its undirected counterpart. If in doubt, note that estimating the SEM in an element-wise fashion, setting up submodels for each of the right hand side probabilities and simultaneously estimating all these submodels, is always appropriate.

In this case, because of the direction of the arrow from A to C, the model in Figure 1.5a cannot be represented by means of one particular loglinear model (see also Hagenaars, 1988). For example, from the model in Figure 1.5a, it follows that A and W are independent of each other, controlling for V, in other words, in table AVW. However in its undirected counterpart (and model {VW,AV,BV,ACW,DW}) A is only independent of W, when controlling for V *and* C (in table AVWC). Therefore, the model in Figure 1.5a must be treated as a SEM for which several loglinear submodels must be defined: saturated submodels for tables VW, AV, BV, and DW and nonsaturated submodel {AW,AC,CW} for table ACW. This model has only four degrees of freedom less than the corresponding model {VW,AV,BV,CW,DW} without the direct effect A-C, but it almost halves the value of G^2 and fits the data in Table 1.2 very well: $G^2 = 45.97$, df = 44, p = .39 (Pearson $X^2 = 44.04$). Adding a similar consistency effect for the relationship B-D does not further improve the fit of the model.

The outcomes for this test-retest model are to a large extent not too different from the corresponding model without the test-retest effect. The same increase of the reliability over time is seen, i.e., the effect of the latent variable on the indicators is larger for wave two than for wave one. Especially variable D turns out to be a more reliable indicator than in the corresponding model without the test-retest effect. Further, the true relationship between Party and Candidate Preference (V-W) is still stronger than the corresponding manifest ones, but somewhat weaker than it was estimated before in model {VW,AV,BV,CW,DW}. The most interesting new finding is of course the nature of the (statistically significant) test-retest effects, shown in Table 1.4.

Table 1.4 Estimated consistency effects A-C

$\hat{\lambda}_{ik}^{AC}$ C / A	1	2	3
1	.683	-.506	-.177
2	-.601	.744	-.142
3	-.082	-.238	.320

The test-retest effect can indeed be interpreted as a consistency effect given the positive parameter estimates on the main diagonal in Table 1.4. The respondents had the tendency to express a preference for a candidate that is in agreement with the preferred party they have just mentioned over and above their true latent candidate preferences.

Even more complicated forms of dependent or systematic misclassifications were needed in the analysis of the data from SIPP – Survey of Income and Program Participation in the USA. The example is taken from Bassi et al. (2000). In SIPP, people were asked about their labor status with categories 1. Employed, 2. Unemployed, 3. Not in the Labor Force. SIPP is a panel study with four rotation panel groups. Each rotation group is interviewed every four months and every month one of the rotation groups is being interviewed. During each interview, information is gathered about the respondents' labor status during the previous four months, called the reference period. The data are presented by Bassi et al. (2000) as monthly figures, where for one reference period trichotomous variable A indicates the observed labor status in the first month of the reference period up to variable D indicating the labor status during the last month of this period.

There were several indications that the manifest data contained errors. As a consequence of the complete SIPP design, the manifest labor status turnover between two consecutive months is observed for three rotation groups within the reference period and for one rotation group from the "seam" between two reference periods. In other words, the data for the turnover table between two consecutive months are based on two different interviews for one rotation group and for the other three groups on data gathered within the same interview, albeit for each of the three groups in a different "phase" of the reference period. Because each rotation group has been drawn as a random sample from the same population, the turnover tables for all four rotation groups should be the same within sampling fluctuations. However, it turned out that the within reference period data always showed less turnover than the between reference periods data. Moreover, the closer the months were to the moment of the interview, the more turnover was observed. Bassi et al. (2000) analyzed these data simultaneously for all four rotation groups and for an extended period of time and they made use of a second dichotomous indicator (Employed or not). They tried to correct for the misclassifications, assuming independent misclassifications for the data coming from different interviews and assuming systematic consistency errors for the within interview data.

For the purposes here, the example is much more modest in scope: only models for the within reference period will be dealt with, using just one (trichotomous) indicator, just one rotation group and only one reference period. The restriction to this simple situation has the big advantage of focusing on the main issue: how to model systematic patterns of misclassifications by means of SEMs. But it has the big disadvantage of discussing models that are as such not identified. Some remarks about identifiability will be made here, but more can be found in the Bassi et al. (2000) article.

Given that there must be misclassifications in the data, the natural point to start accounting for these errors may seem the SEM in Figure 1.6a, after the "graphical simplification", represented by equation (1.10)

$$\pi_{abcdvwyz}^{ABCDVWYZ} = (\pi_{v}^{V}\,\pi_{wv}^{W|V}\,\pi_{yw}^{Y|W}\,\pi_{zy}^{Z|Y})(\pi_{av}^{A|V}\,\pi_{bw}^{B|W}\,\pi_{cy}^{C|Y}\,\pi_{dz}^{D|Z}) \tag{1.10}$$

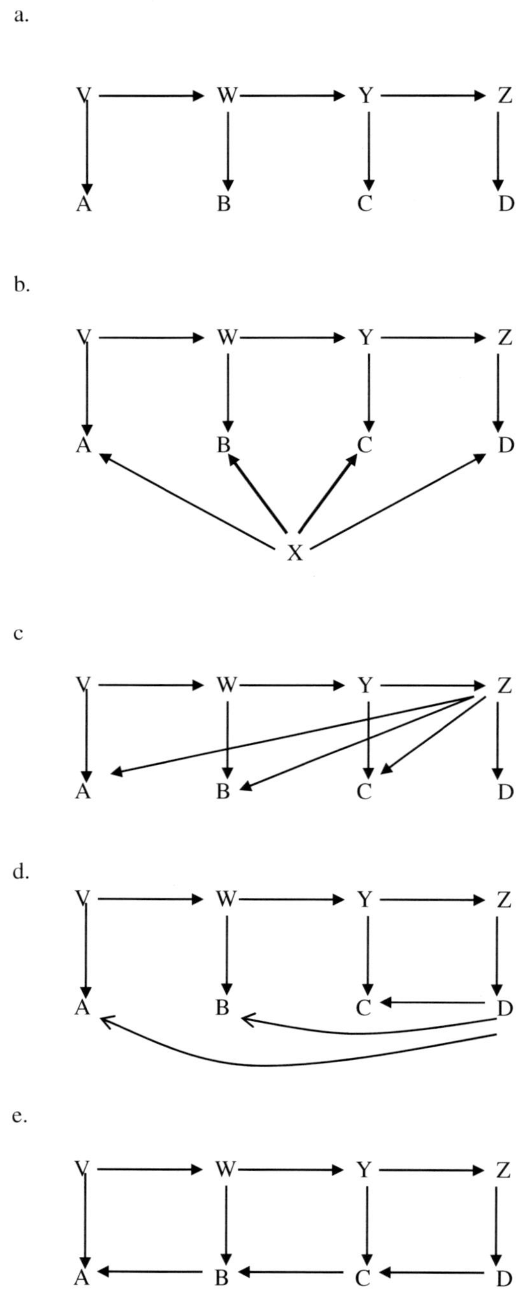

Fig. 1.6 Latent variable models for turnover in labor status.

The trichotomous variables V, W, Y, and Z are the latent analogues of the manifest variables A through D. The true changes are supposed to follow a (latent) first order markov chain in which the true labor status at each particular month within the reference period (e.g., Z), is only influenced directly by the labor status at the immediate previous month (Y) but not by the months (V, W) before that. Higher order markov chains can be defined by introducing extra direct effects from more distant months. The markov chain can be made homogenous by making the turnover tables for two consecutive months equal to each other in terms of the conditional transition probabilities: $\pi_{ij}^{W|V} = \pi_{ij}^{Y|W} = \pi_{ij}^{Z|Y}$ or, a weaker form of homogeneity, in terms of the odds ratios and the loglinear two-variable parameters: $\lambda_{ij}^{VW} = \lambda_{ij}^{WY} = \lambda_{ij}^{YZ}$. Extensive discussions of these and other varieties of latent markov chains are given by Langeheine & van de Pol (2002) and Bergsma, Croon, & Hagenaars (2009). Note that in the model in Figure 1.6a and equation (1.10) the misclassifications are assumed to be independent, as in standard latent class models.

In the simple situation of the model in Figure 1.6a, restrictions on the latent changes or the reliabilities are needed to identify the model's parameters. The latent changes might be expected to follow the first order, homogeneous markovian change indicated above. The reliabilities, that is, the relationships between each latent variable and its indicator can be made equal to each other by imposing appropriate equal response probabilities restrictions or equality restrictions on the relevant loglinear two-variable parameters. Using the complete SIPP data, as Bassi et al. (2000) did, far less restrictive identifying assumptions have to be made.

When Bassi et al. (2000) applied the models with independent misclassifications to the full SIPP data set, the peculiarities in the differences between the outcomes for the four rotation groups that led to the conclusion that there must be classification errors did not disappear. On the contrary, the discrepancies between the rotation groups for the same turnover tables were even larger at the latent level than found at the manifest level. Actually, this was not unexpected. The most natural explanation for the differences among the rotation groups is that the answers (including the misclassifications) given within the same interview, for the same reference period influence each other and are not independent from one another.

A very common and rather general way to account for such extra associations among the answers is to introduce correlated error terms. Correlated error terms are a form of unobserved heterogeneity and indicate essentially that there are extra, unmeasured sources of association between two variables over and above and independent of the measured variables in the model. The model in Figure 1.6b is such a representation of correlated error terms by means of a (n extra) latent variable X. The corresponding equation for this SEM with correlated error terms after the "graphical simplification" is

$$\pi_{abcdxvwyz}^{ABCDXVWYZ} = \pi_x^X \pi_v^V \pi_{wv}^{W|V} \pi_{yw}^{Y|W} \pi_{zy}^{Z|Y} \pi_{avx}^{A|VX} \pi_{bwx}^{B|WX} \pi_{cyx}^{C|YX} \pi_{dzx}^{D|ZX} \qquad (1.11)$$

As usual in correlated error terms models, the extra latent variable X is assumed to be independent from the latent labor status. Element $\pi_x^X \pi_v^V$ appears at the right hand side of equation (1.11) rather than π_{xv}^{XV} or $\pi_{vx}^{V|X}$ and none of the other conditional probabilities for the relationships among the latent variables contains an effect of X, i.e., conditions on X. However, if so wished and largely depending on the substantive interpretation of X, direct relationships with the other latent variables can be added in a straightforward way.

Latent variable X has direct effects on the answers, the manifest variables A through D. Again depending on theoretical considerations, these direct effects can be restricted not to interact with the effects of the latent labor status (as in Figure 1.6b) by imposing the main effects only model {VX,AV,AX} to right hand side element $\pi_{avx}^{A|VX}$ in equation (1.11) and analogous restrictions for the other relevant conditional probabilities. To arrive at an identifiable model more restrictions will be needed. In combination with the earlier mentioned possible restrictions on the latent changes and the reliabilities, it would make sense to impose the restriction that the effects of X on the indicators are the same for all indicators.

Such restrictions are not only useful for identifying the model, but also for substantive reasons. All a variable like "X" contributes in a substantive sense is that there exists another source of association, but the nature of this source is unknown. A model without correlated error terms can almost always made to fit the data by introducing correlated errors in one form or another, but, at the same time, without contributing much to the advancement of our theoretical, substantive knowledge. By trying to specify the substantive meaning of a latent variable in advance and to translate these substantive ideas in the form of empirically testable restricted models, more meaningful results can be expected. For "ordinary" latent variables such as here V, W, etc. these restrictions concern the restricted relations between a latent variable and the manifest variables. For a variable such as X, there are by definition no indicators. Restrictions have to be found in another way.

It starts with the number of categories of X. With a categorical latent variable, this number has to be chosen a priori by the researcher. One might just try it out, say from 2 to 15 latent classes, and see which (identifiable) model fits the data (best). But only rarely will this prove a sound approach. Seeing X here as an extra source of consistency in the answers during one and the same interview suggests three latent classes: the tendency to answer consistently Employed, Unemployed, or Not in the labor respectively. Further restrictions on the direct relations between X and the indicators are possible. Perhaps the only consistency effect is that people avoid saying that they are unemployed when they truly are. In that case one might set, in terms of dummy coding, all λ_{ij}^{AX}'s equal to 0, except λ_{22}^{AX} and similar restrictions concerning the other indicators. This kind of theorizing and modeling may make the interpretation of a latent variable such as X a bit less ex post facto and ad hoc.

Another way of looking at correlated errors is by trying to model the consistency effects more directly rather than in terms of restrictions on (the effects of) X. For example, it might be hypothesized that the extra within reference consistency of the answers is caused by the fact that the respondents adapt all their answers to their true labor status at the moment of interviewing. The best representation of their true status is latent variable Z, which is closest to the moment of interviewing. This particular consistency hypothesis implies that latent variable Z not only has a direct influence on D but also on the other indicators. The model is depicted in Figure 1.6c and in equation form looks like:

$$\pi_{abcdvwyz}^{ABCDVWYZ} = \pi_{v}^{V} \pi_{wv}^{W|V} \pi_{yw}^{Y|W} \pi_{zy}^{Z|Y} \pi_{avz}^{A|VZ} \pi_{bwz}^{B|WZ} \pi_{cyz}^{C|YZ} \pi_{dz}^{D|Z} \quad (1.12)$$

Again, nonsaturated models can be imposed on the right hand side elements as implied by Figure 1.6c.

Still another cause of the extra consistency might be that the respondents tried to be consistent in their given answers, in the sense that they adapt all their answers about their labor status during the reference period to the first given answer (not to their true status) during the interview. If the interviewing process starts within each reference period from the earliest month to the later ones, such an effect can be represented by direct effects of A on B, C, and D. If the interviewing process starts at the moment of interviewing and then goes back in time, this form of consistency implies a direct effect of D on A, B, and C, as is assumed in Figure 1.6d and the following equation:

$$\pi_{abcdvwyz}^{ABCDVWYZ} = \pi_{v}^{V} \pi_{wv}^{W|V} \pi_{yw}^{Y|W} \pi_{zy}^{Z|Y} \pi_{avd}^{A|VD} \pi_{bwd}^{B|WD} \pi_{cyd}^{C|YD} \pi_{dz}^{D|Z} \quad (1.13)$$

Finally, it might be thought, as Bassi et al. (2000) did, that the consistency process might best be represented by an effect of each particular answer on the next one. Again, going back in time, this would lead to Figure 1.6e and the following equation:

$$\pi_{abcdvwyz}^{ABCDVWYZ} = \pi_{v}^{V} \pi_{wv}^{W|V} \pi_{yw}^{Y|W} \pi_{zy}^{Z|Y} \pi_{avb}^{A|VB} \pi_{bwc}^{B|WC} \pi_{cyd}^{C|YD} \pi_{dz}^{D|Z} \quad (1.14)$$

Bassi et al. (2000) also discuss extensively the nature of this last consistency effect and the consequences it has for the true monthly changes (and found that now the "incorrect" observed differences among the rotation groups disappeared at the latent level).

If at all possible, the best way to correct for systematic distortions of the data is to try to measure the nature of these distortions directly, e.g., introduce an explicit measure for "social desirability". However, if this is not possible, as is certainly not the exception, defining precise models for the nature of the classification errors, is a very good and useful second best solution. SEMs are extremely flexible and helpful models in this respect.

1.5 Extensions and Conclusions

The focus in this chapter has been on the basic principles of the application of SEMs to categorical panel data that contain measurement errors. However, the reader should keep in mind that the above indeed only dealt with the basic issues and that a practical research question may need more complicated models and approaches. It is then good to know that many varieties of the basic approach exist (see the literature mentioned throughout this chapter and below) and that the programs mentioned before (LEM, Latent Gold, and Mplus) provide easy access to these more complicated procedures.

These programs also give the user clues about the identifiability of the latent class models (Goodman, 1974b). Not much has been said above about identifiability. A necessary condition for identifiability is that the number of independent unknowns, the parameters to be estimated, do not exceed the number of independent knowns (the number of independent cell frequencies). However, in latent variable models in general and also in latent class models, this is not a sufficient condition. A sufficient condition can be formulated and investigated in terms of the variance-covariance matrix of the estimates (or related matrices) being of full rank. This information is provided by the programs mentioned.

In our examples above, only discrete variables with a few nominal level categories have been dealt with. However from the very beginnings of loglinear modeling, many kinds of models have been developed that take the ordered character of the categories into account, by means of linear or inequality restrictions or estimating the (ordered) scores of the variables; Vermunt (1999) presents an overview; many applications can be found in Hagenaars & McCutcheon (2002).

Starting point of this chapter was the analysis of categorical data and all models treated these data as purely categorical and not as realizations of underlying continuous variables. However, it might well be the case that the research problem involves a data set in which both continuous and discrete variables are present. Such data can be handled within the framework sketched here, depending on the research question and the position of the continuous variables in the models, by imposing linear restrictions on the relationships with these continuous variables and/or by categorizing the continuous variables into five or seven categories. Explicit discussions of combinations of continuous and discrete variables analyses are provided by Dayton & Macready (1988, 2002); Vermunt (2002); van der Heijden, Dessens, & Böckenholt (1996), Mooijaart & van Montfort (2006), among many others; Vermunt & Magidson's Latent Gold and Muthen's Mplus have implemented several models for combinations of discrete and continuous variables.

Panel data often suffer from serious attrition and mortality. It is not rare that only 20% of the potential respondents survives all waves. Loglinear SEMs can easily be extended to include and model response indicators to account for the missing data and the MCAR (missing completely at random), MAR (missing at random) or nonignorable missing data patterns (Fay, 1986; Hagenaars, 1990; Vermunt, 1997).

Sometimes theories require nonstandard, nonrecursive SEMs in which causal loops occur; Koster (1997), Cox & Wermuth (1996), and Hagenaars (1998) discuss several ways of testing such models.

Another way of arriving at a nonstandard, nonrecursive SEM is when restrictions are imposed that involve simultaneously two or more right hand side elements of the basic SEM equation. Such restrictions occurred when the homogenous markov chain was discussed above in the Bassi et al. (2000) example. The program LEM can handle many of these kinds of restrictions. In general, models with simultaneous restrictions on several marginal tables are a form of the marginal modeling approach towards the analysis of panel data. In marginal models, in general, two or more marginal tables formed from the same complete table are analyzed simultaneously. Marginal modeling and conditional modeling in the form of SEM can be combined, e.g., when extra marginal restrictions are imposed on a SEM or when in a panel study the SEM at time one is compared to the similar SEM but then at time two; Bergsma, Croon & Hagenaars (2009) discuss a large number of such combinations for panel data and other kinds of dependent data.

Next to conditional and marginal analyses, subject specific random coefficient models for the analysis of panel data have been mentioned in the beginning of this article. Random effect models can be also defined for latent class and loglinear or logit models (and consequently SEMs) in the form of continuous or discrete random component coefficients, as shown by Vermunt (2003, 2004, 2007). Random effect models can also be regarded as latent variable models to account for unobserved heterogeneity. Categorical SEMs can be easily extended to include possible unknown sources of unobserved heterogeneity, as was essentially done in the discussion above about correlated error terms. In this way, for panel data, the influence of unknown disturbing sources can be corrected for, in much the same vein as is possible for the (first order) fixed effect models for panel data (Hagenaars & McCutcheon, 2002, pp. 345-432; Firebaugh, 2008; Allison, 2009).

With so many possibilities a concluding note of warning is needed. The purpose of an investigation is not just to find a well fitting model. Especially with latent variable models, this is always possible. But the purpose of research is to carefully translate whatever sound theoretical ideas a researcher has into a model that comes as close as possible to these theoretical ideas and then to test this model against sound data. If the model has to be rejected, residuals may play an important role to improve the model, but even more important are theoretical ideas about what might be wrong with the original rejected model. and in any case, whatever comes out of the partially exploratory analyses must be theoretically meaningful and evaluated against additional new evidence. In this process SEMs play an important role, but not the role "blind model fitters" attribute to them.

References

Agresti, A. (1990). *Categorical data analysis*. New York: Wiley.

Allen, M. J., & Yen, W. M. (1979). *Introduction to measurement theory*. Monterey, CA: Brooks/Cole.

Allison , P. D. (2009). *Fixed effects regression models*. QUASS 160. Newbury Park: Sage.

Bassi, F., Hagenaars, J. A., Croon, M. A., & Vermunt, J. K. (2000). Estimating true changes when categorical panel data are affected by uncorrelated and correlated errors. *Sociological Methods and Research, 29*, 230-268.

Bergsma, W. P., Croon M., & Hagenaars, J. A. (2009). *Marginal models for dependent, clustered, and longitudinal categorical data*. Berlin: Springer.

Bishop, Y. M. M., Fienberg, S. E., & Holland, P. W. (1975). *Discrete multivariate analysis: Theory and practice*. Cambridge, Mass: MIT Press.

Carlsson, G. (1972). Lagged structures and cross-sectional methods. *Acta Sociologica, 15*, 323-341.

Clogg C. C. (1995). Latent class models. In G. Arminger, C. C. Clogg, & M. E. Sobel (Eds.), *Handbook of statistical modeling for the social and behavioral sciences* (pp. 311-360). New York: Plenum.

Coleman, J. S. (1964). *Introduction to mathematical sociology.* London: Collier.

Converse, Ph. E. (1964). The nature of belief systems in mass publics. In D. Apter (Ed.), *Ideology and discontent* (pp. 206-261). New York: Free Press.

Converse, Ph. E. (1980). Rejoinder to Judd and Milburn. *American Sociological Review, 45*, 644-646

Cox, D. R., & Wermuth, N. (1996). *Multivariate dependencies: Models, analysis and interpretation.* London: Chapman & Hall.

Dayton, C. M., & Macready, G. B. (1988). Concomitant-variable latent class models. *Journal of the American Statistical Association, 83*, 173-178.

Dayton, C. M., & Macready, G. B. (2002). Use of categorical and continuous covariates in latent class analysis. In J. A. Hagenaars & A. L. McCutcheon (Eds.), *Applied latent class analysis* (pp. 213-233). Cambridge, UK: Cambridge University Press.

Fay, R. E. (1986). Causal models for patterns of nonresponse. *Journal of the American Statistical Association, 81*, 354-365.

Firebaugh, G. (2008). *Seven rules for social research.* Princeton: Princeton University Press.

Goodman, L. A. (1968). The analysis of cross-classified data: Independence, quasi-independence, and interactions in contingency tables with or without missing entries. *Journal of the American Statistical Association, 63*, 1091-1131.

Goodman, L. A. (1970). The multivariate analysis of qualitative data: Interactions among multiple classifications. *Journal of the American Statistical Association, 65*, 225-256.

Goodman, L. A. (1973a). The analysis of multidimensional contingency tables when some variables are posterior to others: A modified path analysis approach. *Biometrika, 60*, 179-192.

Goodman, L. A. (1973b). Causal analysis of data from panel studies and other kinds of surveys. *American Journal of Sociology, 78*, 1135-1191.

Goodman, L. A. (1974a). The analysis of systems of qualitative variables when some of the variables are unobservable. Part I - a modified latent structure approach. *American Journal of Sociology, 79*, 1179-1259.

Goodman, L. A. (1974b). Exploratory latent structure analysis using both identifiable and unidentifiable models. *Biometrika, 61*, 215-231.

Haberman, S. J. (1979). *Analysis of qualitative data: Vol. 2. New developments.* New York: Academic Press.

Hagenaars, J. A. (1988). Latent structure models with direct effects between indicators: Local dependence models. *Sociological Methods and Research, 16*, 379-405.

Hagenaars, J. A. (1990). *Categorical longitudinal data: Log-linear, panel, trend, and cohort analysis.* Newbury Park: Sage.

Hagenaars, J. A. (1993). *Loglinear models with latent variables.* Newbury Park: Sage.

Hagenaars, J. A. (1998). Categorical causal modeling: Latent class analysis and directed loglinear models with latent variables. *Sociological Methods and Research, 26*, 436-486.

Hagenaars, J. A. (2002). Directed loglinear modeling with latent variables: Causal models for categorical data with nonsystematic and systematic measurement errors. In J. A. Hagenaars & A. L. McCutcheon (Eds.), *Applied latent class analysis* (pp. 234-286). Cambridge, UK: Cambridge University Press.

Hagenaars, J. A. (2005). Misclassification phenomena in categorical data analysis: Regression toward the mean and tendency toward the mode. In L. A. van der Ark, M. A. Croon, & K. Sijtsma (Eds.), *New developments in categorical data analysis for the social and behavioral sciences* (pp. 15-39). Mahwah, NJ: Lawrence Erlbaum.

Hagenaars J. A., Heinen, A. G. J., & Hamers P. A. M. (1980). Causale modellen met diskrete latente variabelen: Een variant op de LISREL-benadering. *Methoden en Data Nieuwsbrief, 5*, 38-54. VVS-Vereniging voor Statistiek; SWS-Sociaal-Wetenschappelijke Sectie.

Hagenaars, J. A., & McCutcheon, A. L. (2002). *Applied latent class analysis.* Cambridge, UK: Cambridge University Press.

Heinen, T. (1996). *Latent class and discrete latent trait models: Similarities and differences.* Thousand Oaks, CA: Sage.

Kiiveri, H., & Speed, T. P. (1982). Structural analysis of multivariate data: A review. In S. Leinhardt (Ed.), *Sociological methodology 1982* (pp. 209-289). San Francisco, CA: Jossey-Bass.

Kendall, P. (1954). *Conflict and mood; Factors affecting stability and response.* Glencoe, IL: Free Press.

Kohfeld, C. W., & Salert, B. (1982). Representation of dynamic models. *Political Methodology, 8,* 1-32.

Koster, J. A. (1997). Gibbs and Markov properties of graphs. *Annals of Mathematics and Artificial Intelligence, 21,* 13-26.

Knoke D., & Burke P. J. (1980). *Loglinear models.* Beverly Hills, CA: Sage.

Langeheine, R., & van de Pol, F. (2002). Latent markov chains. In J. A. Hagenaars & A. L. McCutcheon (Eds.), *Applied latent class analysis* (pp. 304-344). Cambridge, UK: Cambridge University Press.

Lazarsfeld, P. F., & Henry, N. W. (1968). *Latent structure analysis.* Boston, MA: Houghton Mifflin.

Lazarsfeld P. F. (1972). The problem of measuring turnover. In P. F. Lazarsfeld, A. K. Pasanella, & M. Rosenberg (Eds.), *Continuities in the language of social research* (pp. 388-398). New York: Free Press.

Lauritzen, S. L. (1996). *Graphical models.* Oxford: Clarendon Press.

Lord, F. M., & Novick, M. R. (1968). *Statistical theories of mental test scores.* Reading, MA: Addison-Wesley.

Molenberghs, G., & Verbeke, G. (2005). *Models for discrete longitudinal data.* New York: Springer-Verlag.

Mooijaart, A., & van Montfort, K. (2006). Latent Markov models for categorical variables and time-dependent covariates. In K. van Montfort, J. Oud, & A. Satorra (Eds.), *Longitudinal models in the behavioral and related sciences* (pp. 1-18). Mahwah, NJ: Lawrence Erlbaum Associates.

Muthén, L. K., & Muthén, B. O. (2006). *Mplus: Statistical analysis with latent variables* (User's guide, 4th ed.). Los Angeles, CA: Muthén and Muthén.

Pearl, J. (2000). *Causality: Models, reasoning, and inference.* Cambridge, UK: Cambridge University Press.

Saris, W. E., & Sniderman P. M. (Eds.) (2004). *Studies in public opinion: Attitudes, nonattitudes, measurement error, and change.* Princeton: Princeton University Press.

Schoenberg, R. (1977.) Dynamic models and cross-sectional data: the consequences of dynamic misspecification. *Social Science Research, 6,* 133-144.

Schuman H., & S. Presser (1981). *Questions and answers in attitude surveys; experiments on question form, wording, and context.* New York: Academic Press.

Sutcliffe, J. P. (1965a). A probability model for errors of classification. I. General considerations. *Psychometrika, 30,* 73-96.

Sutcliffe, J. P. (1965b). A probability model for errors of classification. II. Particular cases. *Psychometrika, 30,* 129-155.

van der Heijden, P. G. M., Dessens, J., & Böckenholt, U. (1996). Estimating the concomitant variable latent class model with the EM algorithm. *Journal of Educational and Behavioral Statistics, 21,* 215-229.

Vermunt, J. K. (1997a). *Log-linear models for event histories.* Thousand Oaks, CA: Sage.

Vermunt, J. K. (1997b). *LEM: A general program for the analysis of categorical data: Users manual*(Tech. Rep.). Tilburg, The Netherlands: Tilburg University.

Vermunt, J. K. (1999). A general class of nonparametric models for ordinal categorical data. *Sociological Methodology 1999, 29,* 187-223. Washington DC: American Sociological Association.

Vermunt, J. K., & Magidson, J. (2002). Latent class cluster analysis. In J. A. Hagenaars & A. McCutcheon (Eds.), *Applied latent class analysis.* (pp. 89-106). Cambridge, UK: Cambridge University Press.

Vermunt, J. K. (2003). Multilevel latent class analysis. In R. M. Stolzenberg (Ed.), *Sociological Methodology 2003.* (pp. 213-240). Washington, DC: American Sociological.

Vermunt, J. K. (2004). An EM algorithm for the estimation of parametric and nonparametric hierarchical nonlinear models. *Statistica Neerlandica, 58*, 220-233.

Vermunt, J. K. (2007). Growth models for categorical response variables: Standard, latent-class, and hybrid approaches. In K. van Montfort, J. Oud & A. Satorra (Eds.), *Longitudinal models in the behavioral and related sciences.* (pp. 139-158). Mahwah, NJ: Erlbaum.

Vermunt, J. K., & Hagenaars, J. A. (2004). Ordinal longitudinal data analysis. In R. C. Hauspie, N. Cameron, & L. Molinari (Eds.), *Methods in human growth research.* (pp. 374-393). Cambridge, UK: Cambridge University Press.

Vermunt, J. K., & Magidson, J. (2005) *Latent GOLD 4.0 user's guide.* Belmont, Massachusetts: Statistical Innovations Inc.

Wiggins, L. M. (1973). *Panel analysis. Latent probability models for attitude and behavior processes.* Amsterdam, NL: Elsevier.

Whittaker, J. W. (1990). *Graphical models in applied multivariate statistics.* New York: Wiley.

Chapter 2
Random Effects Models for Longitudinal Data

Geert Verbeke, Geert Molenberghs, and Dimitris Rizopoulos

Abstract Mixed models have become very popular for the analysis of longitudinal data, partly because they are flexible and widely applicable, partly also because many commercially available software packages offer procedures to fit them. They assume that measurements from a single subject share a set of latent, unobserved, random effects which are used to generate an association structure between the repeated measurements. In this chapter, we give an overview of frequently used mixed models for continuous as well as discrete longitudinal data, with emphasis on model formulation and parameter interpretation. The fact that the latent structures generate associations implies that mixed models are also extremely convenient for the joint analysis of longitudinal data with other outcomes such as dropout time or some time-to-event outcome, or for the analysis of multiple longitudinally measured outcomes. All models will be extensively illustrated with the analysis of real data.

2.1 Introduction

Repeated measures are obtained whenever an outcome is measured repeatedly within a set of units. An array of examples is presented in Section 2.2. The fact that observations from the same unit, in general, will not be independent poses particular challenges to the statistical procedures used for the analysis of such data. In

Geert Verbeke
I-BioStat, Katholieke Universiteit Leuven, Belgium
e-mail: geert.verbeke@med.kuleuven.be

Geert Molenberghs
I-BioStat, Universiteit Hasselt, Diepenbeek, Belgium,
e-mail: geert.molenberghs@uhasselt.be

Dimitris Rizopoulos
Department of Biostatistics, Erasmus University Medical Center, Rotterdam
e-mail: d.rizopoulos@erasmusmc.nl

K. van Montfort et al., *Longitudinal Research with Latent Variables*,
DOI 10.1007/978-3-642-11760-2_2, © Springer-Verlag Berlin Heidelberg 2010

Section 2.3, an overview is presented of the most commonly used method for both Gaussian and non-Gaussian repeated measures.

Given their ubiquity, it is not surprising that methodology for repeated measures has emerged in a variety of fields. For example, Laird and Ware (1982) proposed the so-called linear mixed-effects models in a biometric context, whereas Goldstein (1979) proposed what is termed multilevel modeling in the framework of social sciences. Though the nomenclature is different, the underlying idea is the same: hierarchical data are modeled by introducing random coefficients, constant within a given level but changing across levels. Let us provide two examples. In a longitudinal context, where data are hierarchical because a given subject is measured repeatedly over time, a random effect is one that remains constant within a patient but changes across patients. A typical example of a multilevel setting consists of school children that are nested within classes which are, in turn, nested within schools. Random effects are then introduced to capture class-level as well as school-level variability. Examples abound in other fields as well. Methodology has been developed for continuous, Gaussian data, as well as for non-Gaussian settings, such as binary, count, and ordinal data. Overviews can be found in Verbeke and Molenberghs (2000) for the Gaussian case and in Molenberghs and Verbeke (2005) for the non-Gaussian setting.

In addition, a number of important contemporary extensions and issues will be discussed.

First, it is not uncommon for multiple repeated measures sequences to be recorded and analyzed simultaneously, leading to so-called multivariate longitudinal data. This poses specific methodological and computational challenges, especially when the problem is high-dimensional. An overview is presented in Section 2.4.

Second, it is quite common for longitudinal data to be collected in conjunction with time-to-event outcomes. An overview is presented in Section 2.5. Broadly, there are three main situations where this can occur: (a) The emphasis can be on the survival outcome with the longitudinal outcome(s) acting as a covariate process; (b) interest can be on both simultaneously, such as in the evaluation of surrogate markers in clinical studies, with a longitudinal marker for a time-to-event outcome; (c) the survival process can act, either in discrete or continuous time, as a dropout process on the longitudinal outcome.

The above considerations lead us to include a third main theme, surrogate marker evaluation, in Section 2.6, and a fourth and final theme, incomplete data, in Section 2.7.

2.2 Case Studies

2.2.1 Toenail Data

As a typical longitudinal example, we consider data from a randomized, double blind, parallel group, multicentre study for the comparison of 2 oral treatments (in the sequel coded as *A* and *B*) for toenail dermatophyte onychomycosis (TDO). We refer to De Backer *et al.* (1996) for more details about this study. TDO is a common toenail infection, difficult to treat, affecting more than two percent of the population. Antifungal compounds classically used for treatment of TDO need to be taken until the whole nail has grown out healthy. However, new compounds have reduced the treatment duration to three months. The aim of the present study was to compare the efficacy and safety of two such new compounds, labelled *A* and *B*, and administered during 12 weeks.

Table 2.1 Toenail Data. Number and percentage of patients with severe toenail infection, for each treatment arm separately

	Group A			Group B		
	# severe	# patients	percentage	# severe	# patients	percentage
Baseline	54	146	37.0%	55	148	37.2%
1 month	49	141	34.7%	48	147	32.6%
2 months	44	138	31.9%	40	145	27.6%
3 months	29	132	22.0%	29	140	20.7%
6 months	14	130	10.8%	8	133	6.0%
9 months	10	117	8.5%	8	127	6.3%
12 months	14	133	10.5%	6	131	4.6%

In total, 2×189 patients were randomized, distributed over 36 centres. Subjects were followed during 12 weeks (3 months) of treatment and followed further, up to a total of 48 weeks (12 months). Measurements were taken at baseline, every month during treatment, and every 3 months afterwards, resulting in a maximum of 7 measurements per subject. As a first response, we consider the unaffected nail length (one of the secondary endpoints in the study), measured from the nail bed to the infected part of the nail, which is always at the free end of the nail, expressed in *mm*. Obviously this response will be related to the toe size. Therefore, we will include here only those patients for which the target nail was one of the two big toenails. This reduces our sample under consideration to 146 and 148 subjects respectively. Individual profiles for 30 randomly selected subjects in each treatment group are shown in Figure 2.1. Our second outcome will be severity of the infection, coded as 0 (not severe) or 1 (severe). The question of interest was whether the percentage of severe infections decreased over time, and whether that evolution was different for the two treatment groups. A summary of the number of patients in the study at each time-point, and the number of patients with severe infections is given in Table 2.1.

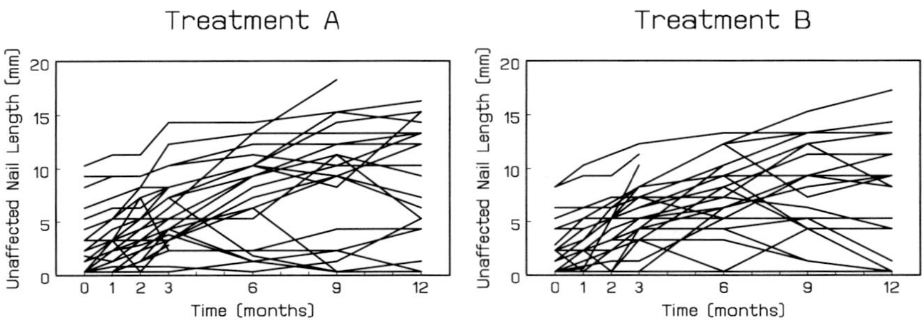

Fig. 2.1 Toenail Data. Individual profiles of 30 randomly selected subjects in each treatment arm.

A key issue in the analysis of longitudinal data is that outcome values measured repeatedly within the same subjects tend to be correlated, and this correlation structure needs to be taken into account in the statistical analysis. This is easily seen with paired observations obtained from, e.g., a pre-test/post-test experiment. An obvious choice for the analysis is the paired t-test, based on the subject-specific difference between the two measurements. While an unbiased estimate for the treatment effect can also be obtained from a two-sample t-test, standard errors and hence also p-values and confidence intervals obtained from not accounting for the correlation within pairs will not reflect the correct sampling variability, and hence still lead to wrong inferences. In general, classical statistical procedures assuming independent observations, cannot be used in the context of repeated measurements. In this chapter, we will give an overview of the most important models useful for the analysis of clinical trial data, and widely available through commercial statistical software packages.

2.2.2 Hearing Data

In a hearing test, hearing threshold sound pressure levels (dB) are determined at different frequencies to evaluate the hearing performance of a subject. A hearing threshold is the lowest signal intensity a subject can detect at a specific frequency. In this study, hearing thresholds measured at eleven different frequencies (125Hz, 250Hz, 500Hz, 750Hz, 1000Hz, 1500Hz, 2000Hz, 3000Hz, 4000Hz, 6000Hz and 8000Hz), obtained on 603 male participants from the Baltimore Longitudinal Study of Aging (BLSA, Shock *et al.* 1984), are considered. Hearing thresholds are measured at the left as well as at the right ear, leading to 22 outcomes measured repeatedly over time. The number of visits per subject varies from 1 to 15 (a median follow-up time of 6.9 years). Visits are unequally spaced. The age at first visit of the participants ranges from 17.2 to 87 years (with a median age at first visit of 50.2 years). Analyses of the hearing data collected in the BLSA study can be found in Brant and Fozard (1990), Morrell and Brant (1991), Pearson *et al.* (1995), Verbeke and Molenberghs (2000), and Fieuws and Verbeke (2006). It is well known that the hearing performance deteriorates as one gets older, which will be reflected by an increase in hearing threshold over time. The aim of our analysis will be to investigate whether this interaction between time and age is frequency related. Also of interest is to study the association between evolutions at different frequencies. Both questions can only be answered using a joint model for all 22 outcomes.

2.2.3 Liver Cirrhosis Data

As an illustrative example for the joint modeling of longitudinal and time-to-event data we consider data on 488 patients with histologically verified liver cirrhosis, collected in Copenhagen from 1962 to 1969 (Andersen *et al.* 1993). Liver cirrhosis is the condition in which the liver slowly deteriorates and malfunctions due to chronic injury. From the 488 patients, 251 were randomly assigned to receive prednisone and 237 placebo. Patients were scheduled to return at 3, 6, and 12 months, and yearly thereafter, and provide several biochemical values related to liver function. Our main research question here is to test for a treatment effect on survival after adjusting for one of these markers namely, the prothrombin index, which is indicative of the severity of liver fibrosis. Since the prothrombin levels are in fact the output of a stochastic process generated by the patients and is only available at the specific visit times the patients came to the study center, it constitutes a typical example of time-dependent covariate measured intermittently and with error.

2.2.4 Orthodontic Growth Data

Consider the orthodontic growth data, introduced by Potthoff and Roy (1964) and used by Jennrich and Schluchter (1986) as well. The data have the typical structure of a clinical trial and are simple yet illustrative. They contain growth measurements for 11 girls and 16 boys. For each subject, the distance from the center of the pituitary to the maxillary fissure was recorded at ages 8, 10, 12, and 14. Figure 2.2 presents the 27 individual profiles. Little and Rubin (2002) deleted 9 of the $[(11+16)\times 4]$ measurements, rendering 9 incomplete subjects which, even though a somewhat unusual practice, has the advantage of allowing a comparison between the incomplete data methods and the analysis of the original, complete data. Deletion is confined to the age 10 measurements and rougly speaking the complete observations at age 10 are those with a higher measurement at age 8. We will put some emphasis on ages 8 and 10, the typical dropout setting, with age 8 fully observed and age 10 partially missing.

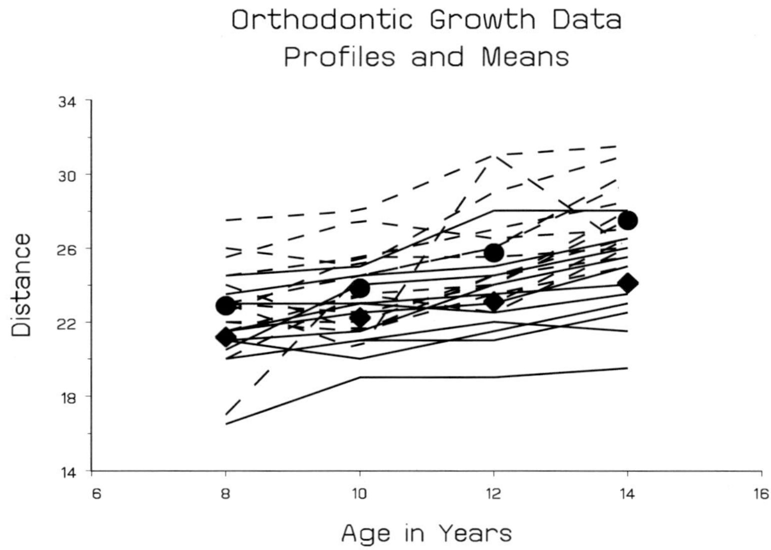

Fig. 2.2 Orthodontic Growth Data. Orthodontic Growth Data. Raw profiles and sample means (girls are indicated with solid lines and diamonds; boys are indicated with dashed lines and bullets).

2.2.5 Age-related Macular Degeneration Trial

These data arise from a randomized multi-center clinical trial comparing an experimental treatment (interferon-α) to a corresponding placebo in the treatment of patients with age-related macular degeneration. In this chapter we focus on the comparison between placebo and the highest dose (6 million units daily) of interferon-α (Z), but the full results of this trial have been reported elsewhere (Pharmacological Therapy for Macular Degeneration Study Group 1997). Patients with macular degeneration progressively lose vision. In the trial, the patients' visual acuity was assessed at different time points (4 weeks, 12 weeks, 24 weeks, and 52 weeks) through their ability to read lines of letters on standardized vision charts. These charts display lines of 5 letters of decreasing size, which the patient must read from top (largest letters) to bottom (smallest letters). The raw patient's visual acuity is the total number of letters correctly read. In addition, one often refers to each line with at least 4 letters correctly read as a 'line of vision.'

Table 2.2 shows the visual acuity (mean and standard error) by treatment group at baseline, at 6 months, and at 1 year. Visual acuity can be measured in several ways. First, one can record the number of letters read. Alternatively, dichotomized versions (at least 3 lines of vision lost, or at least 3 lines of vision lost) can be used as well. Therefore, these data will be useful to illustrate methods for the joint modeling of continuous and binary outcomes, with or without taking the longitudinal nature into account. In addition, though there are 190 subjects with both month 6 and month

Table 2.2 The Age-related Macular Degeneration Trial. Mean (standard error) of visual acuity at baseline, at 6 months and at 1 year according to randomized treatment group (placebo *versus* interferon-α)

Time point	Placebo	Active	Total
Baseline	55.3 (1.4)	54.6 (1.3)	55.0 (1.0)
6 months	49.3 (1.8)	45.5 (1.8)	47.5 (1.3)
1 year	44.4 (1.8)	39.1 (1.9)	42.0 (1.3)

12 measurements available, the total number of longitudinal profiles is 240, but for only 188 of these have the four follow-up measurements been made.

Thus indeed, 50 incomplete subjects could be considered for analysis as well. Both intermittent missingness as well as dropout occurs. An overview is given in Table 2.3. Thus, 78.33% of the profiles are complete, while 18.33% exhibit monotone missingness. Out of the latter group, 2.5% or 6 subjects have no follow-up measurements. The remaining 3.33%, representing 8 subjects, have intermittent missing values. Thus, as in many of the examples seen already, dropout dominates intermediate patterns as the source of missing data

Table 2.3 The Age-related Macular Degeneration Trial. Overview of missingness patterns and the frequencies with which they occur. 'O' indicates observed and 'M' indicates missing

Measurement occasion					
4 wks	12 wks	24 wks	52 wks	Number	%
Completers					
O	O	O	O	188	78.33
Dropouts					
O	O	O	M	24	10.00
O	O	M	M	8	3.33
O	M	M	M	6	2.50
M	M	M	M	6	2.50
Non-monotone missingness					
O	O	M	O	4	1.67
O	M	M	O	1	0.42
M	O	O	O	2	0.83
M	O	M	M	1	0.42

2.3 Modeling Tools for Longitudinal Data

In many branches of science, studies are often designed to investigate changes in a specific parameter which is measured repeatedly over time in the participating subjects. Such studies are called longitudinal studies, in contrast to cross-sectional studies where the response of interest is measured only once for each individual. As pointed out by Diggle *et al.* (2002) one of the main advantages of longitudinal studies is that they can distinguish changes over time within individuals (longitudinal effects) from differences among people in their baseline values (cross-sectional effects).

In randomized clinical trials, for example, where the aim usually is to compare the effect of two (or more) treatments at a specific time-point, the need and advantage of taking repeated measures is at first sight less obvious. Indeed, a simple comparison of the treatment groups at the end of the follow-up period is often sufficient to establish the treatment effect(s) (if any) by virtue of the randomization. However, in some instances, it is important to know how the patients have reached their endpoint, i.e., it is necessary to compare the average profiles (over time) between the treatment groups. Furthermore, longitudinal studies can be more powerful than studies evaluating the treatments at one single time-point. Finally, follow-up studies more often than not suffer from dropout, i.e., some patients leave the study prematurely, for known or unknown reasons. In such cases, a full repeated measures analysis will help in drawing inferences at the end of the study. Given that incompleteness usually occurs for reasons outside of the control of the investigators and may be related to the outcome measurement of interest, it is generally necessary to reflect on the process governing incompleteness. Only in special but important cases is it possible to ignore the missingness process.

When patients are examined repeatedly, missing data can occur for various reasons and at various visits. When missing data result from patient dropout, the missing data pattern is *monotone* pattern. *Non-monotone* missingness occurs when there are intermittent missing values as well. Our focus will be on dropout. We will return to the missing data issue in Section 2.7. We are now in a position to discuss first a key modeling tool for Gaussian longitudinal data, where after we will switch to the non-Gaussian case.

2.3.1 Linear Models for Gaussian Data

With repeated Gaussian data, a general, and very flexible, class of parametric models is obtained from a random-effects approach. Suppose that an outcome Y is observed repeatedly over time for a set of people, and suppose that the individual trajectories are of the type shown in Figure 2.3. Obviously, a linear regression model with intercept and linear time effect seems plausible to describe the data of each person separately. However, different people tend to have different intercepts and different slopes. One can therefore assume that the jth outcome Y_{ij} of subject i ($i = 1,\ldots,N$, $j = 1,\ldots,n_i$), measured at time t_{ij} satisfies $Y_{ij} = \tilde{b}_{i0} + \tilde{b}_{i1}t_{ij} + \varepsilon_{ij}$. Assuming the vector $\tilde{b}_i = (\tilde{b}_{i0}, \tilde{b}_{i1})^\top$ of person-specific parameters to be bivariate normal with mean $(\beta_0, \beta_1)^\top$ and 2×2 covariance matrix D and assuming ε_{ij} to be normal as well, this leads to a so-called linear mixed model. In practice, one will often formulate the model as

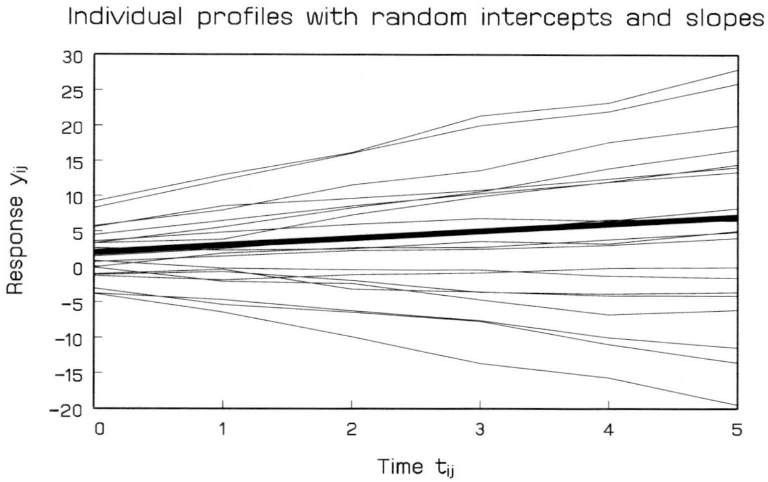

Fig. 2.3 Hypothetical example of continuous longitudinal data which can be well described by a linear mixed model with random intercepts and random slopes. The thin lines represent the observed subject-specific evolutions. The bold line represents the population-averaged evolution. Measurements are taken at six time-points 0, 1, 2, 3, 4, 5.

$$Y_{ij} = (\beta_0 + b_{i0}) + (\beta_1 + b_{i1})t_{ij} + \varepsilon_{ij},$$

with $\tilde{b}_{i0} = \beta_0 + b_{i0}$ and $\tilde{b}_{i1} = \beta_1 + b_{i1}$, and the new random effects $b_i = (b_{i0}, b_{i1})^\top$ are now assumed to have mean zero. The above model is a special case of the general linear mixed model which assumes that the outcome vector Y_i of all n_i outcomes for subject i satisfies

$$Y_i = X_i\beta \; + \; Z_i b_i \; + \; \varepsilon_i, \tag{2.1}$$

in which β is a vector of population-average regression coefficients, called fixed effects, and where b_i is a vector of subject-specific regression coefficients. The b_i are assumed normal with mean vector $\mathbf{0}$ and covariance D, and they describe how the evolution of the ith subject deviates from the average evolution in the population. The matrices X_i and Z_i are $(n_i \times p)$ and $(n_i \times q)$ matrices of known covariates. Note that p and q are the numbers of fixed and subject-specific regression parameters in the model, respectively. The residual components ε_i are assumed to be independent $N(0, \Sigma_i)$, where Σ_i depends on i only through its dimension n_i.

Estimation of the parameters in (2.1) is usually based on maximum likelihood (ML) or restricted maximum likelihood (REML) estimation for the marginal distribution of Y_i which can easily be seen to be

$$Y_i \sim N(X_i\beta, Z_i D Z_i^\top + \Sigma_i). \tag{2.2}$$

Note that model (2.1) implies a model with very specific mean and covariance structures, which may or may not be valid, and hence needs to be checked for every specific data set at hand. Note also that, when $\Sigma_i = \sigma^2 I_{n_i}$, with I_{n_i} equal to the identity matrix of dimension n_i, the observations of subject i are independent conditionally on the random effect b_i. The model is therefore called the conditional-independence model. Even in this simple case, the assumed random-effects structure still imposes a marginal correlation structure for the outcomes Y_{ij}. Indeed, even if all Σ_i equal $\sigma^2 I_{n_i}$, the covariance matrix in (2.2) is not a diagonal matrix, illustrating that, marginally, the repeated measurements Y_{ij} of subject i are not assumed to be uncorrelated. Another special case arises when the random effects are omitted from the model. In that case, the covariance matrix of Y_i is modeled through the residual covariance matrix Σ_i. In the case of completely balanced data, i.e., when n_i is the same for all subjects, and when the measurements are all taken at fixed time points, one can assume all Σ_i to be equal to a general unstructured covariance matrix Σ, which results in the classical multivariate regression model. Inference in the marginal model can be done using classical techniques including approximate Wald tests, t-tests, F-tests, or likelihood ratio tests. Finally, Bayesian methods can be used to obtain 'empirical Bayes estimates' for the subject-specific parameters b_i in (2.1). We refer to Henderson *et al.* (1959), Harville (1974, 1976, 1977), Laird and Ware (1982), Verbeke and Molenberghs (2000), and Fitzmaurice, Laird, and Ware (2004) for more details about estimation and inference in linear mixed models.

2.3.2 Models for Discrete Outcomes

Whenever discrete data are to be analyzed, the normality assumption in the models in the previous section is no longer valid, and alternatives need to be considered. The classical route, in analogy to the linear model, is to specify the full joint distribution for the set of measurements $Y_{ij}, \ldots, Y_{in_i}$ per individual. Clearly, this implies the need to specify all moments up to order n_i. Examples of marginal models can be found in Bahadur (1961), Altham (1978), Efron (1986), Molenberghs and Lesaffre (1994, 1999), Lang and Agresti (1994), and Fahrmeir and Tutz (2001).

Especially for longer sequences and/or in cases where observations are not taken at fixed time points for all subjects, specifying a full likelihood, as well as making inferences about its parameters, traditionally done using maximum likelihood principles, can become very cumbersome. Therefore, inference is often based on a likelihood obtained from a random-effects approach. Associations and all higher-order moments are then implicitly modeled through a random-effects structure. This will be discussed in Section 2.3.2.1. A disadvantage is that the assumptions about all moments are made implicitly, and therefore very hard to check. As a consequence, alternative methods have been in demand, which require the specification of a small number of moments only, leaving the others completely unspecified. In a large number of cases, one is primarily interested in the mean structure, whence only the first moments need to be specified. Sometimes, there is also interest in the association structure, quantified, for example, using odds ratios or correlations. Estimation is then based on so-called generalized estimating equations, and inference no longer directly follows from maximum likelihood theory. This will be explained in Section 2.3.2.2. A comparison of both techniques will be presented in Section 2.3.2.3. In Section 2.3.3, both approaches will be illustrated in the context of the toenail data.

2.3.2.1 Generalized Linear Mixed Models (GLMM)

As discussed in Section 2.3.1, random effects can be used to generate an association structure between repeated measurements. This can be exploited to specify a full joint likelihood in the context of discrete outcomes. More specifically, conditionally on a vector b_i of subject-specific regression coefficients, it is assumed that all responses Y_{ij} for a single subject i are independent, satisfying a generalized linear model with mean $\mu_{ij} = g({x_{ij}}^\top \beta + {z_{ij}}^\top b_i)$ for a pre-specified link function $g(\cdot)$, and for two vectors x_{ij} and z_{ij} of known covariates belonging to subject i at the jth time point. Let $f_{ij}(y_{ij}|b_i)$ denote the corresponding density function of Y_{ij}, given b_i. As for the linear mixed model, the random effects b_i are assumed to be sampled from a normal distribution with mean vector 0 and covariance D. The marginal distribution of Y_i is then given by

$$f(y_i) = \int \prod_{j=1}^{n_i} f_{ij}(y_{ij}|b_i) f(b_i) db_i, \tag{2.3}$$

in which dependence on the parameters β and D is suppressed from the notation. Assuming independence accross subjects, the likelihood can easily be obtained, and maximum likelihood estimation becomes available.

In the linear model, the integral in (2.3) could be worked out analytically, leading to the normal marginal model (2.2). In general, however, this is no longer possible, and numerical approximations are needed. Broadly, we can distinguish between approximations to the integrand in (2.3), and methods based on numerical integration. In the first approach, Taylor series expansions to the integrand are used, simplifying the calculation of the integral. Depending on the order of expansion and the point around which one expands, slightly different procedures are obtained. We refer to Breslow and Clayton (1993), Wolfinger and O'Connell (1993), Molenberghs and Verbeke (2005), and Fitzmaurice, Laird, and Ware (2004) for an overview of estimation methods. In general, such approximations will be accurate whenever the responses y_{ij} are 'sufficiently continuous' and/or if all n_i are sufficiently large. This explains why the approximation methods perform poorly in cases with binary repeated measurements, with a relatively small number of repeated measurements available for all subjects (Wolfinger 1998). Especially in such examples, numerical integration proves very useful. Of course, a wide toolkit of numerical integration tools, available from the optimization literature, can be applied. A general class of quadrature rules selects a set of abscissas and constructs a weighted sum of function evaluations over those. We refer to Hedeker and Gibbons (1994, 1996) and to

Pinheiro and Bates (2000) for more details on numerical integration methods in the context of random-effects models.

2.3.2.2 Generalized Estimating Equations (GEE)

Liang and Zeger (1986) proposed so-called generalized estimating equations (GEE) which require only the correct specification of the univariate marginal distributions provided one is willing to adopt 'working' assumptions about the association structure. More specifically, a generalized linear model (McCullagh and Nelder 1989) is assumed for each response Y_{ij}, modeling the mean μ_{ij} as $g(x_{ij}{}^{\top}\beta)$ for a pre-specified link function $g(\cdot)$, and a vector x_{ij} of known covariates. In case of independent repeated measurements, the classical score equations for the estimation of β are well known to be

$$S(\beta) = \sum_i \frac{\partial \mu_i^{\top}}{\partial \beta} V_i^{-1} (Y_i - \mu_i) = 0, \tag{2.4}$$

where $\mu_i = \mathrm{E}(Y_i)$ and V_i is a diagonal matrix with $v_{ij} = \mathrm{Var}(Y_{ij})$ on the main diagonal. Note that, in general, the mean-variance relation in generalized linear models implies that the elements v_{ij} also depend on the regression coefficients β. Generalized estimating equations are now obtained from allowing non-diagonal 'covariance' matrices V_i in (2.4). In practice, this comes down to the specification of a 'working correlation matrix' which, together with the variances v_{ij}, results in a hypothesized covariance matrix V_i for Y_i.

Solving $S(\beta) = 0$ is done iteratively, constantly updating the working correlation matrix using moment-based estimators. Note that, in general, no maximum likelihood estimates are obtained, since the equations are not first-order derivatives of some log-likelihood function. Still, very similar properties can be derived. More specifically, Liang and Zeger (1986) showed that $\widehat{\beta}$ is asymptotically normally distributed, with mean β and with a covariance matrix that can easily be estimated in practice. Hence, classical Wald-type inferences become available. This result holds provided that the mean was correctly specified, whatever working assumptions were made about the association structure. This implies that, strictly speaking, one can fit generalized linear models to repeated measurements, ignoring the correlation structure, as long as inferences are based on the standard errors that follow from the general GEE theory. However, efficiency can be gained from using a more appropriate working correlation model (Mancl and Leroux 1996).

The original GEE approach focuses on inferences for the first-order moments, considering the association present in the data as nuisance. Later on, extensions have been proposed which also allow inferences about higher-order moments. We refer to Prentice (1988), Lipsitz, Laird and Harrington (1991), and Liang, Zeger and Qaqish (1992) for more details on this.

2.3.2.3 Marginal versus Hierarchical Parameter Interpretation

Comparing the GEE results and the GLMM results in Table 2.4, we observe large differences between the corresponding parameter estimates. This suggests that the parameters in both models have a different interpretation. Indeed, the GEE approach yields parameters with a population-averaged interpretation. Each regression parameter expresses the average effect of a covariate on the probability of having a severe infection. Results from the generalized linear mixed model, however, require an interpretation conditionally on the random effect, i.e., conditionally on the subject. In the context of our toenail example, consider model (2.7) for treatment group A only. The model assumes that the probability of severe infection satisfies a logistic regression model, with the same slope for all subjects, but with subject-specific intercepts. The population-averaged probability of severe infection is obtained from averaging these subject-specific profiles over all subjects. This is graphically presented in Figure 2.4. Clearly, the slope of the average trend is different from the subject-specific slopes, and this effect will be more severe as the subject-specific profiles differ more, i.e., as the random-intercepts variance σ^2 is larger. Formally, the average trend for group A is obtained as

$$
\begin{aligned}
P(Y_i(t)=1) \;=\; E\left[P(Y_i(t)=1|b_i)\right] &= E\left[\frac{\exp(\beta_{A0}+b_i+\beta_{A1}t)}{1+\exp(\beta_{A0}+b_i+\beta_{A1}t)}\right] \\
&\neq E\left[\frac{\exp(\beta_{A0}+\beta_{A1}t)}{1+\exp(\beta_{A0}+\beta_{A1}t)}\right].
\end{aligned}
$$

Hence, the population-averaged evolution is not the evolution for an 'average' subject, i.e., a subject with random effect equal to zero. The second graph in Figure 2.6 shows the fitted profiles for an average subject in each treatment group, and these profiles are indeed very different from the population-averaged profiles shown in the first graph of Figure 2.6 and discussed before. In general, the population-averaged evolution implied by the GLMM is not of a logistic form any more, and the parameter estimates obtained from the GLMM are typically larger in absolute value than their marginal counterparts (Neuhaus, Kalbfleisch, and Hauck 1991). However, one should not refer to this phenomenon as bias given that the two sets of parameters target at different scientific questions. Observe that this difference in parameter interpretation between marginal and random-effects models immediately follows from their non-linear nature, and therefore is absent in the linear mixed model, discussed in Section 2.3.1. Indeed, the regression parameter vector β in the linear mixed model (2.1) is the same as the regression parameter vector modeling the expectation in the marginal model (2.2).

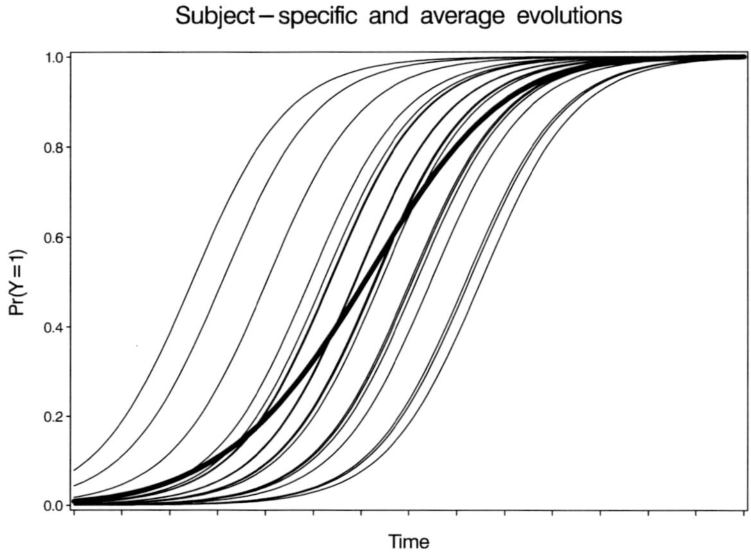

Fig. 2.4 Graphical representation of a random-intercepts logistic model. The thin lines represent the subject-specific logistic regression models. The bold line represents the population-averaged evolution.

2.3.3 Analysis of Toenail Data

As an illustration, we analyze unaffected nail length response in the toenail example. The model proposed by Verbeke, Lesaffre, and Spiessens (2001) assumes a quadratic evolution for each subject, with subject-specific intercepts, and with correlated errors within subjects. More formally, they assume that Y_{ij} satisfies

$$Y_{ij}(t) = \begin{cases} (\beta_{A0} + b_i) + \beta_{A1}t + \beta_{A2}t^2 + \varepsilon_i(t), & \text{in group A} \\ (\beta_{B0} + b_i) + \beta_{B1}t + \beta_{B2}t^2 + \varepsilon_i(t), & \text{in group B,} \end{cases} \tag{2.5}$$

where $t = 0, 1, 2, 3, 6, 9, 12$ is the number of months since randomization. The error components $\varepsilon_i(t)$ are assumed to have common variance σ^2, with correlation of the form $\text{corr}(\varepsilon_i(t), \varepsilon_i(t-u)) = \exp(-\varphi u^2)$ for some unknown parameter φ. Hence, the correlation between within-subject errors is a decreasing function of the time span between the corresponding measurements. Fitted average profiles are shown in Figure 2.5. An approximate F-test shows that, on average, there is no evidence for a treatment effect ($p = 0.2029$). Note that, even when interest would only be in comparing the treatment groups after 12 months, this could still be done based on the above fitted model. The average difference between group A and group B, after 12 months, is given by $(\beta_{A0} - \beta_{B0}) - 12(\beta_{A1} - \beta_{B1}) + 12^2(\beta_{A2} - \beta_{B2})$. The estimate for this difference equals 0.80 mm ($p = 0.0662$). Alternatively, a two-sample t-test could be performed based on those subjects that have completed the study. This yields an

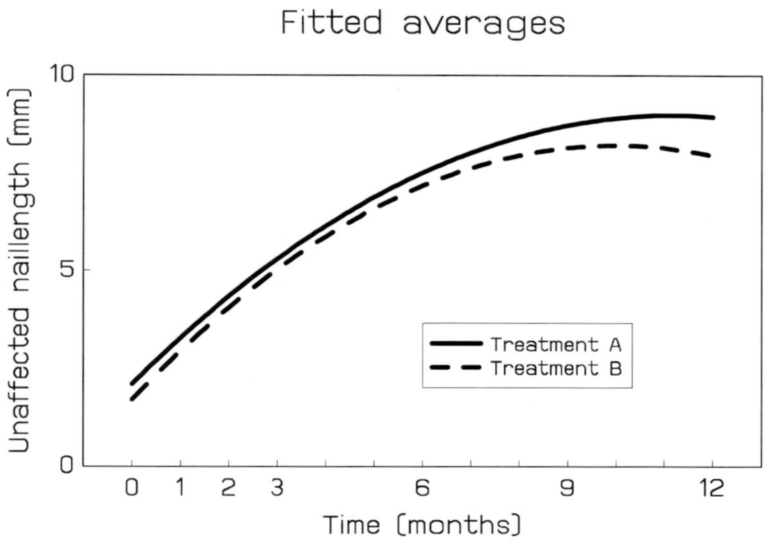

Fig. 2.5 Toenail Data. Fitted average profiles based on model (2.5).

estimated treatment effect of 0.77 *mm* ($p = 0.2584$) illustrating that modeling the whole longitudinal sequence also provides more efficient inferences at specific time-points.

As an illustration of GEE and GLMM, we analyze the binary outcome 'severity of infection' in the toenail study. We will first apply GEE, based on the marginal logistic regression model

$$\log\left[\frac{P(Y_i(t)=1)}{1-P(Y_i(t)=1)}\right] = \begin{cases} \beta_{A0}+\beta_{A1}t, & \text{in group A} \\ \beta_{B0}+\beta_{B1}t, & \text{in group B.} \end{cases} \tag{2.6}$$

Furthermore, we use an unstructured 7×7 working correlation matrix. The results are reported in Table 2.4, and the fitted average profiles are shown in the top graph of Figure 2.6. Based on a Wald-type test we obtain a significant difference in the average slope between the two treatment groups ($p = 0.0158$).

Table 2.4 Toenail Data. Parameter estimates (standard errors) for a generalized linear mixed model (GLMM) and a marginal model (GEE)

	GLMM	GEE
Parameter	Estimate (s.e.)	Estimate (s.e.)
Intercept group A (β_{A0})	−1.63 (0.44)	−0.72 (0.17)
Intercept group B (β_{B0})	−1.75 (0.45)	−0.65 (0.17)
Slope group A (β_{A1})	−0.40 (0.05)	−0.14 (0.03)
Slope group B (β_{B1})	−0.57 (0.06)	−0.25 (0.04)
Random intercepts s.d. (σ)	4.02 (0.38)	

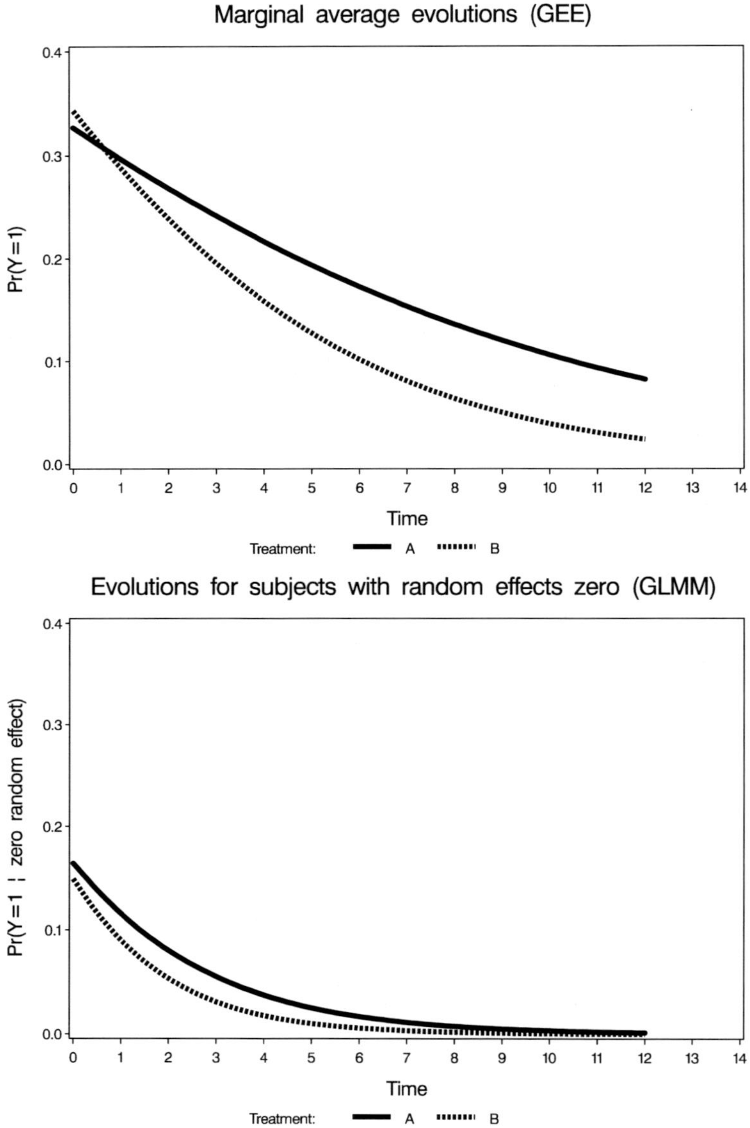

Fig. 2.6 Toenail Data. Treatment-specific evolutions. (a) Marginal evolutions as obtained from the marginal model (2.6) fitted using GEE, (b) Evolutions for subjects with random effects in model (2.7) equal to zero.

Alternatively, we consider a generalized linear mixed model, modeling the association through the inclusion of subject-specific, i.e., random, intercepts. More specifically, we will now assume that

$$\log\left[\frac{P(Y_i(t)=1|b_i)}{1-P(Y_i(t)=1|b_i)}\right] = \begin{cases} \beta_{A0}+b_i+\beta_{A1}t, & \text{in group A} \\ \beta_{B0}+b_i+\beta_{B1}t, & \text{in group B} \end{cases} \tag{2.7}$$

with b_i normally distributed with mean 0 and variance σ^2. The results, obtained using numerical integration methods, are also reported in Table 2.4. As before, we obtain a significant difference between β_{A1} and β_{B1} ($p = 0.0255$).

2.4 Multivariate Longitudinal Data

So far, we have considered a single, repeatedly measured outcome. However, often one observes more than one outcome at the same time, which is essentially known as multivariate outcomes. These can all be of the same data type, e.g., all Gaussian or all binary, or of a mixed type, e.g., when the outcome vector is made up of continuous and binary components. Statistical problems where various outcomes of a mixed nature are observed have been around for about half a century and are rather common at present. Many research questions can often only fully be addressed in a joint analysis of all outcomes simultaneously. For example, the association structure can be of direct scientific relevance.

It is definitely possible for all of these features to occur simultaneously, whereby a multivariate outcome vector, possible of a mixed nature, is measured repeatedly over time. An array of research questions can then be addressed in this way. A possible question might be how the association between outcomes evolves over time or how outcome-specific evolutions are related to each other (Fieuws and Verbeke 2004). Another example is discriminant analysis based on multiple, longitudinally measured, outcomes. Third, interest may be in the comparison of average trends for different outcomes. As an example, consider testing the difference in evolution between many outcomes or joint testing of a treatment effect on a set of outcomes. All of these situations require a joint model for all outcomes.

Let us focus, for a moment, on the combined analysis of a continuous and a discrete outcome. There then broadly are three approaches. The first one postulates a marginal model for the binary outcome and then formulates a conditional model for the continuous outcome, given the categorical one. For the former, one can use logistic regression, whereas for the latter conditional normal models are a straightforward choice, i.e., a normal model with the categorical outcome used as a covariate (Tate 1954). The second family starts from the reverse factorization, combining a marginal model for the continuous outcome with a conditional one for the categorical outcome. Conditional models have been discussed by Cox and Wermuth (1992, 1994a, 1994b), Krzanowski (1988), and Little and Schluchter (1985). Schafer (1997) presents a so-called *general location model* where a number of continuous and binary outcomes can be modeled together. The third model family directly formulates a joint model for the two outcomes. In this context, one often starts from a bivariate continuous variable, one component of which is explicitly observed and the other one observed in dichotomized, or generally discretized, version only

(Tate 1955). Molenberghs, Geys, and Buyse (2001) presented a model based on a Plackett-Dale approach, where a bivariate Plackett distribution is assumed, of which one margin is directly observed and the other one only after dichotomization. General multivariate exponential family based models have been proposed by Prentice and Zhao (1991), Zhao, Prentice, and Self (1992), and Sammel, Ryan, and Legler (1997).

Of course, these developments have not been limited to bivariate joint outcomes. One can obviously extend these ideas and families to a multivariate continuous outcome and/or a multivariate categorical outcome. For the first and second families, one then starts from conditional and marginal multivariate normal and appropriately chosen multinomial models. Such a model within the first family has been formulated by Olkin and Tate (1961). Within the third family, models were formulated by Hannan and Tate (1965) and Cox (1974) for a multivariate normal with a univariate bivariate or discrete variable.

As alluded to before, apart from an extension from the bivariate to the multivariate case, one can introduce other hierarchies as well. We will now assume that each of the outcomes may be measured repeatedly over time, and there could even be several repeated outcomes in both the continuous and the categorical subgroup. A very specific hierarchy stems from clustered data, where a continuous and a categorical, or several of each, are observed for each member of a family, a household, a cluster, etc. For the specific context of developmental toxicity studies, often conducted in rats and mice, a number of developments have been made. An overview of such methods, together with developments for probit-normal and Plackett-Dale based models, was presented in Regan and Catalano (2002). Catalano and Ryan (1992) and Fitzmaurice and Laird (1995) propose models for a combined continuous and discrete outcome, but differ in the choice of which outcome to condition on the other one. Both use generalized estimating equations to allow for clustering. Catalano (1997) extended the model by Catalano and Ryan (1992) to accommodate ordinal variables. An overview can be found in Aerts *et al* (2002).

Regan and Catalano (1999a) proposed a probit-type model to accommodate joint continuous and binary outcomes in a clustered data context, thus extending the correlated probit model for binary outcomes (Ochi and Prentice 1984) to incorporate continuous outcomes. Molenberghs, Geys, and Buyse (2001) used a Plackett latent variable to the same effect, extending the bivariate version proposed by Molenberghs, Geys, and Buyse (2001). Estimation in such hierarchical joint models can be challenging. Regan and Catalano (1999a) proposed maximum likelihood, but considered GEE as an option too (Regan and Catalano 1999b). Geys, Molenberghs, and Ryan (1999) made use of pseudo-likelihood. Ordinal extensions have been proposed in Regan and Catalano (2000).

Thus, many applications of this type of joint models can already be found in the statistical literature. For example, the approach has been used in a non-longitudinal setting to validate surrogate endpoints in meta-analyses (Buyse *et al.* 2000, Burzykowski *et al.* 2001) or to model multivariate clustered data (Thum 1997). Gueorguieva (2001) used the approach for the joint modeling of a continuous and a binary outcome measure in a developmental toxicity study on mice.

Also in a longitudinal setting, Chakraborty *et al.* (2003) obtained estimates of the correlation between blood and semen HIV-1 RNA by using a joint random-effects model. Other examples with longitudinal studies can be found in MacCallum *et al.* (1997), Thiébaut *et al.* (2002) and Shah, Laird, and Schoenfeld (1997). All of these examples refer to situations where the number of different outcomes is relatively low. Although the model formulation can be done irrespective of the number of outcomes to be modeled jointly, standard fitting procedures, such as maximum likelihood estimation, is only feasible when the dimension is sufficiently low or if one is willing to make a priori strong assumptions about the association between the various outcomes. An example of the latter can be found in situations where the corresponding random effects of the various outcomes are assumed to be perfectly correlated (Oort 2001, Sivo 2001, Roy and Lin 2000, and Liu and Hedeker 2006). Fieuws and Verbeke (2006) have developed a model-fitting procedure that is applicable, irrespective of the dimensionality of the problem. This is the route that will be followed in the next sections.

2.4.1 A Mixed Model for Multivariate Longitudinal Outcomes

A flexible joint model that can handle any number of outcomes measured longitudinally, without any restriction to the nature of the outcomes can be obtained by modeling each outcome separately using a mixed model (linear, generalized linear, or non-linear), by assuming that, conditionally on these random effects, the different outcomes are independent, and by imposing a joint multivariate distribution on the vector of all random effects. This approach has many advantages and is applicable in a wide variety of situations. First, the data can be highly unbalanced. For example, it is not necessary that all outcomes are measured at the same time points. Moreover, the approach is applicable for combining linear mixed models, non-linear mixed models, or generalized linear mixed models. The procedure also allows the combination of different types of mixed models, such as a generalized linear mixed model for a discrete outcome and a non-linear mixed model for a continuous outcome.

Let m be the dimension of the problem, i.e., the number of outcomes that need to be modeled jointly. Further, let Y_{rij} denote the jth measurement taken on the ith subject, for the rth outcome, $i = 1,\ldots,N$, $r = 1,\ldots,m$, and $j = 1,\ldots,n_{ri}$. Note that we do not assume that the same number of measurements is available for all subjects, nor for all outcomes. Let Y_{ri} be the vector of n_{ri} measurements taken on subject i, for outcome r. Our model assumes that each Y_{ri} satisfies a mixed model. Let $f_{ri}(y_{ri}|b_{ri},\theta_r)$ be the density of Y_{ri}, conditional on a q_r-dimensional vector b_{ri} of random effects for the rth outcome on subject i. The vector θ_r contains all fixed effects and possibly also a scale parameter needed in the model for the rth outcome. Note that we do not assume the same type of model for all outcomes: A combination of linear, generalized linear, and non-linear mixed models is possible. It is also not assumed that the same number q_r of random effects is used for all m outcomes. Finally, the model is completed by assuming that the vector b_i of all random effects for subject i is multivariate normal with mean zero and covariance D, i.e.,

$$b_i = \begin{pmatrix} b_{1i} \\ b_{2i} \\ \vdots \\ b_{mi} \end{pmatrix} \sim N\left[\begin{pmatrix} 0 \\ 0 \\ \vdots \\ 0 \end{pmatrix}, \begin{pmatrix} D_{11} & D_{12} & \cdots & D_{1m} \\ D_{21} & D_{22} & \cdots & D_{2m} \\ \vdots & \vdots & \ddots & \vdots \\ D_{m1} & D_{m2} & \cdots & D_{mm} \end{pmatrix}\right].$$

The matrices D_{rs} represent the covariances between b_{ri} and b_{si}, $r,s = 1,\ldots,m$. Finally, D is the matrix with blocks D_{rs} as entries.

A special case of the above model is the so-called shared-parameter model, which assumes the same set of random effects for all outcomes. This clearly can be obtained as a special case of the above model by assuming perfect correlation between some of the random effects. The advantage of such shared-parameter models is the relatively low dimension of the random-effects distribution, when compared to the above model. The dimension of the random effects in shared parameter models does not increase with the number of outcomes to be modeled. In the above model, each new outcome added to the model introduces new random effects, thereby increasing the dimension of b_i. Although the shared-parameter models can reasonably easily be fitted using standard software, this is no longer the case for the model considered here. Estimation and inference under the above model will require specific procedures, which will be discussed in Section 2.4.2. A disadvantage of the shared-parameter model is that it is based on much stronger assumptions about the association between the outcomes, which may not be valid, especially in high-dimensional settings as considered in this chapter. Note also that, joining valid univariate mixed models does not necessarily lead to a correct joint model. Fieuws and Verbeke (2004) illustrate this in the context of linear mixed models for two continuous outcomes. It is shown how the joint model may imply association structures between the two sets of longitudinal profiles that may strongly depend on the actual parameterization of the individual models and that are not necessarily valid.

2.4.2 A Pairwise Model-fitting Approach

Whereas the modeling approach from the previous setting is rather versatile, it might become computationally cumbersome for high-dimensional applications. It is therefore useful to consider the approach of Fieuws and Verbeke (2006), when a large number of repeated sequences are to be analyzed simultaneously. The general idea is that all parameters in the full multivariate model can be identified from all pairwise models, i.e., all bivariate models for each pair of outcomes. Therefore, using pseudo-likelihood ideas, also termed pairwise or composite likelihood (Molenberghs and Verbeke 2005), fitting the full model is replaced by maximum likelihood estimation of each bivariate model separately. This can be done using standard statistical software. Afterwards, all results are appropriately combined, and Wald-type inferences become available from noticing that the pairwise fitting approach is equivalent to maximizing the sum of all the log-likelihoods from all fitted pairs. This sum can be interpreted as a pseudo-log-likelihood function, and inferences then immediately follow from the general pseudo-likelihood theory, as will now be explained in the following sections.

2.4.2.1 Pairwise Fitting

Let Ψ^* be the vector of all parameters in the multivariate joint mixed model for $(Y_1, Y_2, \ldots, Y_m)$. The pairwise fitting approach starts from fitting all $m(m-1)/2$ bivariate models, i.e., all joint models for all possible pairs

$$(Y_1,Y_2),(Y_1,Y_3),\ldots,(Y_1,Y_m),(Y_2,Y_3),\ldots,(Y_2,Y_m),\ldots,(Y_{m-1},Y_m)$$

of the outcomes $Y_1, Y_2, \ldots, Y_m$. Let the log-likelihood function corresponding to the pair (r,s) be denoted by $\ell(y_r, y_s|\Psi_{rs})$, and let Ψ_{rs} be the vector containing all parameters in the bivariate model for pair (r,s).

Let Ψ now be the stacked vector combining all $m(m-1)/2$ pair-specific parameter vectors Ψ_{rs}. Estimates for the elements in Ψ are obtained by maximizing each of the $m(m-1)/2$ log-likelihoods $\ell(y_r, y_s|\Psi_{rs})$ separately. It is important to realize that the parameter vectors Ψ and Ψ^* are not equivalent. Indeed, some parameters in Ψ^* will have a single counterpart in Ψ, e.g., the covariances between random effects of different outcomes. Other elements in Ψ^* will have multiple counterparts in Ψ, e.g., fixed effects from one single outcome. In the latter case, a single estimate for the corresponding parameter in Ψ^* is obtained by averaging all corresponding pair-specific estimates in $\widehat{\Psi}$. Standard errors of the so-obtained estimates clearly cannot be obtained from averaging standard errors or variances. Indeed, two pair-specific estimates corresponding to two pairwise models with a common outcome are based on overlapping information and hence correlated. This correlation should also be accounted for in the sampling variability of the combined estimates in $\widehat{\Psi}^*$. Correct asymptotic standard errors for the parameters in $\widehat{\Psi}$, and consequently in $\widehat{\Psi}^*$, can be obtained from pseudo-likelihood ideas.

2.4.2.2 Inference for Ψ

Fitting all bivariate models is equivalent to maximizing the function

$$p\ell(\Psi) \equiv p\ell(y_{1i}, y_{2i}, \ldots, y_{mi}|\Psi) = \sum_{r<s} \ell(y_r, y_s|\Psi_{rs}), \tag{2.8}$$

ignoring the fact that some of the vectors Ψ_{rs} have common elements, i.e., assuming that all vectors Ψ_{rs} are completely distinct. The function in (2.8) can be considered a pseudo-likelihood function, maximization of which leads to so-called pseudo-likelihood estimates, with well-known asymptotic statistical properties. We refer to Arnold and Strauss (1991) and Geys, Molenberghs, and Ryan (1997) for more details. Our application of pseudo-likelihood methodology is different from most other applications in the sense that the same parameter vector is usually present in the different parts of the pseudo-likelihood function. Here, the set of parameters in Ψ_{rs} is treated pair-specific, which allows separate maximization of each term in the pseudo log-likelihood function (2.8). In Section 2.4.2.3, we will account for the fact that Ψ_{rs} and $\Psi_{rs'}$, $s \neq s'$, are not completely distinct, as they share the parameters referring to the rth outcome.

It now follows directly from the general pseudo-likelihood theory that $\widehat{\Psi}$ asymptotically satisfies

$$\sqrt{N}(\widehat{\Psi} - \Psi) \approx N(0, I_0^{-1} I_1 I_0^{-1})$$

in which $I_0^{-1} I_1 I_0^{-1}$ is a 'sandwich-type' robust variance estimator, and where I_0 and I_1 can be constructed using first- and second-order derivatives of the components in (2.8). Strictly speaking, I_0 and I_1 depend on the unknown parameters in Ψ, but these are traditionally replaced by their estimates in $\widehat{\Psi}$.

2.4.2.3 Combining Information: Inference for Ψ^*

In a final step, estimates for the parameters in Ψ^* can be calculated, as suggested before, by taking averages of all the available estimates for that specific parameter. Obviously, this implies that $\widehat{\Psi}^* = A^\top \widehat{\Psi}$ for an appropriate weight matrix A. Hence, inference for the elements in $\widehat{\Psi}^*$ will be based on

$$\sqrt{N}(\widehat{\Psi}^* - \Psi^*) = \sqrt{N}(A^\top \widehat{\Psi} - A^\top \Psi) \approx N(0, A^\top I_0^{-1} I_1 I_0^{-1} A).$$

It can be shown that pseudo-likelihood estimates are less efficient than the full maximum likelihood estimates (Arnold and Strauss 1991). However, these results refer to efficiency for the elements in Ψ, not directly to the elements in Ψ^*. In general, the degree of loss of efficiency depends on the context, but Fieuws and Verbeke (2006) have presented evidence for only very small losses in efficiency in the present context of the pairwise fitting approach for multivariate random-effects models.

2.4.3 Analysis of the Hearing Data

Let $Y_{r,i}(t)$ denote the rth hearing threshold for subject i taken at time t, $r = 1, \ldots, 11$ for the right ear, and $r = 12, \ldots, 22$ for the left ear. Morrell and Brant (1991), and Pearson *et al.* (1995) have proposed the following linear mixed model to analyze the evolution of the hearing threshold for a single frequency:

$$\begin{aligned} Y_{r,i}(t) = &(\beta_{r,1} + \beta_{r,2}\text{Age}_i + \beta_{r,3}\text{Age}_i^2 + a_{r,i}) + \\ &+ (\beta_{r,4} + \beta_{r,5}\text{Age}_i + b_{r,i})t + \beta_{r,6}V_i(t) + \varepsilon_{r,i}(t). \end{aligned} \tag{2.9}$$

The time t is expressed in years from entry in the study and Age_i equals the age of subject i at the time of entry in the study. The binary time-varying covariate V_i represents a learning effect from the first to the subsequent visits. Finally, the $a_{r,i}$ are random intercepts, the $b_{r,i}$ are the random slopes for time, and the $\varepsilon_{r,i}$ represent the usual error components. The regression coefficients $\beta_{r,1},\ldots,\beta_{r,6}$ are fixed, unknown parameters. The 44 random effects $a_{1,i},a_{2,i},\ldots,a_{22,i},b_{1,i},b_{2,i},\ldots,b_{22,i}$ are assumed to follow a joint zero-mean normal distribution with covariance matrix D. At each

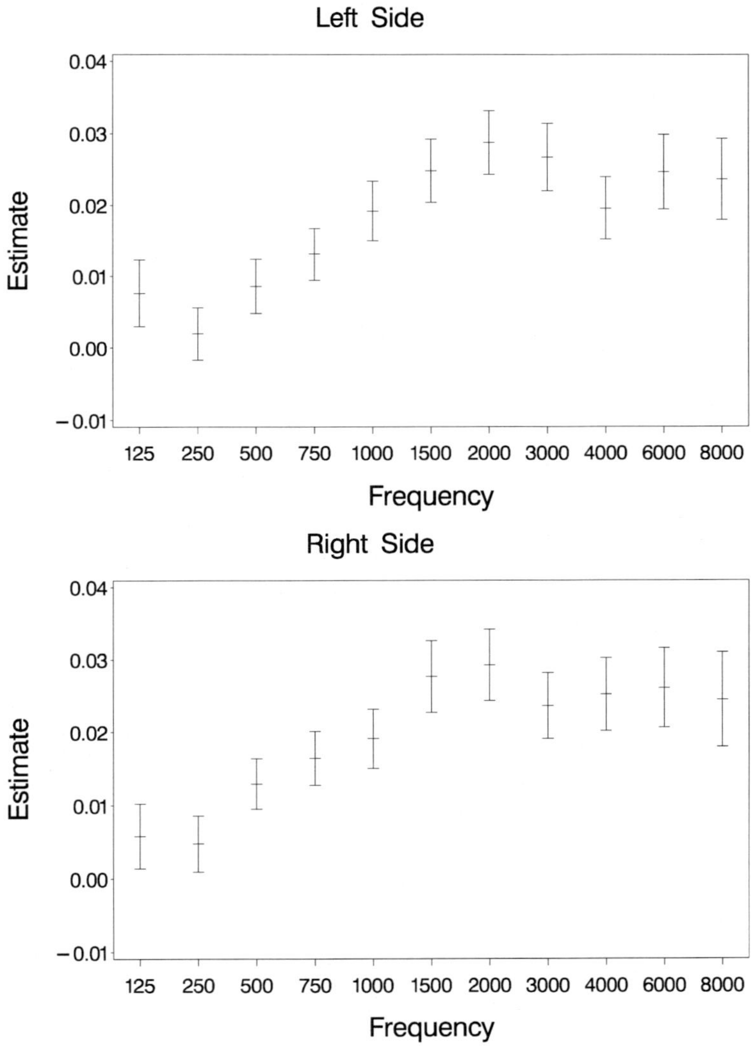

Fig. 2.7 Hearing Data. Estimates $\widehat{\beta}_{r,5}$ with associated 95% confidence intervals, for the measurements from left and right ear separately.

time point t, the error components $\varepsilon_{1,i},\ldots,\varepsilon_{22,i}$ follow a 22-dimensional zero-mean normal distribution with covariance matrix R. The total number of parameters in D and R equals $990+253=1243$.

We applied the pairwise approach to fit model (2.9) to the Hearing data introduced in Section 2.2.2. As discussed before, one of the key research questions is whether the deterioration of hearing ability with age is different for different frequencies, because this would yield evidence for selective deterioration. Formally, this requires testing the null-hypotheses $H_0: \beta_{1,5}=\beta_{2,5}=\ldots=\beta_{11,5}$ for the right side, and $H_0: \beta_{12,5}=\beta_{13,5}=\ldots=\beta_{22,5}$ for the left side. Figure 2.7 shows all estimates $\widehat{\beta}_{r,5}$ with associated 95% confidence intervals, for the left and right ear separately. We clearly observe an increasing trend implying that age accelerates hearing loss, but that this is more severe for higher frequencies. Wald-type tests indicate that these estimates are significantly different between the outcomes, at the left side ($\chi^2_{10}=90.4$, $p<0.0001$) as well as at the right side ($\chi^2_{10}=110.9$, $p<0.0001$).

2.4.4 Some Reflections

The advantage of this technique is that all implied univariate models belong to the well-known mixed model family. This implies that one can first model each outcome separately (with separate data exploration and model building), before joining the univariate models into the full multivariate model. Moreover, the parameters in the multivariate model keep their interpretation from the separate univariate models. Finally, this approach is sufficiently flexible to allow for different types of models for the different outcomes (linear, non-linear, generalized linear).

A disadvantage of the approach is that, when the number of outcomes becomes large, the dimension of the random effects can become too large to fit the full multivariate model using standard software for mixed models. Using results of Fieuws and Verbeke (2006), and Fieuws *et al.* (2006), we have explained how all parameters in the multivariate model can be estimated from fitting the model to all possible pairs of outcomes. Inferences follow from pseudo-likelihood theory. Although the estimates obtained from the pairwise approach do not maximize the full multivariate likelihood, they still have similar asymptotic properties, with no or only marginal loss of efficiency when compared to the maximum likelihood estimates. It should be emphasized that we do not advocate fitting multivariate models in order to gain efficiency for parameters in single univariate models. As long as no inferences are needed for combinations of parameters from different outcomes, and if no outcomes share the same parameters, univariate mixed models are by far the preferred tools for the analysis.

Fitting of the models can usually be done using standard software for the linear, non-linear, and generalized linear mixed models. Software is available from `http://med.kuleuven.be/biostat/software/software.htm/`and several examples from the book website `http://www.econ.upf.edu /~satorra/longitudinallatent/readme.html`. Calculation of the standard errors, however, requires careful data manipulation. In case all univariate mixed models are of the linear type (e.g., our model for the Hearing Data example), a SAS macro can be used.

2.5 Joint Models for Longitudinal and Time-to-Event Data

As we have seen earlier in this chapter, it is very common in longitudinal studies to collect measurements on several types of outcomes. In this section we focus on settings in which the outcomes recorded on the subjects simultaneously include a set of repeated measurements and the time at which an event of particular interest occurs, for instance, death, development of a disease or dropout from the study. Typical areas where such studies are encountered encompass HIV/AIDS and cancer studies. In HIV studies, seropositive patients are monitored until they develop AIDS or die, and they are regularly measured for the condition of their immune system using markers such as the CD4 lymphocyte count, the estimated viral load, or whether viral load is below detectable limits. Similarly, in cancer trials the event outcome is death or metastasis, while patients also provide longitudinal measurements of antibody levels or of other markers of carcinogenesis, such as the prostate specific antigen levels for prostate cancer.

Depending on the research questions, these two outcomes can be analyzed either separately or jointly. Here, we will focus on situations in which a joint analysis is required. This is typically the case when interest is on the event time and one wishes to account for the effect of the longitudinal outcome as a time-dependent covariate. Traditional approaches for analyzing time-to-event data, such as the partial likelihood for the Cox proportional hazards models, assume that the time-dependent covariate is a predictable process; that is, the value of this covariate at time point t is not affected by the occurrence of an event at time point u, with $t > u$ (Therneau and Grambsch, 2000, Sect. 1.3). For instance, age can be included as predictable time-dependent covariate in a standard analysis, because if we know the age of a subject at baseline, we can 'predict' her age at every time point without error. However, the type of time-dependent covariates encountered in longitudinal studies are often not predictable. In particular, they are the output of a stochastic process generated at the level of the subject, and it is directly related to the failure mechanism. The stochastic nature of these covariates complicates matters in two ways. First, we do not actually observe the 'true' values for these covariates, owing to the fact that the longitudinal responses usually contain measurement error. Second, we are only able to observe the, error-contaminated, values intermittently at the specific time points at which we have collected measurements and not at any time point t. These special features complicate analysis with the traditional partial likelihood approaches (Tsiatis, DeGruttola, and Wolfsohn 1995, Wulfsohn and Tsiatis 1997). Hence, to produce valid inferences, a model for the joint distribution of the longitudinal and survival outcomes is required instead.

Early attempts to tackle such problems considered using the last available value of the longitudinal outcome for each subject as a representative value for the complete longitudinal history. This method is also known as 'Last Value or Last Observation Carried Forward' (LVCF or LOCF, Molenberghs and Kenward 2007). Even though the simplicity of such an approach is apparent, Prentice (1982) showed that it leads to severe bias in the estimation of the model parameters. Later approaches (Self and Pawitan, 1992; Tsiatis, DeGruttola, and Wulfsohn 1995) focused on joint models with a survival sub-model for the time-to-event and a longitudinal sub-model for the longitudinal process, in which so-called two-stage procedures have been proposed to derive estimates of the model parameters. In particular, at a first stage, the longitudinal model is estimated ignoring the survival outcome, and at the second stage a survival model is fitted using the subject-specific predictions of time-dependent covariates based on the longitudinal model. Such approaches were shown to reduce bias compared to the naive LVCF without completely eliminating it. This persistent bias prompted a turn of focus to full maximum likelihood methods. A fully parametric approach was proposed by DeGruttola and Tu (1994) who postulated a log-normal sub-model for the time-to-event and a linear mixed model for the longitudinal responses, respectively. Later, Wulfsohn and Tsiatis (1997) extended this work by assuming a relative risk model for the survival times with an unspecified baseline risk function. Excellent overviews of the joint modeling literature are given by Tsiatis and Davidian (2004) and Yu *et al.* (2004). In the rest of this section we will present the basics of the joint modeling framework and provide a perspective on its features.

2.5.1 Joint Modeling Framework

To introduce joint models for longitudinal and time-to-event data, we need to adapt and extend the notation introduced so far in this chapter. In particular, for the time-to-event outcome we denote by T_i the observed failure time for the ith subject ($i = 1, \dots, n$), which is taken as the minimum of the true event time T_i^* and the censoring time C_i, i.e., $T_i = \min(T_i^*, C_i)$. Furthermore, we define the event indicator as $\delta_i = I(T_i^* \le C_i)$, where $I(\cdot)$ is the indicator function that takes the value 1 if the condition $T_i^* \le C_i$ is satisfied, and 0 otherwise. Thus, the observed data for the time-to-event outcome consist of the pairs $\{(T_i, \delta_i), i = 1, \dots, n\}$. For the longitudinal responses, we let $y_i(t)$ to denote the value of the longitudinal outcome at time point t for the ith subject. However, we do not actually observe $y_i(t)$ at all time points but only at very specific occasions t_{ij} at which measurements were taken. Thus, the observed longitudinal data consist of the measurements $y_{ij} = \{y_i(t_{ij}), j = 1, \dots, n_i\}$. As noted above, this feature of the longitudinal outcome is one of the main reasons why it cannot be simply included as a standard time-dependent covariate in a survival model.

In survival analysis, relative risk models have traditionally been used to quantify effects of both time-independent and time-dependent covariates on the risk of an event (Therneau and Grambsch, 2000). In our setting, we introduce the term $m_i(t)$ that denotes the *true* and *unobserved* value of the longitudinal outcome at time t, which is included as a time-dependent covariate in a relative risk model:

$$\begin{aligned} h_i(t \mid \mathcal{M}_i(t), w_i) &= \lim_{dt \to 0} P\{t \le T_i^* < t + dt \mid T_i^* \ge t, \mathcal{M}_i(t), w_i\}/dt \\ &= h_0(t) \exp\{\gamma^\top w_i + \alpha m_i(t)\}, \end{aligned} \tag{2.10}$$

where $\mathcal{M}_i(t) = \{m_i(u), 0 \le u < t\}$ denotes the history of the true unobserved longitudinal process up to time point t, $h_0(\cdot)$ denotes the baseline risk function, and w_i a vector of baseline covariates, such as a treatment indicator, history of diseases, etc., with a corresponding vector of regression coefficients γ. Similarly, parameter α quantifies the effect of the underlying longitudinal outcome to the risk for an event. For instance, in the AIDS example introduced in Section 2.5, α measures the effect of the number of CD4 cells to the risk for death. An important note regarding Model (2.10) is that the risk for an event at time t is assumed to depend on the longitudinal history $\mathcal{M}_i(t)$ only through the current value of the time-dependent covariate $m_i(t)$; on the contrary, survival probabilities depend on the whole history via:

$$\begin{aligned} \mathcal{S}_i(t \mid \mathcal{M}_i(t), w_i) &= P(T_i^* > t \mid \mathcal{M}_i(t), w_i) \\ &= \exp\left(- \int_0^t h_0(s) \exp\{\gamma^\top w_i + \alpha m_i(s)\}\, ds \right), \end{aligned} \tag{2.11}$$

which implies that a correct specification of $\mathcal{M}_i(t)$ is required to produce valid estimates of $\mathcal{S}_i(t \mid \mathcal{M}_i(t), w_i)$. To complete the specification of the survival model, we need to specify the baseline risk function. Within the joint modeling framework, $h_0(t)$ is typically left unspecified (Wulfsohn and Tsiatis 1997). However, Hsieh, Tseng, and Wang (2006) have recently noted that leaving this function completely unspecified leads to an underestimation of the standard errors of the parameter estimates. In particular, problems arise stemming from the fact that the non-parametric maximum likelihood estimate for this function cannot be obtained explicitly under the random-effects structure. To avoid this problem, we could either opt for a standard survival distribution on the one hand, such as the Weibull or Gamma distributions, or for more flexible models on the other, in which $h_0(t)$ is sufficiently well approximated using step functions or spline-based approaches.

So far, in the definition of the survival model we have assumed that the true underlying longitudinal covariate $m_i(t)$ is available at any time point t. Nevertheless, longitudinal information is actually collected intermittently for each subject at a few time points t_{ij}. Therefore, our aim is to estimate $m_i(t)$ and successfully reconstruct the complete longitudinal history, using the available measurements $y_{ij} = \{y_i(t_{ij}), j = 1, \dots, n_i\}$ of each subject and a set of modeling assumptions. For the remainder of this section, we will focus on normal data and postulate a linear mixed effects model, as in Section 2.3.1, to describe the subject-specific longitudinal evolutions. Here we make explicit the model's time-dependent nature,

$$y_i(t) = m_i(t) + \varepsilon_i(t) = x_i^\top(t)\beta + z_i^\top(t) b_i + \varepsilon_i(t), \quad \varepsilon_i(t) \sim N(0, \sigma^2), \tag{2.12}$$

where β denotes the vector of the unknown fixed effects parameters, $x_i(t)$ and $z_i(t)$ denote row vectors of the design matrices for the fixed and random effects, respectively, and $\varepsilon_i(t)$ is the measurement error term, which is assumed independent of b_i, and with variance σ^2. As we have seen above, the survival function is a function of the complete longitudinal history, and therefore, it is important to adequately specify $x_i(t)$ and $z_i(t)$ to capture interesting characteristics of the data and produce a good estimate of $\mathscr{M}_i(t)$. For instance, in applications in which subjects show highly non-linear longitudinal trajectories, it is advisable to consider flexible representations for $x_i(t)$ and $z_i(t)$ using a possibly high-dimensional vector of functions of time t, expressed in terms of high-order polynomials or splines (Ding and Wang 2008, Brown, Ibrahim, and DeGruttola 2005).

An alternative approach is to consider correlated error terms. Joint models with such error structures have been proposed by Wang and Taylor (2001), who postulated an integrated Ornstein-Uhlenbeck process, and by Henderson, Diggle, and Dobson (2000), who considered a latent Gaussian stochastic process shared by both the longitudinal and event processes. We should note, however, that there is a conflict for information between the random-effects structure and a measurement error structure that assumes correlated errors, given that both aim at modeling the marginal correlation in the data. Thus, depending on the features of the data at hand, it is advisable to either opt for an elaborate random-effects structure (using e.g., splines in the design matrix $z_i(t)$) or for correlated error terms, but not for both. For an enlightening discussion on the philosophical differences between these two approaches, we refer to Tsiatis and Davidian (2004, Sect. 2.2).

Finally, a suitable distributional assumption for the random-effects component is required to complete the specification of the joint model. So far, in this chapter, we have relied on standard parametric assumptions for this distribution, with a typical choice being the multivariate normal distribution with mean zero and covariance matrix D. However, within the joint modeling framework and mainly for two reasons, there is the concern that relying on standard distributions may influence the derived inferences. First, the random effects have a more prominent role in joint models, because on the one hand they capture the correlations between the repeated measurements in the longitudinal outcome and on the other they associate the longitudinal outcome with the event process. Second, joint models belong to the general family of shared parameter models, and correspond to a non-random dropout mechanism. We wil return to this in Section 2.7. As is known from the missing-data literature, handling dropout can be highly sensitive to modeling assumptions. These features motivated Song, Davidian, and Tsiatis (2002) to explore the need for a more flexible model for the distribution of the random effects, especially in the joint modeling framework. However, the findings of these authors suggested that parameter estimates and standard errors were rather robust to misspecification. This feature has been further theoretically corroborated by Rizopoulos, Verbeke, and Molenberghs (2008), who showed that, as the number of repeated measurements per subject n_i increases, misspecification of the random-effects distribution has a minimal effect in parameter estimators and standard errors.

2.5.2 Likelihood and Estimation

The main estimation methods that have been proposed for joint models are (semi-parametric) maximum likelihood (Hsieh, Tseng, and Wang 2006, Henderson, Diggle, and Dobson 2000, Wulfsohn and Tsiatis 1997) and Bayes using MCMC techniques (Chi and Ibrahim 2006, Brown and Ibrahim 2003, Wang and Taylor 2001, Xu and Zeger 2001). Moreover, Tsiatis and Davidian (2001) have proposed a conditional score approach in which the random effects are treated as nuisance parameters, and they developed a set of unbiased estimating equations that yields consistent and asymptotically normal estimators. Here, we review the basics of the maximum likelihood method for joint models as one of the more traditional approaches.

Maximum likelihood estimation for joint models is based on the maximization of the log-likelihood corresponding to the joint distribution of the time-to-event and longitudinal outcomes $\{T_i, \delta_i, y_i\}$. To define this joint distribution, we will assume that the vector of time-independent random effects b_i underlies both the longitudinal and survival processes. This means that these random effects account for both the association between the longitudinal and event outcomes, and the correlation between the repeated measurements in the longitudinal process. Formally, we have that,

$$f(T_i, \delta_i, y_i \mid b_i; \theta) = f(T_i, \delta_i \mid b_i; \theta)\, f(y_i \mid b_i; \theta), \tag{2.13}$$

$$f(y_i \mid b_i; \theta) = \prod_j f\{y_i(t_{ij}) \mid b_i; \theta\}, \tag{2.14}$$

where θ is the parameter vector, y_i is the $n_i \times 1$ vector of longitudinal responses of the ith subject, and $f(\cdot)$ denotes an appropriate probability density function. Under this conditional independence assumption we can now define separate models for the longitudinal responses and the event time data by conditioning on the shared random effects. Under the modeling assumptions presented in the previous section and the conditional independence assumptions (2.13) and (2.14), the joint likelihood contribution for the ith subject can be formulated as

$$f(T_i, \delta_i, y_i; \theta) = \int f(T_i, \delta_i \mid b_i; \theta) \Big[\prod_j f\{y_i(t_{ij}) \mid b_i; \theta\}\Big] f(b_i; \theta)\, db_i, \tag{2.15}$$

where the likelihood of the survival part is written as

$$f(T_i, \delta_i \mid b_i; \theta) = \{h_i(T_i \mid b_i; \theta)\}^{\delta_i} \mathscr{S}_i(T_i \mid b_i; \theta), \tag{2.16}$$

with $h_i(\cdot)$ and $\mathscr{S}_i(\cdot)$ are given by (2.10) and (2.11), respectively, $f\{y_i(t_{ij}) \mid b_i; \theta\}$ is the univariate normal density for the longitudinal responses, and $f(b_i; \theta)$ is the multivariate normal density for the random effects. A further implicit assumption in the above definition of the likelihood is that both the censoring mechanism and the visiting process (i.e., the stochastic mechanism that generates the time points at which the longitudinal measurements are collected) are non-informative, and thus they can be ignored. This non-informativeness assumption is similar in spirit to the missing at random (MAR) assumption in the missing data framework (see also Section 2.7), and in particular, it is assumed that the probabilities of visiting and censoring at time point t depend only on the observed longitudinal history but not on the event times and future longitudinal measurements themselves. As observed longitudinal history we define all available information for the longitudinal process prior to time point t, i.e., $\mathscr{Y}_i(t) = \{y_i(u), 0 \leq u < t\}$; note that this is different from $\mathscr{M}_i(t)$, which denotes the history of the true unobserved longitudinal outcome $m_i(t)$. In practice, this assumption is valid when the decision on whether a subject withdraws from the study or appears at the study center for the scheduled visit to provide a longitudinal measurement at time t, depends only on $\mathscr{Y}_i(t)$ (and possibly on baseline covariates), but there is no additional dependence on future longitudinal responses and the underlying random effects b_i. Unfortunately, the observed data do not often contain enough information to corroborate these assumptions, and therefore, it is essential to use external information from subject-matter experts as to their validity.

Maximization of the log-likelihood function corresponding to (2.15) with respect to θ is a computationally challenging task, because it requires a combination of numerical integration and optimization algorithms. Numerical integration is required, owing to the fact that neither the integral with respect to the random effects in (2.15), nor the integral of the risk function in (2.11) allow for an analytical solution, except in very special cases. Standard numerical integration techniques, such as Gaussian quadrature and Monte Carlo have been successfully applied in the joint modelling framework (Song, Davidian, and Tsiatis 2002, Henderson, Diggle, and Dobson 2000, Wulfsohn and Tsiatis 1997). Furthermore, Rizopoulos, Verbeke, and Lesaffre (2009b) have recently discussed the use of Laplace approximations for joint models, that can be especially useful in high-dimensional random-effects settings (e.g., when splines are used in random-effects design matrix). For the maximization of the approximated log-likelihood the EM algorithm has been traditionally used in which the random effects are treated as 'missing data'. The main motivation for using this algorithm is the closed-form M-step updates for certain parameters of the joint model. However, a serious drawback of the EM algorithm is its linear convergence rate that results in slow convergence especially near the maximum. Nonetheless, Rizopoulos, Verbeke, and Lesaffre (2009b) have noted that a direct maximization of the observed data log-likelihood, using for instance, a quasi-Newton algorithm (Lange 2004), requires very similar computations to the EM algorithm. Therefore hybrid optimization approaches that start with EM and then continue with direct maximization can be easily employed.

2.5.3 Analysis of Liver Cirrhosis Data

To illustrate the virtues of the joint modeling approach, we will start with a 'naive' analysis, in which we ignore the special characteristics of the prothrombin index and we fit a Cox model that includes treatment indicator and prothrombin as an ordinary time-dependent covariate. The results are presented in Table 2.5. We observe

Table 2.5 Liver Cirrhosis Data. Parameter estimates with standard errors in parenthesis. For the longitudinal process 'a:b' denotes the interaction term between covariates 'a' and 'b'. For the random effects σ_{b1} denotes the standard deviation of the random intercepts term, σ_{b2} the standard deviation of the random slopes term, ρ_{b12} the correlation between the random intercepts and random slopes, and σ the measurement error standard deviation

	Survival Process		Longitudinal Process		Variance Comp.	
Model	Parameter	Estimate (s.e.)	Effect	Est. (s.e.)	Param.	Est.
Naive	prednisone	0.054 (0.130)				
Cox	prothrombin	−0.032 (0.003)				
Joint	prednisone	−0.214 (0.140)	intercept	70.49 (1.36)	σ_{b1}	18.51
Model	prothrombin	−0.040 (0.004)	prednisone	11.10 (1.96)	σ_{b2}	4.22
			baseline	−1.49 (1.35)	ρ_{b12}	0.04
			baseline:prednisone	−11.20 (1.89)	σ	16.86
			time	0.40 (0.39)		
			time:prednisone	−1.05 (0.68)		

that, after adjusting for prothrombin in the Cox model, there is no statistical evidence for a treatment effect. We proceed by specifying and fitting a joint model that explicitly postulates a linear mixed effects model for the prothrombin index. In particular, in the longitudinal sub-model, we include fixed effects of time, treatment, and an indicator for the baseline measurement at $t = 0$, as well as the interactions of treatment with time and treatment with the baseline indicator. In the random-effects design matrix, we include an intercept and a time term. For the survival sub-model and similarly to the Cox model above we include the treatment effect and as time-dependent covariate the true underlying effect of prothrombin as estimated from the longitudinal model. The baseline risk function is assumed piecewise constant

$$h_0(t) = \sum_{q=1}^{Q} \xi_q I(v_{q-1} < t \leq v_q),$$

where $0 = v_0 < v_1 < \cdots < v_Q$ denotes a split of the time scale, with v_Q being larger than the largest observed time, and ξ_q denotes the value of the hazard in the interval $(v_{q-1}, v_q]$. For the internal knots $v_1, \ldots, v_{Q-1}$ we use equally spaced percentiles of the observed survival times T_i.

The parameter estimates and standard errors from the joint model fit are also shown in Table 2.5. For the treatment effect, we arrive at a similar conclusion as with the standard analysis, that is, there is no clear evidence that prednisone decreases the risk for an event. However, a comparison between the standard time-dependent Cox model with the joint model reveals some interesting features. In particular, we observe that the estimated treatment effect from the joint model is much bigger in size and on the opposite direction compared to the time-dependent Cox model, with a standard error of the same magnitude in both models. Similarly, the effect of the prothrombin index from the joint model is about 2.5 standard errors larger compared to the same effect from the Cox model. These comparisons convincingly demonstrate the degree of attenuation in the regression coefficients of the standard analysis due to the measurement error in the prothrombin levels.

2.5.4 *Some Reflections*

Joint modeling of longitudinal and time-to-event data is one of the most rapidly evolving areas of current biostatistics research, with several extensions of the standard joint model that we have presented here already proposed in the literature. These include, among others, handling multiple failure types (Elashoff and Li 2008), considering categorical longitudinal outcomes (Faucett, Schenker, and Elashoff 1998), assuming that several longitudinal outcomes affect the time-to-event (Chi and Ibrahim 2006, Brown and Ibrahim 2003), replacing the relative risk model by an accelerated failure time model (Tseng, Hsieh, and Wang 2005), and associating the two outcomes via latent classes instead of random effects (Proust-Lima *et al.* 2009, Lin *et al.* 2002). Even though there has been considerable work on such extensions, little attention has been given to the development of diagnostic and model-assessment tools for these models. The main problem of using standard diagnostic tools, such as residuals, is the nonrandom dropout caused by the occurrence events. To this end, Dobson and Henderson (2003) defined residuals conditional on the dropout times and recommended plotting these residuals per dropout pattern. Another, more recent proposal by Rizopoulos, Verbeke, and Molenberghs (2009a) takes dropout into account by multiply imputing the longitudinal responses that would have been observed had the event not occurred, and use afterwards standard residuals plots.

Finally, one of the main practical limitations for joint modeling finding its way into the tool box of modern statisticians was the lack of free and reliable software. The R package **JM** has been developed to fill this gap to some extent. **JM** can be freely downloaded from the CRAN website at `http://cran.r-project.org/` with more information at `http://wiki.r-project.org/rwiki/doku.php?id=packages:cran:jm/` or from the book website at `http://www.econ.upf.edu/~satorra/longitudinallatent/readme.html`. **JM** has a user-friendly interface to fit joint models and also provides several supporting functions that extract or calculate various quantities based on the fitted model (e.g., residuals, fitted values, empirical Bayes estimates, various plots, and others).

2.6 The Use in Surrogate Markers

Over the years, longitudinal data models, survival analysis tools, and the combination thereof, have been used in the so-called validation of surrogate endpoints in clinical studies. Reviews can be found in Burzykowski, Molenberghs, and Buyse (2005), Molenberghs *et al.* (2008, 2009). We provide a bird's eye perspective on these developments and their extensions towards information theory.

The field is interesting in its own right, because the use of surrogate endpoints in the development of new therapies has always been very controversial, partly owing to a number of unfortunate historical instances where treatments showing a highly positive effect on a surrogate endpoints were ultimately shown to be detrimental to the subjects' clinical outcome, and conversely, some instances of treatments conferring clinical benefit without measurable impact on presumed surrogates (Fleming and DeMets 1996). For example, in cardiovascular disease, the unsettling discovery that the two major anti arrhythmic drugs encanaide and flecanaide reduced arrhythmia but caused a more than 3-fold increase in overall mortality stressed the need for caution in using non-validated surrogate markers in the evaluation of the possible clinical benefits of new drugs (CAST 1989). On the other hand, the dramatic surge of the AIDS epidemic, the impressive therapeutic results obtained early on with zidovudine, and the pressure for an accelerated evaluation of new therapies, have all led to the use of CD4 blood count and later of viral load as endpoints that replaced time to clinical events and overall survival (DeGruttola and Tu 1994), in spite of serious concerns about their limitations as surrogate markers for clinically relevant endpoints (Lagakos and Hoth 1992). Loosely speaking, a *surrogate endpoint* is a biomarker that is intended to substitute for a clinical endpoint. A surrogate endpoint is expected to predict clinical benefit, harm, or lack thereof.

One important reason for the present interest in surrogate endpoints is the advent of a large number of biomarkers that closely reflect the disease process. An increasing number of new drugs have a well-defined mechanism of action at the molecular level, allowing drug developers to measure the effect of these drugs on the relevant biomarkers (Ferentz 2002). There is increasing public pressure for new, promising drugs to be approved for marketing as rapidly as possible, and such approval will have to be based on biomarkers rather than on some long-term clinical endpoint (Lesko and Atkinson 2001). If the approval process is shortened, there will be a corresponding need for earlier detection of safety signals that could point to toxic problems with new drugs. It is a safe bet, therefore, that the evaluation of tomorrow's drugs will be based primarily on biomarkers, rather than on the longer-term, harder clinical endpoints that have dominated the development of new drugs until now. It is therefore imperative to use *validated* surrogates, though one needs to reflect on the precise meaning and extent of validation (Schatzkin and Gail 2002).

2.6.1 A Meta-analytic Framework for Normally Distributed Outcomes

Several methods have been suggested for the formal evaluation of surrogate markers, some based on a single trial with others, currently gaining momentum, of a meta-analytic nature. The first formal single trial approach to validate markers is due to Prentice (1989), who gave a definition of the concept of a surrogate endpoint, followed by a series of operational criteria. Freedman, Graubard, and Schatzkin (1992) augmented Prentice's hypothesis-testing based approach, with the estimation paradigm, through the so-called *proportion of treatment effect explained.* In turn, Buyse and Molenberghs (1998) added two further measures: the *relative effect* and the *adjusted association.* All of these proposals are hampered by the fact that they are single-trial based, in which there evidently is replication at the patient level, but not at the level of the trial.

Although the single trial based methods are relatively easy in terms of implementation, they are surrounded with the difficulties stated before. Therefore, several authors, such as Daniels and Hughes (1997), Buyse *et al.* (2000), and Gail *et al.* (2000) have introduced the meta-analytic approach. This section briefly outlines the methodology.

The meta-analytic approach was formulated originally for two continuous, normally distributed outcomes, and extended in the meantime to a large collection of outcome types, ranging from continuous, binary, ordinal, time-to-event, and longitudinally measured outcomes (Burzykowski, Molenberghs, and Buyse 2005). First, we focus on the continuous case, where the surrogate and true endpoints are jointly normally distributed.

The method is based on the linear mixed model of Section 2.3.1. Both a fixed-effects and a random-effects view can be taken. Let T_{ij} and S_{ij} be the random variables denoting the true and surrogate endpoints for the jth subject in the ith trial, respectively, and let Z_{ij} be the indicator variable for treatment. First, consider the following fixed-effects models:

$$S_{ij} = \mu_{Si} + \alpha_i Z_{ij} + \varepsilon_{Sij}, \tag{2.17}$$

$$T_{ij} = \mu_{Ti} + \beta_i Z_{ij} + \varepsilon_{Tij}, \tag{2.18}$$

where μ_{Si} and μ_{Ti} are trial-specific intercepts, α_i and β_i are trial-specific effects of treatment Z_{ij} on the endpoints in trial i, and ε_{Si} and ε_{Ti} are correlated error terms, assumed to be zero-mean normally distributed with covariance matrix

$$\Sigma = \begin{pmatrix} \sigma_{SS} & \sigma_{ST} \\ & \sigma_{TT} \end{pmatrix}. \tag{2.19}$$

In addition, we can decompose

$$\begin{pmatrix} \mu_{Si} \\ \mu_{Ti} \\ \alpha_i \\ \beta_i \end{pmatrix} = \begin{pmatrix} \mu_S \\ \mu_T \\ \alpha \\ \beta \end{pmatrix} + \begin{pmatrix} m_{Si} \\ m_{Ti} \\ a_i \\ b_i \end{pmatrix}, \tag{2.20}$$

where the second term on the right hand side of (2.20) is assumed to follow a zero-mean normal distribution with covariance matrix

$$D = \begin{pmatrix} d_{SS} & d_{ST} & d_{Sa} & d_{Sb} \\ & d_{TT} & d_{Ta} & d_{Tb} \\ & & d_{aa} & d_{ab} \\ & & & d_{bb} \end{pmatrix}. \tag{2.21}$$

A classical hierarchical, random-effects modeling strategy results from the combination of the above two steps into a single one:

$$S_{ij} = \mu_S + m_{Si} + \alpha Z_{ij} + a_i Z_{ij} + \varepsilon_{Sij}, \tag{2.22}$$

$$T_{ij} = \mu_T + m_{Ti} + \beta Z_{ij} + b_i Z_{ij} + \varepsilon_{Tij}. \tag{2.23}$$

Here, μ_S and μ_T are fixed intercepts, α and β are fixed treatment effects, m_{Si} and m_{Ti} are random intercepts, and a_i and b_i are random treatment effects in trial i for the surrogate and true endpoints, respectively. The random effects $(m_{Si}, m_{Ti}, a_i, b_i)$ are assumed to be mean-zero normally distributed with covariance matrix (2.21). The error terms ε_{Sij} and ε_{Tij} follow the same assumptions as in the fixed effects models.

After fitting the above models, surrogacy is captured by means of two quantities: trial-level and individual-level coefficients of determination. The former quantifies the association between the treatment effects on the true and surrogate endpoints at the trial level, while the latter measures the association at the level of the individual patient, after adjustment for the treatment effect. The former is given by:

$$R^2_{\text{trial}} = R^2_{b_i|m_{Si},a_i} = \frac{\begin{pmatrix} d_{Sb} \\ d_{ab} \end{pmatrix}^{\top} \begin{pmatrix} d_{SS} & d_{Sa} \\ d_{Sa} & d_{aa} \end{pmatrix}^{-1} \begin{pmatrix} d_{Sb} \\ d_{ab} \end{pmatrix}}{d_{bb}}. \tag{2.24}$$

The above quantity is unitless and, at the condition that the corresponding variance-covariance matrix is positive definite, lies within the unit interval.

Apart from estimating the strength of surrogacy, the above model can also be used for prediction purposes. To this end, observe that $(\beta + b_0 | m_{S0}, a_0)$ follows a normal distribution with mean and variance:

$$E(\beta + b_0 | m_{S0}, a_0) = \beta + \begin{pmatrix} d_{Sb} \\ d_{ab} \end{pmatrix}^{\top} \begin{pmatrix} d_{SS} & d_{Sa} \\ d_{Sa} & d_{aa} \end{pmatrix}^{-1} \begin{pmatrix} \mu_{S0} - \mu_S \\ \alpha_0 - \alpha \end{pmatrix}, \tag{2.25}$$

$$\text{Var}(\beta + b_0 | m_{S0}, a_0) = d_{bb} - \begin{pmatrix} d_{Sb} \\ d_{ab} \end{pmatrix}^{\top} \begin{pmatrix} d_{SS} & d_{Sa} \\ d_{Sa} & d_{aa} \end{pmatrix}^{-1} \begin{pmatrix} d_{Sb} \\ d_{ab} \end{pmatrix}. \tag{2.26}$$

A prediction can be made using (2.25), with prediction variance (2.26). Of course, one has to properly acknowledge the uncertainty resulting from the fact that parameters are not known but merely estimated.

Though the above hierarchical modeling is elegant, it often poses a considerable computational challenge (Burzykowski, Molenberghs, and Buyse 2005). To address this problem, Tibaldi *et al.* (2003) suggested several simplifications.

2.6.2 Non-Gaussian Endpoints

Statistically speaking, the surrogate endpoint and the clinical endpoint are realizations of random variables. As will be clear from the formalism in Section 2.6.1, one is in need of the joint distribution of these variables. The easiest, but not the only, situation is where both are Gaussian random variables, but one also encounters binary (e.g., CD4+ counts over 500/mm3, tumor shrinkage), categorical (e.g., cholesterol levels <200 mg/dl, 200-299 mg/dl, 300+ mg/dl, tumor response as complete response, partial response, stable disease, progressive disease), censored continuous (e.g., time to undetectable viral load, time to cardiovascular death), longitudinal (e.g., CD4+ counts over time, blood pressure over time), and multivariate longitudinal (e.g., CD4+ and viral load over time jointly, various dimensions of quality of life over time) endpoints. The models used to validate a surrogate for a clinical endpoint will depend on the type of variables observed in the problem at hand. Table 2.6 shows some examples of potential surrogate endpoints in various diseases. In what follows, we will briefly discuss the settings of binary endpoints, failure-time endpoints, the combination of an ordinal and a survival endpoint, and longitudinal endpoints.

Table 2.6 Examples of possible surrogate endpoints in various diseases (Abbreviations: AIDS = acquired immune deficiency syndrome; ARMD = age-related macular degeneration; HIV = human immunodeficiency virus)

Disease	Surrogate Endpoint	Type	Final Endpoint	Type
Resectable solid tumor	Time to recurrence	Censored	Survival	Censored
Advanced cancer	Tumor response	Binary	Time to progression	Censored
Osteoporosis	Bone mineral density	Longitudinal	Fracture	Binary
Cardiovascular disease	Ejection fraction	Continuous	Myocardial infraction	Binary
Hypertension	Blood pressure	Longitudinal	Coronary heart disease	Binary
Arrhythmia	Arrhythmic episodes	Longitudinal	Survival	Censored
ARMD	6-month visual acuity	Continuous	24-month visual acuity	Continuous
Glaucoma	Intraoccular pressure	Continuous	Vision loss	Censored
Depression	Biomarkers	Multivariate	Depression scale	Continuous
HIV infection	CD4 counts + viral load	Multivariate	Progression to AIDS	Censored

2.6.2.1 Binary Endpoints

Renard *et al.* (2002) have shown that extension to this situation is easily done using a latent variable formulation. That is, one posits the existence of a pair of continuously distributed latent variable responses $(\widetilde{S}_{ij}, \widetilde{T}_{ij})$ that produce the actual values of (S_{ij}, T_{ij}). These unobserved variables are assumed to have a joint normal distribution and the realized values follow by double dichotomization. On the latent-variable scale, we obtain a model similar to (2.17)–(2.18) and in the matrix (2.19) the variances are set equal to unity in order to ensure identifiability. This leads to the following model:

$$\begin{cases} \Phi^{-1}(P[S_{ij}=1|Z_{ij},m_{S_i},a_i,m_{T_i},b_i]) = \mu_S + m_{S_i} + (\alpha + a_i)Z_{ij}, \\ \Phi^{-1}(P[T_{ij}=1|Z_{ij},m_{S_i},a_i,m_{T_i},b_i]) = \mu_T + m_{T_i} + (\beta + b_i)Z_{ij}, \end{cases}$$

where Φ denotes the standard normal cumulative distribution function. Renard *et al.* (2002) used pseudo-likelihood methods to estimate the model parameters. Similar ideas have been used in the case one of the endpoints is continuous, with the other one binary or categorical (Burzykowski, Molenberghs, and Buyse 2005, Ch. 6).

2.6.2.2 Two Failure-time Endpoints

Assume now that S_{ij} and T_{ij} are failure-time endpoints. Model (2.17)–(2.18) is replaced by a model for two correlated failure-time random variables. Burzykowski *et al.* (2001) used copulas to this end (Clayton 1978, Hougaard 1986). Precisely, one assumes the joint survivor function of (S_{ij}, T_{ij}) is written as:

$$F(s,t) = P(S_{ij} \geq s, T_{ij} \geq t) = K_\xi\{F_{Sij}(s), F_{Tij}(t)\}, \qquad s,t \geq 0, \tag{2.27}$$

where (F_{Sij}, F_{Tij}) denote marginal survivor functions and K_ξ is a copula, i.e., a distribution function on $[0,\,1]^2$ with ξ taking values on the real line.

When the hazard functions are specified, estimates of the parameters for the joint model can be obtained using maximum likelihood. Shih and Louis (1995) discuss alternative estimation methods. The association parameter is generally hard to interpret. However, it can be shown (Genest and McKay 1986) that there is a link with Kendall's τ:

$$\tau = 4\int_0^1\int_0^1 K_\xi(u,v)K_\xi(du,dv) - 1,$$

providing an easy measure of surrogacy at the individual level. At the second stage R^2_{trial} can be computed based on the pairs of treatment effects estimated at the first stage.

2.6.2.3 An Ordinal Surrogate and a Survival Endpoint

Assume that T is a failure-time random variable and S is a categorical variable with K ordered categories. To propose validation measures, similar to those introduced in the previous section, Burzykowski *et al.* (2004) also used bivariate copulas, combining ideas of Molenberghs, Geys, and Buyse (2001) and Burzykowski *et al.* (2001). One marginal distribution is a proportional odds logistic regression, while the other is a proportional hazards model. The Plackett copula (Dale 1986) was chosen to capture the association between both endpoints. The ensuing global odds ratio is relatively easy to interpret.

2.6.2.4 Longitudinal Endpoints

Most of the previous work focuses on univariate responses. Alonso *et al* (2003) showed that going from a univariate setting to a multivariate framework represents new challenges. The R^2 measures proposed by Buyse *et al* (2000), are no longer applicable. Alonso *et al* (2003) based their calculations of surrogacy measures on a two-stage approach rather than a full random-effects approach. They assume that information from $i = 1, \ldots, N$ trials is available, in the ith of which, $j = 1, \ldots, n_i$ subjects are enrolled and they denoted the time at which subject j in trial i is measured as t_{ijk}. If T_{ijk} and S_{ijk} denote the associated true and surrogate endpoints, respectively, and Z_{ij} is a binary indicator variable for treatment then along the ideas of Galecki (1994), they proposed the following joint model, at the first stage, for both responses

$$\begin{cases} T_{ijk} = \mu_{Ti} + \beta_i Z_{ij} + g_{Tij}(t_{ijk}) + \varepsilon_{Tijk}, \\ S_{ijk} = \mu_{Si} + \alpha_i Z_{ij} + g_{Sij}(t_{ijk}) + \varepsilon_{Sijk}, \end{cases} \tag{2.28}$$

where μ_{Ti} and μ_{Si} are trial-specific intercepts, β_i and α_i are trial-specific effects of treatment Z_{ij} on the two endpoints and g_{Tij} and g_{Sij} are trial-subject-specific time functions that can include treatment-by-time interactions. They also assume that the vectors, collecting all information over time for patient j in trial i, $\widetilde{\varepsilon}_{T_{ij}}$ and $\widetilde{\varepsilon}_{S_{ij}}$ are correlated error terms, following a mean-zero multivariate normal distribution with covariance matrix

$$\Sigma_i = \begin{pmatrix} \Sigma_{TTi} & \Sigma_{TSi} \\ \Sigma_{TSi}^{\top} & \Sigma_{SSi} \end{pmatrix} = \begin{pmatrix} \sigma_{TTi} & \sigma_{TSi} \\ \sigma_{TSi} & \sigma_{SSi} \end{pmatrix} \otimes R_i. \tag{2.29}$$

Here, R_i is a correlation matrix for the repeated measurements.

If treatment effect can be assumed constant over time, then (2.24) can still be useful to evaluate surrogacy at the trial level. However, at the individual level the situation is totally different, the R^2_{ind} no longer being applicable, and new concepts are needed.

Using multivariate ideas, Alonso *et al* (2003) proposed the *variance reduction factor* (VRF) to capture individual-level surrogacy in this more elaborate setting. They quantified the relative reduction in the true endpoint variance after adjustment by the surrogate as

$$VRF_{\text{ind}} = \frac{\sum_i \{\text{tr}(\Sigma_{TTi}) - \text{tr}(\Sigma_{(T|S)i})\}}{\sum_i \text{tr}(\Sigma_{TTi})}, \tag{2.30}$$

where $\Sigma_{(T|S)_i}$ denotes the conditional variance-covariance matrix of $\widetilde{\varepsilon}_{T_{ij}}$ given $\widetilde{\varepsilon}_{S_{ij}}$: $\Sigma_{(T|S)i} = \Sigma_{TTi} - \Sigma_{TSi}\Sigma_{SSi}^{-1}\Sigma_{TSi}^{\top}$. Here, Σ_{TTi} and Σ_{SSi} are the variance-covariance matrices associated with the true and surrogate endpoint respectively and Σ_{TSi} contains the covariances between the surrogate and the true endpoint. Alonso *et al* (2003) showed that the VRF_{ind} ranges between zero and one, and that $VRF_{\text{ind}} = R^2_{\text{ind}}$ when the endpoints are measured only once.

An alternative proposal is

$$\theta_p = \sum_i \frac{1}{Np_i} \text{tr}\left\{ \left(\Sigma_{TTi} - \Sigma_{(T|S)i}\right) \Sigma_{TTi}^{-1} \right\}. \tag{2.31}$$

Structurally, both VRF and θ_p are similar, the difference being the reversal of summing the trace and calculating the ratio. In spite of this strong structural similarity the VRF is not symmetric in S and T and it is only invariant with respect to linear orthogonal transformations, whereas θ_p is both symmetric and invariant with respect to the broader class of linear bijective transformations.

A common problem of all previous proposals is that they are strongly based on the normality assumption and extensions to non-normal settings are difficult. To overcome this limitation, Alonso *et al* (2005), introduced a new parameter, the so-called R^2_Λ, to evaluate surrogacy at the individual level when both responses are measured over time or in general when multivariate or repeated measures are available

$$R^2_\Lambda = \frac{1}{N} \sum_i (1 - \Lambda_i), \tag{2.32}$$

where: $\Lambda_i = \dfrac{|\Sigma_i|}{|\Sigma_{TTi}|\,|\Sigma_{SSi}|}$. This parameter not only allows the detection of more general patterns of association but can also be extended to more general settings that those defined by the normal distribution. They proved that R^2_Λ ranges between zero and one, and that in the cross-sectional case $R^2_\Lambda = R^2_{\text{ind}}$. These authors have shown that $R^2_\Lambda = 1$ whenever there is a deterministic relationship between two linear combinations of both endpoints, allowing the detection of strong associations in cases where the VRF or θ_p would fail in doing so.

2.6.3 Towards a Unified Approach

The longitudinal method of the previous section, while elegant, hinges upon normality. First using the likelihood reduction factor (Section 2.6.3.1) and then an information-theoretic approach (Section 2.6.3.2), extension, and therefore unification, will be achieved.

2.6.3.1 The Likelihood Reduction Factor

Estimating individual-level surrogacy, as the previous developments clearly show, has frequently been based on a variance-covariance matrix coming from the distribution of the residuals. However, if we move away from the normal distribution, it is not always clear how to quantify the association between both endpoints after adjusting for treatment and trial effect. To address this problem, Alonso *et al* (2004) considered the following generalized linear models in the ith trial

$$g_T(T_{ij}) = \mu_{T_i} + \beta_i Z_{ij}, \tag{2.33}$$

$$g_T(T_{ij}) = \theta_{0i} + \theta_{1i} Z_{ij} + \theta_{2i} S_{ij}. \tag{2.34}$$

The longitudinal case would be covered by considering particular functions of time in (2.33) and (2.34). Consider G_i^2 as the log-likelihood ratio test statistics to compare (2.33) with (2.34) in trial i, and quantify the association between both endpoints at the individual level using a scaled likelihood reduction factor (LRF)

$$\text{LRF} = 1 - \frac{1}{N} \sum_i \exp\left(-\frac{G_i^2}{n_i}\right). \tag{2.35}$$

Alonso *et al.* (2004) established a number of properties for LRF, in particular its ranging in the unit interval, and its reduction to R_Λ^2 in the longitudinal and to R_{ind}^2 in the cross-sectional case.

2.6.3.2 An Information-theoretic Unification

This proposal avoids the needs for a joint, hierarchical model, and allows for unification across different types of endpoints. The entropy of a random variable (Shannon 1948), a good measure of randomness or uncertainty, is defined in the following way for the case of a discrete random variable Y, taking values $\{k_1, k_2, \ldots, k_m\}$, and with probability function $P(Y = k_i) = p_i$:

$$H(Y) = \sum_i p_i \log\left(\frac{1}{p_i}\right). \tag{2.36}$$

The differential entropy $h_d(X)$ of a continuous variable X with density $f_X(x)$ and support S_{f_X} equals

$$h_d(Y) = -E[\log f_X(X)] = -\int_{S_{f_X}} f_X(x) \log f_X(x) dx. \tag{2.37}$$

The joint and conditional (differential) entropies are defined in an analogous fashion. Defining the information of a single event as $I(A) = \log p_A$, the entropy is $H(A) = -I(A)$. No information is gained from a totally certain event, $p_A \approx 1$, so $I(A) \approx 0$), while an improbable event is informative.

$H(Y)$ is the average uncertainty associated with P. Entropy is always non-negative, satisfies $H(Y|X) \leq H(Y)$ for any pair of random variables, with equality holding under independence, and is invariant under a bijective transformation (Cover and Tomas 1991). Differential entropy enjoys some but not all properties of entropy: it can be infinitely large, negative, or positive, and is coordinate dependent. For a bijective transformation $Y = y(X)$, it follows $h_d(Y) = h_d(X) - \mathrm{E}_Y\left(\log\left|\frac{dx}{dy}(y)\right|\right)$.

We can now quantify the amount of uncertainty in Y, expected to be removed if the value of X were known, by $I(X,Y) = h_d(Y) - h_d(Y|X)$, the so-called *mutual information*. It is always non-negative, zero if and only if X and Y are independent, symmetric, invariant under bijective transformations of X and Y, and $I(X,X) = h_d(X)$. The mutual information measures the information of X, shared by Y.

We will now introduce the entropy-power (Shannon 1948) for comparison of continuous random variables. Let X be a continuous n-dimensional random vector. The entropy-power of X is

$$\mathrm{EP}(X) = \frac{1}{(2\pi e)^n} e^{2h(X)}. \tag{2.38}$$

The differential entropy of a continuous normal random variable is $h(X) = \frac{1}{2}\log\left(2\pi\sigma^2\right)$, a simple function of the variance and, on the natural logarithmic scale: $\mathrm{EP}(X) = \sigma^2$. In general, $\mathrm{EP}(X) \leq \mathrm{Var}(X)$ with equality if and only if X is normally distributed.

We can now define an information-theoretic measure of association (Schemper and Stare 1996):

$$R_h^2 = \frac{\mathrm{EP}(Y) - \mathrm{EP}(Y|X)}{\mathrm{EP}(Y)}, \tag{2.39}$$

which ranges in the unit interval, equals zero if and only if (X,Y) are independent, is symmetric, is invariant under bijective transformation of X and Y, and, when $R_h^2 \to 1$ for continuous models, there is usually some degeneracy appearing in the distribution of (X,Y). There is a direct link between R_h^2 and the mutual information: $R_h^2 = 1 - e^{-2I(X,Y)}$. For Y discrete: $R_h^2 \leq 1 - e^{-2H(Y)}$, implying that R_h^2 then has an upper bound smaller than 1; we then redefine

$$R_h^2\mathrm{max} = \frac{R_h^2}{1 - e^{-2H(Y)}},$$

reaching 1 when both endpoints are deterministically related.

We can now redefine surrogacy, while preserving previous proposals as special cases. While we will focus on individual-level surrogacy, all results apply to the trial level too. Let $Y = T$ and $X = S$ be the true and surrogate endpoints, respectively. We consider S a good surrogate for T at the individual (trial) level, if a "large" amount of uncertainty about T (the treatment effect on T) is reduced when S (the treatment effect on S) is known. Equivalently, we term S a good surrogate for T at the individual level, if our lack of knowledge about the true endpoint is substantially reduced when the surrogate endpoint is known.

A meta-analytic framework, with N clinical trials, produces N_q different R_{hi}^2, and hence we propose a meta-analytic R_h^2:

$$R_h^2 = \sum_{i=1}^{N_q} \alpha_i R_{hi}^2 = 1 - \sum_{i=1}^{N_q} \alpha_i e^{-2I_i(S_i,T_i)},$$

where $\alpha_i > 0$ for all i and $\sum_{i=1}^{N_q} \alpha_i = 1$. Different choices for α_i lead to different proposals, producing an uncountable family of parameters. This opens the additional issue of finding an *optimal* choice. In particular, for the cross-sectional normal-normal case, Alonso and Molenberghs (2007) have shown that $R_h^2 = R_{\text{ind}}^2$. The same holds for R_Λ^2, defined in (2.28) for the longitudinal case. Finally, when the true and surrogate endpoints have distributions in the exponential family, then LRF $\xrightarrow{P} R_h^2$ when the number of subjects per trial goes to infinity.

2.6.3.3 Fano's Inequality and the Theoretical Plausibility of Finding a Good Surrogate

Fano's inequality shows the relationship between entropy and prediction:

$$\mathrm{E}\left[(T - g(S))^2\right] \geq \mathrm{EP}(T)(1 - R_h^2) \tag{2.40}$$

where $\mathrm{EP}(T) = \dfrac{1}{2\pi e} e^{2h(T)}$. Note that nothing has been assumed about the distribution of our responses and no specific form has been considered for the prediction function g. Also, (2.40) shows that the predictive quality strongly depends on the characteristics of the endpoint, specifically on its power-entropy. Fano's inequality states that the prediction error increases with $\mathrm{EP}(T)$ and therefore, if our endpoint has a large power-entropy then a surrogate should produce a large R_h^2 to have some predictive value. This means that, for some endpoints, the search for a good surrogate can be a dead end street: the larger the entropy of T the more difficult it is to predict. Studying the the power-entropy before trying to find a surrogate is therefore advisable.

2.7 Incomplete Data

When referring to the missing-value, or non-response, process we will use the terminology of Little and Rubin (2002). A non-response process is said to be *missing completely at random* (MCAR), if the missingness is independent of both unobserved and observed data, and *missing at random* (MAR), if, conditional on the observed data, the missingness is independent of the unobserved measurements. A process that is neither MCAR nor MAR is termed *non-random* (MNAR). In the context of likelihood inference, and when the parameters describing the measurement process are functionally independent of the parameters describing the missingness process, MCAR and MAR are *ignorable,* while a non-random process is non-ignorable. Thus, under ignorable dropout, one can literally ignore the

Table 2.7 *Overview of missing data mechanisms.*

Acronym	Description	Likelih./Bayesian	Frequentist
MCAR	missing completely at random	ignorable	ignorable
MAR	missing at random	ignorable	non-ignorable
MNAR	missing not at random	non-ignorable	non-ignorable

missingness process and nevertheless obtain valid estimates of, say, the treatment. Above definitions are conditional on including the correct set of covariates into the model. An overview of the various mechanisms, and their (non-)ignorability under likelihood, Bayesian, or frequentist inference, is given in Table 2.7.

Let us first consider the case where one follow-up measurement per patient is made. When dropout occurs, and hence there are no follow-up measurements, one usually is forced to discard such a patient from analysis, thereby violating the intention to treat (ITT) principle which stipulates that all randomized patients should be included in the primary analysis and according to the randomisation scheme. Of course, the effect of treatment can be investigated under extreme assumptions, such as, for example, a worst case and a best case scenario, but such scenarios are most often not really helpful.

Early work regarding missingness focused on the consequences of the induced lack of balance of deviations from the study design (Afifi and Elashoff 1966, Hartley and Hocking 1971). Later, algorithmic developments took place, such as the expectation-maximization algorithm (EM; Dempster, Laird, and Rubin 1977) and multiple imputation (Rubin 1987). These have brought likelihood-based ignorable analysis within reach of a large class of designs and models. However, they usually require extra programming in addition to available standard statistical software.

In the meantime, however, clinical trial practice has put a strong emphasis on methods such as *complete case analysis* (CC) and *last observation carried forward* (LOCF) or other simple forms of imputation. Claimed advantages include computational simplicity, no need for a full longitudinal model analysis (e.g., when the scientific question is in terms of the last planned measurement occasion only) and, for LOCF, compatibility with the ITT principle. However, a CC analysis assumes MCAR and the LOCF analysis makes peculiar assumptions on the (unobserved) evolution of the response, underestimates the variability of the response and ignores the fact that imputed values are no real data.

On the other hand, a likelihood-based longitudinal analysis requires only MAR, uses all data (obviating the need for both deleting and filling in data) and is also consistent with the ITT principle. Further, it can be shown that also the incomplete sequences contribute to estimands of interest (treatment effect at the end of the study), even early dropouts. For continuous responses, the linear mixed model is quite popular and is a direct extension of ANOVA and MANOVA approaches, but more broadly valid in incomplete data settings. For categorical responses and count data, so-called marginal (e.g., generalized estimating equations, GEE) and random-effects (e.g., generalized linear mixed-effects models, GLMM) approaches are in use. While GLMM parameters can be fitted using maximum likelihood, the same is not true for the frequentist GEE method but modifications have been proposed to accommodate the MAR assumption (Robins, Rotnitzky, and Zhao 1995).

Finally, MNAR missingness can never be fully ruled out based on the observed data only. It is argued that, rather than going either for discarding MNAR models entirely or for placing full faith on them, a sensible compromise is to make them a component of a sensitivity analysis.

2.7.1 Direct Likelihood Analysis

For continuous outcomes, Verbeke and Molenberghs (2000) describe likelihood-based mixed-effects models, in the spirit of Section 2.3.1, that are valid under the MAR assumption. Indeed, for longitudinal studies, where missing data are involved, a mixed model only requires that missing data are MAR. As opposed to the traditional techniques, mixed-effects models permit the inclusion of subjects with missing values at some time points (both dropout and intermittent missingness).

This likelihood-based MAR analysis is also termed likelihood-based ignorable analysis, or, as we will be using in the remainder of this section, a *direct likelihood analysis*. In such a direct likelihood analysis, the observed data are used without deletion nor imputation. In doing so, appropriate adjustments are made to parameters at times when data are incomplete, due to the within-patient correlation.

Thus, even when interest lies, for example, in a comparison between the two treatment groups at the last occasion, such a full longitudinal analysis is a good approach, since the fitted model can be used as the basis for inference at the last occasion.

In many clinical trials, the repeated measures are balanced in the sense that a common (and often limited) set of measurement times is considered for all subjects, which allows the a priori specification of a "saturated" model. For example, a full group-by-time interaction for the fixed effects combined with an unstructured covariance matrix. A direct-likelihood analysis is equivalent to a classical MANOVA analysis when data are complete. This is a strong answer to the common criticism that a direct likelihood method is making strong assumptions. Indeed, its coincidence with MANOVA for data without missingness shows that the assumptions made are very mild. However, when data are incomplete, one should be aware that MANOVA and comparisons per time point are only valid under MCAR and less efficient compared to a likelihood analysis; this was also noted in Section 2.3.3, where the t-test for treatment differences at month 12 for the toenail data was found less efficient than the linear mixed effects model. On the other hand, under MAR, both MANOVA and comparisons per time point will not only be less efficient, but more importantly, they will produced biased results, because they do not take into account that the observed data no longer constitute a random sample from the target population. Therefore, the full likelihood analysis constitutes a very promising alternative to CC and LOCF. When a relatively large number of measurements is made within a single subject, the full power of random-effects modeling can be used (Verbeke and Molenberghs 2000). The practical implication is that a software module with likelihood estimation facilities and with the ability to handle incompletely observed subjects, manipulates the correct likelihood, providing valid parameter estimates and likelihood ratio values.

A few cautionary remarks are warranted. First, when at least part of the scientific interest is directed towards the nonresponse process, obviously both processes need to be considered. Under MAR, both questions can be answered separately. This implies that a conventional method can be used to study questions in terms of the the outcomes of interest, such as treatment effect and time trend, whereafter a separate model can be considered to study missingness. Second, likelihood inference is often surrounded with references to the sampling distribution (e.g., to construct measures of precision for estimators and for statistical hypothesis tests, Kenward and Molenberghs 1998). However, the practical implication is that standard errors and associated tests, when based on the observed rather than the expected information matrix and given that the parametric assumptions are correct, are valid. Thirdly, it may be hard to rule out the operation of an MNAR mechanism. This point was brought up in the introduction and will be discussed further in Section 2.7.4.

2.7.2 Illustration: Orthodontic Growth Data

The simple methods and direct likelihood method are now compared using the growth data of Section 2.2.4. For this purpose, a linear mixed model is used, assuming unstructured mean, i.e., assuming a separate mean for each of the eight age$\times$sex combinations, together with an unstructured covariance structure, and using maximum likelihood (ML) as well as restricted maximum likelihood (REML). The mean profiles of the linear mixed model using maximum likelihood for all four data sets, for boys, are given in Figure 2.8. The girls' profiles are similar and hence not shown.

Next to this longitudinal approach, we will consider a full MANOVA analysis and a univariate ANOVA analysis, i.e., one per time point. For all of these analyses, Table 2.8 shows the estimates and standard errors for boys at ages 8 and 10, for the original data and all available incomplete data, as well as for the CC and the LOCF data.

First, we consider the group means for the boys in the original data set in Figure 2.8, i.e., we observe relatively a straight line. Clearly, there seems to be a linear trend in the mean profile.

In a complete case analysis of the growth data, the 9 subjects which lack one measurement are deleted, resulting in a working data set with 18 subjects. This implies that 27 available measurements will not be used for analysis, a quite severe penalty on a relatively small data set. Observing the profiles for the CC data set in

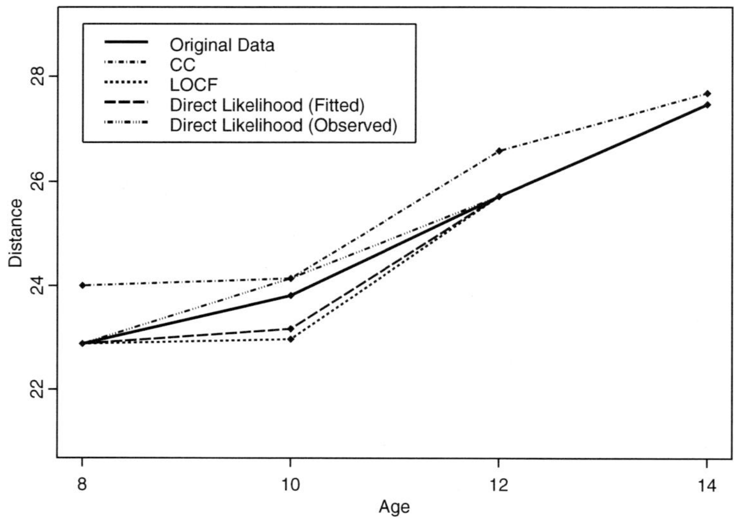

Fig. 2.8 Orthodontic Growth Data. Profiles for the original data, CC, LOCF, and direct likelihood for boys.

Figure 2.8, all group means increased relative to the original data set but mostly so at age 8. The net effect is that the profiles overestimate the average length.

For the LOCF data set, the 9 subjects that lack a measurement at age 10 are completed by imputing the age 8 value. It is clear that this procedure will affect the apparently increasing linear trend found for the original data set. Indeed, the imputation procedure forces the means at ages 8 and 10 to be more similar, thereby destroying the linear relationship. Hence, a simple, intuitively appealing interpretation of the trends is made impossible.

In case of direct likelihood, we now see two profiles. One for the observed means and one for the fitted means. These two coincide at all ages except age 10. As mentioned earlier, the complete observations at age 10 are those with a higher measurement at age 8. Due to the within-subject correlation, they are the ones with a higher measurement at age 10 as well, and therefore the fitted model corrects in the appropriate direction. The consequences of this are very important. While we are inclined to believe that the fitted means do not follow the observed means all that well, this nevertheless is precisely what we should observe. Indeed, since the observed means are based on a non-random subset of the data, the fitted means take into account all observed data points, as well as information on the observed data at age 8, through the measurements that have been taken for such children, at different time points.

As an aside to this, note that, in case of direct likelihood, the observed average at age 10 coincides with the CC average, while the fitted average does not coincide with anything else. Indeed, if the model specification is correct, then a direct likelihood analysis produces a consistent estimator for the average profile, as if nobody had dropped out. Of course, this effect might be blurred in relatively small

Table 2.8 Orthodontic Growth Data. Comparison of analyses based on means at (completely observed age 8 and incompletely observed age 10 measurement)

Method	Boys at Age 8	Boys at Age 10
Original Data		
Direct likelihood, ML	22.88 (0.56)	23.81 (0.49)
Direct likelihood, REML	22.88 (0.58)	23.81 (0.51)
MANOVA	22.88 (0.58)	23.81 (0.51)
ANOVA per time point	22.88 (0.61)	23.81 (0.53)
All Available Incomplete Data		
Direct likelihood, ML	22.88 (0.56)	23.17 (0.68)
Direct likelihood, REML	22.88 (0.58)	23.17 (0.71)
MANOVA	24.00 (0.48)	24.14 (0.66)
ANOVA per time point	22.88 (0.61)	24.14 (0.74)
Complete Case Analysis		
Direct likelihood, ML	24.00 (0.45)	24.14 (0.62)
Direct likelihood, REML	24.00 (0.48)	24.14 (0.66)
MANOVA	24.00 (0.48)	24.14 (0.66)
ANOVA per time point	24.00 (0.51)	24.14 (0.74)
Last Observation Carried Forward Analysis		
Direct likelihood, ML	22.88 (0.56)	22.97 (0.65)
Direct likelihood, REML	22.88 (0.58)	22.97 (0.68)
MANOVA	22.88 (0.58)	22.97 (0.68)
ANOVA per time point	22.88 (0.61)	22.97 (0.72)

data sets due to small-sample variability. Irrespective of the small-sample behavior encountered here, the validity under MAR and the ease of implementation are good arguments that favor this direct likelihood analysis over other techniques.

Let us now compare the different methods by means of Table 2.8, which shows the estimates and standard errors for boys at age 8 and 10, for the original data and all available incomplete data, as well as for the CC data and the LOCF data.

Table 2.8 shows some interesting features. In all four cases, a CC analysis gives an upward biased estimate, for both age groups. This is obvious, since the complete observations at age 10 are those with a higher measurement at age 8, as we have seen before. The LOCF analysis gives a correct estimate for the average outcome for boys at age 8. This is not surprising since there were no missing observations at this age. As noted before, the estimate for boys of age 10 is biased downwards. When the incomplete data are analyzed, we see from Table 2.8 that direct likelihood produces good estimates. The MANOVA and ANOVA per time point analyses give an overestimation of the average of age 10, like in the CC analysis. Further, the MANOVA analysis also yields an overestimation of the average at age 8, again the same as in the CC analysis.

Thus, direct likelihood shares the elegant and appealing features of ANOVA and MANOVA for fully observed data, but is superior with incompletely observed profiles.

2.7.3 Incompleteness and Estimating Equations

2.7.3.1 Weighted Generalized Estimating Equations

As Liang and Zeger (1986) pointed out, GEE-based inferences are valid only under MCAR, due to the fact that they are based on frequentist considerations. An important exception, mentioned by these authors, is the situation where the working correlation structure (discussed in the previous section) happens to be correct, since then the estimates and model-based standard errors are valid under the weaker MAR. This is because then, the estimating equations can be interpreted as likelihood equations. In general, of course, the working correlation structure will not be correctly specified. The ability to do so is the core motivation of the method, and therefore Robins, Rotnitzky, and Zhao (1995) proposed a class of *weighted estimating equations* to allow for MAR, extending GEE.

The idea is to weight each subject's contribution in the GEEs by the inverse probability that a subject drops out at the time he dropped out. This can be calculated, for example, as

$$\begin{aligned} \nu_{id_i} \equiv P[D_i = d_i] = & \prod_{k=2}^{d_i-1} (1 - P[R_{ik} = 0 | R_{i2} = \ldots = R_{i,k-1} = 1]) \times \\ & P[R_{id_i} = 0 | R_{i2} = \ldots = R_{i,d_i-1} = 1]^{I\{d_i \le T\}}. \end{aligned}$$

Recall that we partitioned Y_i into the unobserved components Y_i^m and the observed components Y_i^o. Similarly, we can make the exact same partition of μ_i into $\mu_i{}^m$ and $\mu_i{}^o$. In the weighted GEE approach, which is proposed to reduce possible bias of $\hat{\beta}$, the score equations to be solved when taking into account the correlation structure are:

$$\begin{aligned} S(\beta) = & \sum_{i=1}^{N} \frac{1}{\nu_{id_i}} \frac{\partial \mu_i}{\partial \beta^\top} (A_i^{1/2} C_i A_i^{1/2})^{-1} (y_i - \mu_i) \; = \; 0 \\ = & \sum_{i=1}^{N} \sum_{d=2}^{n+1} \frac{I(D_i = d)}{\nu_{id}} \frac{\partial \mu_i}{\partial \beta^\top}(d) (A_i^{1/2} C_i A_i^{1/2})^{-1}(d)(y(d) - \mu_i(d)) \; = \; 0, \end{aligned} \qquad (2.41)$$

where $y_i(d)$ and $\mu_i(d)$ are the first $d-1$ elements of y_i and μ_i respectively. We define $\frac{\partial \mu_i}{\partial \beta^\top}(d)$ and $(A_i^{1/2} C_i A_i^{1/2})^{-1}(d)$ analogously.

It is worthwhile to note that the recently proposed so-called *doubly robust* methods (van der Laan and Robins 2002) is more efficient and robust to a wider class of deviations. However, it is harder to implement than the original proposal.

An alternative mode of analysis, generally overlooked but proposed by Schafer (2003), would consist in multiply imputing the missing outcomes using a parametric model, e.g., of a random-effects or conditional type, followed by conventional GEE and conventional multiple-imputation inference on the so-completed sets of data. This approach is discussed in Beunckens, Sotto, and Molenberghs (2007).

2.7.3.2 Analysis of the Age-related Macular Degeneration Trial

We compare analyses performed on the completers only (CC), on the LOCF imputed data, as well as on the observed data. For the observed, partially incomplete data, GEE is supplemented with WGEE. The GEE analyses are reported in Table 2.9. A working exchangeable correlation matrix is considered. The model has four intercepts and four treatment effects. Precisely, the marginal regression model takes the form

$$\text{logit}[P(Y_{ij} = 1|T_i)] = \beta_{j1} + \beta_{j2}T_i,$$

where $j = 1,\ldots,4$ refers to measurement occasion, T_i is the treatment assignment for subject $i = 1,\ldots,240$ and Y_{ij} is the indicator for whether or not 3 lines of vision have been lost for subject i at time j. The advantage of having separate treatment effects at each time is that particular attention can be given at the treatment effect assessment at the last planned measurement occasion, i.e., after one year. From Table 2.9 it is clear that the model-based and empirically corrected standard errors agree extremely well. This is due to the unstructured nature of the full time by treatment mean structure. However, we do observe differences in the WGEE analyses. Not only are the parameter estimates mildly different between the two GEE versions, there is a dramatic difference between the model-based and empirically corrected standard errors. Nevertheless, the two sets of empirically corrected standard errors agree very closely, which is reassuring.

When comparing parameter estimates across CC, LOCF, and observed data analyses, it is clear that LOCF has the effect of artificially increasing the correlation between measurements. The effect is mild in this case. The parameter estimates of the observed-data GEE are close to the LOCF results for earlier time points and close to CC for later time points. This is to be expected, as at the start of the study the LOCF and observed populations are virtually the same, with the same holding between CC and observed populations near the end of the study. Note also that the treatment effect under LOCF, especially at 12 weeks and after 1 year, is biased downward in comparison to the GEE analyses. To properly use the information in the missingness process, WGEE can be used. To this end, a logistic regression for dropout, given covariates and previous outcomes, needs to be fitted. Parameter estimates and standard errors are given in Table 2.10. Intermittent missingness will be ignored. Covariates of importance are treatment assignment, the level of lesions

Table 2.9 Age-related Macular Degeneration Trial. Parameter estimates (model-based standard errors; empirically corrected standard errors) for the marginal models: GEE on the CC and LOCF population, and on the observed data. In the latter case, also WGEE is used

Effect	Par.	CC	LOCF	Observed data	
				Unweighted	WGEE
Int.4	β_{11}	-1.01(0.24;0.24)	-0.87(0.20;0.21)	-0.87(0.21;0.21)	-0.98(0.10;0.44)
Int.12	β_{21}	-0.89(0.24;0.24)	-0.97(0.21;0.21)	-1.01(0.21;0.21)	-1.78(0.15;0.38)
Int.24	β_{31}	-1.13(0.25;0.25)	-1.05(0.21;0.21)	-1.07(0.22;0.22)	-1.11(0.15;0.33)
Int.52	β_{41}	-1.64(0.29;0.29)	-1.51(0.24;0.24)	-1.71(0.29;0.29)	-1.72(0.25;0.39)
Tr.4	β_{12}	0.40(0.32;0.32)	0.22(0.28;0.28)	0.22(0.28;0.28)	0.80(0.15;0.67)
Tr.12	β_{22}	0.49(0.31;0.31)	0.55(0.28;0.28)	0.61(0.29;0.29)	1.87(0.19;0.61)
Tr.24	β_{32}	0.48(0.33;0.33)	0.42(0.29;0.29)	0.44(0.30;0.30)	0.73(0.20;0.52)
Tr.52	β_{42}	0.40(0.38;0.38)	0.34(0.32;0.32)	0.44(0.37;0.37)	0.74(0.31;0.52)
Corr.	ρ	0.39	0.44	0.39	0.33

at baseline (a four-point categorical variable, for which three dummies are needed), and time at which dropout occurs. For the latter covariates, there are three levels, since dropout can occur at times 2, 3, or 4. Hence, two dummy variables are included. Finally, the previous outcome does not have a significant impact, but will be kept in the model nevertheless. In spite of there being no strong evidence for MAR, the results between GEE and WGEE differ quite a bit. It is noteworthy that at 12 weeks, a treatment effect is observed with WGEE which goes unnoticed with the other marginal analyses. This finding is mildly confirmed by the random-intercept model, when the data as observed are used.

Table 2.10 Age-related Macular Degeneration Trial. Parameter estimates (standard errors) for a logistic regression model to describe dropout

Effect	Parameter	Estimate (s.e.)
Intercept	ψ_0	0.14 (0.49)
Previous outcome	ψ_1	0.04 (0.38)
Treatment	ψ_2	-0.86 (0.37)
Lesion level 1	ψ_{31}	-1.85 (0.49)
Lesion level 2	ψ_{32}	-1.91 (0.52)
Lesion level 3	ψ_{33}	-2.80 (0.72)
Time 2	ψ_{41}	-1.75 (0.49)
Time 3	ψ_{42}	-1.38 (0.44)

2.7.4 Sensitivity Analysis

When there is residual doubt about the plausibility of MAR, one can conduct a sensitivity analysis. While many proposals have been made, this is still a very active area of research. Obviously, a number of MNAR models can be fitted, provided one is prepared to approach formal aspects of model comparison with due caution. Such analyses can be complemented with appropriate (global and/or local) influence analyses (Verbeke *et al.* 2001). Another route is to construct pattern-mixture models, where the measurement model is considered, conditional upon the observed dropout pattern, and to compare the conclusions with those obtained from the selection model framework, where the reverse factorization is used (Michiels *et al.* 2002, Thijs *et al.* 2002). Alternative sensitivity analyses frameworks are provided by Robins, Rotnitzky, and Scharfstein (1998), Forster and Smith (1998) who present a Bayesian sensitivity analysis, and Raab and Donnelly (1999). A further paradigm, useful for sensitivity analysis, are so-called shared parameter models, where common latent or random-effects drive both the measurement process as well as the process governing missingness.

Nevertheless, ignorable analyses may provide reasonably stable results, even when the assumption of MAR is violated, in the sense that such analyses constrain the behavior of the unseen data to be similar to that of the observed data. A discussion of this phenomenon in the survey context has been given in Rubin, Stern, and Vehovar (1995). These authors firstly argue that, in well conducted experiments (some surveys and many confirmatory clinical trials), the assumption of MAR is often to be regarded as a realistic one. Secondly, and very important for confirmatory trials, a MAR analysis can be specified *a priori* without additional work relative to a situation with complete data. Thirdly, while MNAR models are more general and explicitly incorporate the dropout mechanism, the inferences they produce are typically highly dependent on the untestable and often implicit assumptions built in regarding the distribution of the unobserved measurements given the observed ones. The quality of the fit to the observed data need not reflect at all the appropriateness of the implied structure governing the unobserved data. Based on these considerations, we recommend, for primary analysis purposes, the use of ignorable likelihood-based methods or appropriately modified frequentist methods. To explore the impact of deviations from the MAR assumption on the conclusions, one should ideally conduct a sensitivity analysis (Verbeke and Molenberghs 2000).

2.7.5 The Link Between Joint Modeling and Incomplete Data

In Section 2.5, the main research interest was in the time-to-event outcome, and we have motivated joint modeling approaches in order to adequately take into account in our analysis the effect of a time-dependent covariate measured with error. However, joint modeling may be also required when interest is in the longitudinal outcome. In particular, the occurrence of events causes dropout due to the fact that no longitudinal measurements are usually available at and after the event (e.g., death). As we have seen earlier in this section, if the probability of dropout depends on unobserved longitudinal components, i.e., is MNAR, then the dropout process must be explicitly taken into account in order to produce valid inferences for the longitudinal model. One of the modeling frameworks that has been proposed in the missing data literature to handle nonrandom dropout is the shared parameter models (Wu and Carroll 1988, Follmann and Wu 1995). These models posit a survival sub-model for the time-to-dropout and a mixed effects sub-model for the longitudinal responses, and therefore, they belong in fact to same family as the joint model (2.15). When approached from the missing-data point of view, the basic assumption behind these models is that the probability of dropout at time t depends on values of the longitudinal outcome at both past and future time points, through a set of random effects. To show this more clearly, we define for each subject the observed and missing part of the longitudinal response vector. The observed part $y_i^o = \{y_i(t_{ij}) : t_{ij} < T_i, j = 1,\ldots,n_i\}$ contains all observed longitudinal measurements of the ith subject before the observed event time, whereas the missing part $y_i^m = \{y_i(t_{ij}) : t_{ij} \geq T_i, j = 1,\ldots,n_i'\}$ contains the longitudinal measurements that would have been taken until the end of the study, had the event not occurred. Under these definitions, we can derive the dropout mechanism, which is the conditional distribution of the time-to-dropout given the complete vector of longitudinal responses (y_i^o, y_i^m),

$$f(T_i^* \mid y_i^o, y_i^m; \theta) = \int f(T_i^* \mid b_i; \theta) f(b_i \mid y_i^o, y_i^m; \theta)\, db_i, \qquad (2.42)$$

which states that the time-to-dropout depends on y_i^m through the posterior distribution of the random effects $f(b_i \mid y_i^o, y_i^m; \theta)$. In practice, this implies that such models are most meaningful when subjects who experience the event sooner, are the ones that show steeper evolutions in their longitudinal profiles.

2.8 Software Considerations

Let us provide a brief overview of useful software tools, relative to the methodology described and exemplified in this chapter.

Linear mixed models can be fitted using the SAS procedures MIXED, GLIMMIX, and NLMIXED, and the R packages nlme and lme4.

Generalized linear mixed models have been implemented in the SAS procedures GLIMMIX and NLMIXED; they can also be fitted using the R packages lme4, glmmML, MCMCglmm among others.

GEE can be fitted using the SAS procedure GENMOD and the R packages gee and geepack.

When incomplete data are analyzed using multiple imputation, the SAS procedures MI and MIANALYZE apply. Likewise, a suite of R functions is available in packages mice, mitools and Hmisc. For direct-likelihood analysis, simply the aforementioned SAS and R tools apply. Weighted estimating equations require user-defined software.

User-defined software is also needed for the validation of surrogate markers, for high-dimensional data, and for joint modeling.

The authors and their collaborators have developed a variety of software tools, made available via their web sites.

2.9 Concluding Remarks

Models for the analysis of longitudinal and otherwise hierarchical data are omnipresent these days throughout empirical research. Indeed, models and analysis techniques for longitudinal data, be it for Gaussian or non-Gaussian outcomes, are showing up in biometry, medical statistics, epidemiology, psychometry, econometrics, social science, and survey applications. The models are appealing for the intuition behind their formulation. Inferential apparatus is now well developed, and many methods have been implemented in standard software packages.

In this chapter, we have presented basic methodology for Gaussian and non-Gaussian longitudinal data, including the linear and generalized linear mixed model and generalized estimating equations. We also placed a strong emphasis on the use of these methods in conjunction with a time-to-event outcome, also known as joint modeling. Furthermore, we have indicated how models for longitudinal data are playing a role in the validation of surrogate markers.

Finally, we have placed some emphasis on the problem of incomplete data, and how likelihood-based or Bayesian analysis of incomplete longitudinal data can be performed easily when data are not fully observed, given the missing data are missing at random. Related to this, we have indicated how the joint modeling framework can play a role when the missing data are not missing at random.

Acknowledgements The authors gratefully acknowledge support from IAP research Network P6/03 of the Belgian Government (Belgian Science Policy).

References

Aerts, M., Geys, H., Molenberghs, G., & Ryan, L. (2002). *Topics in modelling of clustered data*. London: Chapman & Hall.

Afifi, A., & Elashoff, R. (1966). Missing observations in multivariate statistics I: Review of the literature. *Journal of the American Statistical Association, 61*, 595-604.

Alonso, A., Geys, H., Molenberghs, G., & Vangeneugden, T. (2003). Validation of surrogate markers in multiple randomized clinical trials with repeated measurements. *Biometrical Journal, 45*, 931-945.

Alonso, A., & Molenberghs, G. (2007). Surrogate marker evaluation from an information theory perspective. *Biometrics, 63*, 180-186.
Alonso, A., Molenberghs, G., Geys, H., & Buyse, M. (2005). A unifying approach for surrogate marker validation based on Prentice's criteria. *Statistics in Medicine, 25*, 205-211.
Alonso, A., Molenberghs, G., Burzykowski, T., Renard, D., Geys, H., Shkedy, Z., Tibaldi, F., Abrahantes, J., & Buyse, M. (2004). Prentice's approach and the meta analytic paradigm: a reflection on the role of statistics in the evaluation of surrogate endpoints. *Biometrics, 60*, 724-728.
Altham, P. M. E. (1978). Two generalizations of the binomial distribution. *Applied Statistics, 27*, 162-167.
Andersen, P., Borgan, O., Gill, R., & Keiding, N. (1993). *Statistical models based on counting processes*. New York: Springer.
Arnold, B. C., & Strauss, D. (1991). Pseudolikelihood estimation: some examples. *Sankhya: The Indian Journal of Statistics - Series B, 53*, 233-243.
Bahadur, R. R. (1961). A representation of the joint distribution of responses to *n* dichotomous items. In H. Solomon (Ed.), *Studies in item analysis and prediction, Stanford Mathematical Studies in the Social Sciences VI*. Stanford, CA: Stanford University Press.
Beunckens, C., Sotto, C., & Molenberghs, G. (2007). A simulation study comparing weighted estimating equations with multiple imputation based estimating equations for longitudinal binary data. *Computational Statistics and Data Analysis, 52*, 1533-1548.
Brant, L. J., & Fozard, J. L. (1990). Age changes in pure-tone hearing thresholds in a longitudinal study of normal human aging. *Journal of the Acoustical Society of America, 88*, 813-820.
Breslow, N. E., & Clayton, D. G. (1993). Approximate inference in generalized linear mixed models. *Journal of the American Statistical Association, 88*, 9-25.
Brown, E., & Ibrahim, J. (2003). A Bayesian semiparametric joint hierarchical model for longitudinal and survival data. *Biometrics, 59*, 221-228.
Brown, E., Ibrahim, J., & DeGruttola, V. (2005). A flexible B-spline model for multiple longitudinal biomarkers and survival. *Biometrics, 61*, 64-73.
Burzykowski, T., Molenberghs, G., & Buyse, M. (2004). The validation of surrogate endpoints using data from randomized clinical trials: a case-study in advanced colorectal cancer. *Journal of the Royal Statistical Society, Series A, 167*, 103-124.
Burzykowski, T., Molenberghs, G., & Buyse, M. (2005). *The evaluation of surrogate endpoints*. New York: Springer.
Burzykowski, T., Molenberghs, G., Buyse, M., Geys, H., & Renard, D. (2001). Validation of surrogate endpoints in multiple randomized clinical trials with failure time end points. *Applied Statistics, 50*, 405-422.
Buyse, M., & Molenberghs, G. (1998). The validation of surrogate endpoints in randomized experiments. *Biometrics, 54*, 1014-1029.
Buyse, M., Molenberghs, G., Burzykowski, T., Renard, D., & Geys, H. (2000). The validation of surrogate endpoints in meta-analyses of randomized experiments. *Biostatistics, 1*, 49-67.
Cardiac Arrhythmia Suppression Trial (CAST) Investigators (1989). Preliminary report: effect of encainide and flecainide on mortality in a randomized trial of arrhythmia suppression after myocardial infraction. *New England Journal of Medicine, 321*, 406-412.
Catalano, P. J. (1997). Bivariate modelling of clustered continuous and ordered categorical outcomes. *Statistics in Medicine, 16*, 883-900.
Catalano, P. J., & Ryan, L. M. (1992). Bivariate latent variable models for clustered discrete and continuous outcomes. *Journal of the American Statistical Association, 87*, 651-658.
Chakraborty, H., Helms, R. W., Sen, P. K., & Cohen, M. S. (2003). Estimating correlation by using a general linear mixed model: Evaluation of the relationship between the concentration of HIV-1 RNA in blood and semen. *Statistics in Medicine, 22*, 1457-1464.
Chi, Y.-Y., & Ibrahim, J. (2006). Joint models for multivariate longitudinal and multivariate survival data. *Biometrics, 62*, 432-445.

Clayton, D. G. (1978). A model for association in bivariate life tables and its application in epidemiological studies of familial tendency in chronic disease incidence. *Biometrika, 65*, 141-151.

Cover, T., & Tomas, J. (1991). *Elements of information theory.* New York: Wiley.

Cox, N. R. (1974). Estimation of the correlation between a continuous and a discrete variable. *Biometrics, 30*, 171-178.

Cox, D. R., & Wermuth, N. (1992). Response models for mixed binary and quantitative variables. *Biometrika, 79*, 441-461.

Cox, D. R., & Wermuth, N. (1994a). A note on the quadratic exponential binary distribution. *Biometrika, 81*, 403-408.

Cox, D. R., & Wermuth, N. (1994b). *Multivariate dependencies: Models, analysis and interpretation.* London: Chapman & Hall.

Dale, J. R. (1986). Global cross ratio models for bivariate, discrete, ordered responses. *Biometrics, 42*, 909-917.

Daniels, M. J., & Hughes, M. D. (1997). Meta-analysis for the evaluation of potential surrogate markers. *Statistics in Medicine, 16*, 1515-1527.

De Backer, M., De Keyser, P., De Vroey, C., & Lesaffre, E. (1996). A 12-week treatment for dermatophyte toe onychomycosis: terbinafine 250mg/day vs. itraconazole 200mg/day–a double-blind comparative trial. *British Journal of Dermatology, 134*, 16-17.

DeGruttola, V., & Tu, X. (1994). Modeling progression of CD-4 lymphocyte count and its relationship to survival time. *Biometrics, 50*, 1003-1014.

Dempster, A. P., Laird, N. M., & Rubin, D. B. (1977). Maximum likelihood from incomplete data via the EM algorithm (with discussion). *Journal of the Royal Statistical Society, Series B, 39*, 1-38.

Diggle, P. J., Heagerty, P., Liang, K.-Y., & Zeger, S. L. (2002). *Analysis of longitudinal data.* New York: Oxford University Press.

Ding, J., & Wang, J.-L. (2008). Modeling longitudinal data with nonparametric multiplicative random effects jointly with survival data. *Biometrics, 64*, 546-556.

Dobson, A., & Henderson, R. (2003). Diagnostics for joint longitudinal and dropout time modeling. *Biometrics, 59*, 741-751.

Efron, B. (1986). Double exponential families and their use in generalized linear regression. *Journal of the American Statistical Association, 81*, 709-721.

Elashoff, R., Li, G., & Li, N. (2008). A joint model for longitudinal measurements and survival data in the presence of multiple failure types. *Biometrics, 64*, 762-771.

Fahrmeir, L., & Tutz, G. (2001). *Multivariate statistical modelling based on generalized linear models.* Heidelberg: Springer.

Faucett, C., Schenker, N., & Elashoff, R. (1998). Analysis of censored survival data with intermittently observed time-dependent binary covariates. *Journal of the American Statistical Association, 93*, 427-437.

Ferentz, A. E. (2002). Integrating pharmacogenomics into drug development. *Pharmacogenomics, 3*, 453-467.

Fieuws, S., & Verbeke, G. (2004). Joint modelling of multivariate longitudinal profiles: Pitfalls of the random-effects approach. *Statistics in Medicine, 23*, 3093-3104.

Fieuws, S., & Verbeke, G. (2006). Pairwise fitting of mixed models for the joint modelling of multivariate longitudinal profiles. *Biometrics, 62*, 424-431.

Fieuws, S., Verbeke, G., Boen, F., & Delecluse, C. (2006). High-dimensional multivariate mixed models for binary questionnaire data. *Applied Statistics, 55*, 1-12.

Fitzmaurice, G. M., & Laird, N. M. (1995). Regression models for a bivariate discrete and continuous outcome with clustering. *Journal of the American Statistical Association, 90*, 845-852.

Fitzmaurice, G. M., Laird, N. M., & Ware, J. H. (2004). *Applied longitudinal analysis.* New York: John Wiley & Sons.

Fleming, T. R., & DeMets, D. L. (1996). Surrogate endpoints in clinical trials: are we being misled? *Annals of Internal Medicine, 125*, 605-613.

Folk, V. G., & Green, B. F. (1989). Adaptive estimation when the unidimensionality assumption of IRT is violated. *Applied Psychological Measurement, 13*, 373-389.

Follmann, D., & Wu, M. (1995). An approximate generalized linear model with random effects for informative missing data. *Biometrics, 51*, 151-168.

Forster, J. J., & Smith, P. W. (1998). Model-based inference for categorical survey data subject to non-ignorable non-response. *Journal of the Royal Statistical Society, Series B, 60*, 57-70.

Freedman, L. S., Graubard, B. I., & Schatzkin, A. (1992). Statistical validation of intermediate endpoints for chronic diseases. *Statistics in Medicine, 11*, 167-178.

Gail, M. H., Pfeiffer, R., van Houwelingen, H. C., & Carroll, R. J. (2000). On meta-analytic assessment of surrogate outcomes. *Biostatistics, 1*, 231-246.

Galecki, A. (1994). General class of covariance structures for two or more repeated factors in longitudinal data analysis. *Communications in Statistics: Theory and Methods, 23*, 3105-3119.

Genest, C., & McKay, J. (1986). The joy of copulas: bivariate distributions with uniform marginals. *American Statistician, 40*, 280-283.

Geys, H., Molenberghs, G., & Ryan, L. M. (1997). Pseudo-likelihood inference for clustered binary data. *Communications in Statistics: Theory and Methods, 26*, 2743-2767.

Geys, H., Molenberghs, G., & Ryan, L. (1999). Pseudolikelihood modeling of multivariate outcomes in developmental toxicology. *Journal of the American Statistical Association, 94*, 734-745.

Gueorguieva, R. (2001). A multivariate generalized linear mixed model for joint modelling of clustered outcomes in the exponential family. *Statistical Modelling, 1*, 177-193.

Goldstein, H. (1979). *The design and analysis of longitudinal studies*. London: Academic Press.

Hartley, H. O., & Hocking, R. (1971). The analysis of incomplete data. *Biometrics, 27*, 7783-7808.

Harville, D. A. (1974). Bayesian inference for variance components using only error contrasts. *Biometrika, 61*, 383-385.

Harville, D. A. (1976). Extension of the Gauss-Markov theorem to include the estimation of random effects. *The Annals of Statistics, 4*, 384-395.

Harville, D. A. (1977). Maximum likelihood approaches to variance component estimation and to related problems. *Journal of the American Statistical Association, 72*, 320-340.

Hedeker, D., & Gibbons, R. D. (1994). A random-effects ordinal regression model for multilevel analysis. *Biometrics, 50*, 933-944.

Hedeker, D., & Gibbons, R. D. (1996). MIXOR: A computer program for mixed-effects ordinal regression analysis. *Computer Methods and Programs in Biomedicine, 49*, 157-176.

Henderson, R., Diggle, P., & Dobson, A. (2000). Joint modelling of longitudinal measurements and event time data. *Biostatistics, 1*, 465-480.

Henderson, C. R., Kempthorne, O., Searle, S. R., & Von Krosig, C. N. (1959). Estimation of environmental and genetic trends from records subject to culling. *Biometrics, 15*, 192-218.

Hougaard, P. (1986). Survival models for heterogeneous populations derived from stable distributions. *Biometrika, 73*, 387-396.

Hsieh, F., Tseng, Y.-K., & Wang, J.-L. (2006). Joint modeling of survival and longitudinal data: likelihood approach revisited. *Biometrics, 62*, 1037-1043.

Jennrich, R. I., & Schluchter, M. D. (1986). Unbalanced repeated measures models with structured covariance matrices. *Biometrics, 42*, 805-820.

Kenward, M. G., & Molenberghs, G. (1998). Likelihood based frequentist inference when data are missing at random. *Statistical Science, 12*, 236-247.

Krzanowski, W. J. (1988). *Principles of multivariate analysis*. Oxford: Clarendon Press.

Lagakos, S. W., & Hoth, D. F. (1992). Surrogate markers in AIDS: Where are we? Where are we going? *Annals of Internal Medicine, 116*, 599-601.

Laird, N. M., & Ware, J. H. (1982). Random effects models for longitudinal data. *Biometrics, 38*, 963-974.

Lang, J. B., & Agresti, A. (1994). Simultaneously modeling joint and marginal distributions of multivariate categorical responses. *Journal of the American Statistical Association, 89*, 625-632.

Lange, K. (2004). *Optimization*. New York: Springer.

Lesko, L. J., & Atkinson, A. J. (2001). Use of biomarkers and surrogate endpoints in drug development and regulatory decision making: criteria, validation, strategies. *Annual Review of Pharmacological Toxicology, 41*, 347-366.

Liang, K.-Y., & Zeger, S. L. (1986). Longitudinal data analysis using generalized linear models. *Biometrika, 73*, 13-22.

Liang, K.-Y., Zeger, S.L., & Qaqish, B. (1992). Multivariate regression analyses for categorical data. *Journal of the Royal Statistical Society, Series B, 54*, 3-40.

Lin, H., Turnbull, B., McCulloch, C., & Slate, E. (2002). Latent class models for joint analysis of longitudinal biomarker and event process data: Application to longitudinal prostate-specific antigen readings and prostate cancer. *Journal of the American Statistical Association, 97*, 53-65.

Lipsitz, S. R., Laird, N. M., & Harrington, D. P. (1991). Generalized estimating equations for correlated binary data: using the odds ratio as a measure of association. *Biometrika, 78*, 153-160.

Little, R. J. A., & Rubin, D. B. (2002). *Statistical analysis with missing data*. New York: John Wiley & Sons.

Little, R. J. A., & Schluchter, M. D. (1985). Maximum likelihood estimation for mixed continuous and categorical data with missing values. *Biometrika, 72*, 497-512.

Liu, L. C., & Hedeker, D. (2006). A mixed-effects regression model for longitudinal multivariate ordinal data. *Biometrics, 62*, 261-268.

MacCallum, R., Kim, C., Malarkey, W., & Kiecolt-Glaser, J. (1997). Studying multivariate change using multilevel models and latent curve models. *Multivariate Behavioral Research, 32*, 215-253.

Mancl, L. A., & Leroux, B. G. (1996). Efficiency of regression estimates for clustered data. *Biometrics, 52*, 500-511.

McCullagh, P., & Nelder, J. A. (1989). *Generalized linear models*. London: Chapman & Hall/CRC.

Michiels, B., Molenberghs, G., Bijnens, L., Vangeneugden, T., & Thijs, H. (2002). Selection models and pattern-mixture models to analyze longitudinal quality of life data subject to dropout. *Statistics in Medicine, 21*, 1023-1041.

Molenberghs, G., Burzykowski, T., Alonso, A., Assam, P., Tilahun, A., & Buyse, M. (2008). The meta-analytic framework for the evaluation of surrogate endpoints in clinical trials. *Journal of Statistical Planning and Inference, 138*, 432-449.

Molenberghs, G., Burzykowski, T., Alonso, A., Assam, P., Tilahun, A., & Buyse, M. (2009). A unified framework for the evaluation of surrogate endpoints in clinical trials. *Statistical Methods in Medical Research, 00*, 000-000.

Molenberghs, G., Geys, H., & Buyse, M. (2001). Evaluation of surrogate end-points in randomized experiments with mixed discrete and continuous outcomes. *Statistics in Medicine, 20*, 3023-3038.

Molenberghs, G., & Kenward, M. G. (2007). *Missing data in clinical studies*. Chichester: John Wiley & Sons.

Molenberghs, G., & Lesaffre, E. (1994). Marginal modelling of correlated ordinal data using a multivariate Plackett distribution. *Journal of the American Statistical Association, 89*, 633-644.

Molenberghs, G., & Lesaffre, E. (1999). Marginal modelling of multivariate categorical data. *Statistics in Medicine, 18*, 2237-2255.

Molenberghs, G., & Verbeke, G. (2005). *Models for discrete longitudinal data*. New York: Springer.

Morrell, C. H., & Brant, L. J. (1991). Modelling hearing thresholds in the elderly. *Statistics in Medicine, 10*, 1453-1464.

Neuhaus, J. M., Kalbfleisch, J. D., & Hauck, W. W. (1991). A comparison of cluster-specific and population-averaged approaches for analyzing correlated binary data. *International Statistical Review, 59*, 25-30.

Ochi, Y., & Prentice, R. L. (1984). Likelihood inference in a correlated probit regression model. *Biometrika, 71*, 531-543.

Olkin, I., & Tate, R. F. (1961). Multivariate correlation models with mixed discrete and continuous variables. *Annals of Mathematical Statistics, 32*, 448-465 (with correction in *36*, 343-344).

Oort, F. J. (2001). Three-mode models for multivariate longitudinal data. *British Journal of Mathematical and Statistical Psychology, 54*, 49-78.

Pearson, J. D., Morrell, C. H., Gordon-Salant, S., Brant, L. J., Metter, E. J., Klein, L. L., & Fozard, J. L. (1995). Gender differences in a longitudinal study of age-associated hearing loss. *Journal of the Acoustical Society of America, 97*, 1196-1205.

Pharmacological Therapy for Macular Degeneration Study Group (1997). Interferon α-IIA is ineffective for patients with choroidal neovascularization secondary to age-related macular degeneration. Results of a prospective randomized placebo-controlled clinical trial. *Archives of Ophthalmology, 115*, 865-872.

Pinheiro, J. C., & Bates, D. M. (2000). *Mixed effects models in S and S-Plus*. New York: Springer.

Prentice, R. L., & Zhao, L. P. (1991). Estimating equations for parameters in means and covariances of multivariate discrete and continuous responses. *Biometrics, 47*, 825-839.

Potthoff, R. F., & Roy, S. N. (1964). A generalized multivariate analysis of variance model useful especially for growth curve problems. *Biometrika, 51*, 313-326.

Prentice, R. (1982). Covariate measurement errors and parameter estimates in a failure time regression model. *Biometrika, 69*, 331-342.

Prentice, R. L. (1988). Correlated binary regression with covariates specific to each binary observation. *Biometrics, 44*, 1033-1048.

Prentice, R. L. (1989). Surrogate endpoints in clinical trials: definitions and operational criteria. *Statistics in Medicine, 8*, 431-440.

Proust-Lima, C., Joly, P., Dartigues, J. F., & Jacqmin-Gadda, H. (2009). Joint modelling of multivariate longitudinal outcomes and a time-to-event: a nonlinear latent class approach. *Computational Statistics and Data Analysis, 53*, 1142-1154.

Raab, G. M., & Donnelly, C. A. (1999). Information on sexual behaviour when some data are missing. *Applied Statistics, 48*, 117-133.

Regan, M. M., & Catalano, P. J. (1999a). Likelihood models for clustered binary and continuous outcomes: Application to developmental toxicology. *Biometrics, 55*, 760-768.

Regan, M. M., & Catalano, P. J. (1999b). Bivariate dose-response modeling and risk estimation in developmental toxicology. *Journal of Agricultural, Biological and Environmental Statistics, 4*, 217-237.

Regan, M. M., & Catalano, P. J. (2000). Regression models for mixed discrete and continuous outcomes with clustering. *Risk Analysis, 20*, 363-376.

Regan, M. M., & Catalano, P. J. (2002). Combined continuous and discrete outcomes. In M. Aerts, H. Geys, G. Molenberghs, & L. Ryan (Eds.), *Topics in modelling of clustered data*. London: Chapman & Hall.

Renard, D., Geys, H., Molenberghs, G., Burzykowski, T., & Buyse, M. (2002). Validation of surrogate endpoints in multiple randomized clinical trials with discrete outcomes. *Biometrical Journal, 44*, 1-15.

Rizopoulos, D., Verbeke, G., & Molenberghs, G. (2009a). Multiple-imputation-based residuals and diagnostic plots for joint models of longitudinal and survival outcomes. *Biometrics*, to appear. doi: 10.1111/j.1541-0420.2009.01273.x

Rizopoulos, D., Verbeke, G., & Lesaffre, E. (2009b). Fully exponential Laplace approximation for the joint modelling of survival and longitudinal data. *Journal of the Royal Statistical Society, Series B, 71*, 637-654.

Rizopoulos, D., Verbeke, G., & Molenberghs, G. (2008). Shared parameter models under random effects misspecification. *Biometrika, 95*, 63-74.

Robins, J. M., Rotnitzky, A., & Scharfstein, D. O. (1998). Semiparametric regression for repeated outcomes with non-ignorable non-response. *Journal of the American Statistical Association, 93*, 1321-1339.

Robins, J. M., Rotnitzky, A., & Zhao, L. P. (1995). Analysis of semiparametric regression models for repeated outcomes in the presence of missing data. *Journal of the American Statistical Association, 90*, 106-121.

Roy, J., & Lin, X. (2000). Latent variable models for longitudinal data with multiple continuous outcomes. *Biometrics, 56*, 1047-1054.

Rubin, D. B. (1987). *Multiple imputation for nonresponse in surveys*. New York: John Wiley & Sons.

Rubin, D. B., Stern, H. S., & Vehovar, V. (1995). Handling "don't know" survey responses: The case of the Slovenian plebiscite. *Journal of the American Statistical Association, 90*, 822-828.

Sammel, M. D., Ryan, L. M., & Legler, J. M. (1997). Latent variable models for mixed discrete and continuous outcomes. *Journal of the Royal Statistical Society, Series B, 59*, 667-678.

Schafer J. L. (1997). *Analysis of incomplete multivariate data*. London: Chapman & Hall.

Schafer, J. L. (2003). Multiple imputation in multivariate problems when the imputation and analysis models differ. *Statistica Neerlandica, 57*, 19-35.

Schatzkin, A., & Gail, M. (2002). The promise and peril of surrogate end points in cancer research. *Nature Reviews Cancer, 2*, 19-27.

Schemper, M., & Stare, J. (1996). Explained variation in survival analysis. *Statistics in Medicine, 15*, 1999-2012.

Self, S., & Pawitan, Y. (1992). Modeling a marker of disease progression and onset of disease. In N.P. Jewell, K. Dietz, & V.T. Farewell (Eds.), *AIDS epidemiology: Methodological issues*. Boston: Birkhauser.

Shah, A., Laird, N., & Schoenfeld, D. (1997). A random-effects model for multiple characteristics with possibly missing data. *Journal of the American Statistical Association, 92*, 775-779.

Shannon, C. (1948). A mathematical theory of communication. *Bell System Technical Journal, 27*, 379-423 and 623-656.

Shock, N. W., Greullich, R. C., Andres, R., Arenberg, D., Costa, P. T., Lakatta, E. G., & Tobin, J. D. (1984). Normal human aging: The Baltimore Longitudinal Study of Aging. *National Institutes of Health publication 84-2450*.

Shih, J. H., & Louis, T. A. (1995). Inferences on association parameter in copula models for bivariate survival data. *Biometrics, 51*, 1384-1399.

Sivo, S. A. (2001). Multiple indicator stationary time series models. *Structural Equation Modeling, 8*, 599-612.

Song, X., Davidian, M., & Tsiatis, A. (2002). A semiparameteric likelihood approach to joint modeling of longitudinal and time-to-event data. *Biometrics, 58*, 742-753.

Tate, R. F. (1954). Correlation between a discrete and a continuous variable. *Annals of Mathematical Statistics, 25*, 603-607.

Tate, R.F. (1955). The theory of correlation between two continuous variables when one is dichotomized. *Biometrika, 42*, 205-216.

Thijs, H., Molenberghs, G., Michiels, B., Verbeke, G., & Curran, D. (2002). Strategies to fit pattern-mixture models. *Biostatistics, 3*, 245-265.

Therneau, T., & Grambsch, P. (2000). *Modeling survival data: Extending the Cox Model*. New York: Springer.

Thiébaut, R., Jacqmin-Gadda, H., Chêne, G., Leport, C., & Commenges, D. (2002). Bivariate linear mixed models using SAS PROC MIXED. *Computer Methods and Programs in Biomedicine, 69*, 249-256.

Thum, Y. M. (1997). Hierarchical linear models for multivariate outcomes. *Journal of Educational and Behavioral Statistics, 22*, 77-108.

Tibaldi, F. S, Cortiñas Abrahantes, J., Molenberghs, G., Renard, D., Burzykowski, T., Buyse, M., Parmar, M., Stijnen, T., & Wolfinger, R. (2003). Simplified hierarchical linear models for the evaluation of surrogate endpoints. *Journal of Statistical Computation and Simulation, 73*, 643-658.

Tseng, Y.-K., Hsieh, F., & Wang, J.-L. (2005). Joint modelling of accelerated failure time and longitudinal data. *Biometrika, 92*, 587-603.

Tsiatis, A., & Davidian, M. (2001). A semiparametric estimator for the proportional hazards model with longitudinal covariates measured with error. *Biometrika, 88*, 447-458.

Tsiatis, A., & Davidian, M. (2004). Joint modeling of longitudinal and time-to-event data: An overview. *Statistica Sinica, 14*, 809-834.

Tsiatis, A., DeGruttola, V., & Wulfsohn, M. (1995). Modeling the relationship of survival to longitudinal data measured with error: applications to survival and CD4 counts in patients with AIDS. *Journal of the American Statistical Association, 90*, 27-37.

Van der Laan, M. J., & Robins, J. M. (2002). *Unified methods for censored longitudinal data and causality*. New York: Springer.

Verbeke, G., Lesaffre, E., & Spiessens, B. (2001). The practical use of different strategies to handle dropout in longitudinal studies. *Drug Information Journal, 35*, 419-434.

Verbeke, G., & Molenberghs, G. (2000). *Linear mixed models for longitudinal data*. New York: Springer.

Verbeke, G., Molenberghs, G., Thijs, H., Lesaffre, E., & Kenward, M. G. (2001). Sensitivity analysis for non-random dropout: A local influence approach. *Biometrics, 57*, 7-14.

Wang, Y., & Taylor, J. (2001). Jointly modeling longitudinal and event time data with application to acquired immunodeficiency syndrome. *Journal of the American Statistical Association, 96*, 895-905.

Wolfinger, R. D. (1998). Towards practical application of generalized linear mixed models. In B. Marx & H. Friedl (Eds.), *Proceedings of the 13th International Workshop on Statistical Modeling* (pp. 388-395). New Orleans, Louisiana, USA.

Wolfinger, R., & O'Connell, M. (1993). Generalized linear mixed models: a pseudo-likelihood approach. *Journal of Statistical Computation and Simulation, 48*, 233-243.

Wu, M., & Carroll, R. (1988). Estimation and comparison of changes in the presence of informative right censoring by modeling the censoring process. *Biometrics, 44*, 175-188.

Wulfsohn, M., & Tsiatis, A. (1997). A joint model for survival and longitudinal data measured with error. *Biometrics, 53*, 330-339.

Xu, J., & Zeger, S. (2001). Joint analysis of longitudinal data comprising repeated measures and times to events. *Applied Statistics, 50*, 375-387.

Yu, M., Law, N., Taylor, J., & Sandler, H. (2004). Joint longitudinal-survival-cure models and their application to prostate cancer. *Statistica Sinica, 14*, 835-832.

Zhao, L. P., Prentice, R. L., & Self, S. G. (1992). Multivariate mean parameter estimation by using a partly exponential model. *Journal of the Royal Statistical Society B, 54*, 805-811.

Chapter 3
Multivariate and Multilevel Longitudinal Analysis

Nicholas T. Longford

Abstract This chapter presents a review of perspectives and methods for analysis of longitudinal data on several related variables. A connection is made with multilevel analysis in which the longitudinal and multivariate dimensions of the data can naturally be subsumed. With the focus on large-scale longitudinal studies of human subjects who are in general disinterested in and not highly motivated by the agenda of the study, methods for dealing with nonresponse are an essential addendum to the analytical equipment.

3.1 Introduction

Modern practice of data collection from human subjects is highly aware of the costs and difficulties in retaining survey respondents, especially in longitudinal studies in which survey subjects are to be contacted on several occasions, sometimes over a long period of time. One reaction to these pressures is to collect more complete information from complying subjects, so that the resulting data would be well suited for a wider analytical agenda within the remit of the survey. In particular, it would enable us to study the associations of several variables, and how these associations are altered over time.

In this perspective, it is more appropriate to consider as an elementary data item the value of a vector $\mathbf{X}^{(t)}$ observed on a (single) occasion t. Any one component of $\mathbf{X}^{(t)}$ offers little information without the other components of $\mathbf{X}^{(t)}$. However, the vector $\mathbf{X}^{(t)}$ offers only a snapshot of a social, economic or epidemiological development in the studied population so, for any single t, $\mathbf{X}^{(t)}$ is also a much poorer source of information than the sequence $\mathbf{X}^{(1)}, \ldots, \mathbf{X}^{(T)}$. Such a sequence can be presented as a random matrix, and data for a random sample from the population as a

Nicholas T. Longford
SNTL, Barcelona, Spain
e-mail: NTL@sntl.co.uk

K. van Montfort et al., *Longitudinal Research with Latent Variables*,
DOI 10.1007/978-3-642-11760-2_3,

three-dimensional array $\mathbf{X}$ composed of vectors $\mathbf{x}_i^{(t)}$ for subjects $i = 1, \ldots, n$ at time points $t = 1, \ldots, T$.

We assume that the goal of an analysis is inference about a particular finite but large population $\mathscr{P}$, and that this population is represented by a sample $\mathscr{S}$ drawn from $\mathscr{P}$ by a simple random sampling design. We assume that the time points $1, \ldots, T$, at which the values of a vector of variables $\mathbf{X}$ are observed, are selected noninformatively, without regard for any of the values $\mathbf{X}$ of the units in the sample.

We have two perspectives which lead to diverging approaches to inference. In the *sampling-design based perspective*, there is a finite set of units $1, \ldots, N$ with fixed (unchanging) values of $\mathbf{X}^{(t)}$ for every time point t. In a replication of the study, a different set of units would be selected into the sample, but if a unit i happened to be selected in both samples, its values of $\mathbf{X}^{(t)}$ would be the same in the two replications. In this view, sampling is the only source of variation, and the sampling design provides its complete description.

In the *model-based perspective*, the values of $\mathbf{X}^{(t)}$, $t = 1, \ldots, T$, are generated by a particular stochastic process, the definition of which (or, in most practical settings, an approximation to it) is the analyst's responsibility. Inferences are made assuming this model, but the analysis is accompanied by a careful diagnosis that searches for contradictions of the data with the assumptions made. This approach is much more common nowadays because it is more flexible, with a greater variety of software tools that have the necessary elements for its implementation.

The two perspectives are not completely separated. Dealing with nonresponse is a notable concern that they have in common. Even in the sampling-design based perspective, a model has to be posited for how the missing data are related to the recorded data (Little and Rubin, 2002); without a model the analysis would be at a dead end. In contrast, the model-based perspective ignores all the units with empty records (no data available); in many analyses no information is available about the units that were selected into the sample but nothing was recorded about them. The perspective is, however, concerned about making use of the information in incomplete records for which some, but not all, values are recorded. The concern about good representation of a population often appears out of place because no reference population is defined, or the model is specified in such a way that it implies or generates an impression of universality; that, within some reasonable bounds, it applies to *any* population.

Our view is that this perspective is constructive but not valid. We qualify this view by adding that we do not regard model validity as an imperative for a respectable analysis. We illustrate this on a simple example of a growth model

$$\mathbf{y}_i = \mathbf{Z}_i\beta + \varepsilon_i\,, \tag{3.1}$$

where $\mathbf{y}_i$ are $T \times 1$ (column) vectors of outcomes for units $i = 1, \ldots, n$, $\mathbf{Z}_i$ is the regression matrix for unit i, β is the vector of regression parameters, and ε_i are a random sample from a centred multivariate normal distribution, $\mathscr{N}(\mathbf{0}, \Sigma)$. We may regard $\mu_i = \mathbf{Z}_i\beta$ as the growth for a typical unit, but deviations from μ_i, unless they are extreme, cannot be regarded as anything untypical. The vector of deviations ε_i

does not represent any errors, because deviation from μ_i is not necessarily a sign of anything that has gone wrong. Usually incorrect is the central assumption of the functional form of μ_i — an unavoidable 'error' committed by the analyst against the environment (nature) in which units are exposed to a multitude of influences, many of them continual, the impact of which defies both our understanding and any neat algebraic summarisation. The choice of the variables in **Z** is governed by analytical pragmatism, attempting to capture the most important features of the studied phenomenon. With more extensive data (more observed units), we can capture finer detail and include more variables in **Z**. When more variables are recorded, we have a wider choice of variables in **Z**. Validity of a model, defined as a collection of distributions according to one of which the data is generated, is an unattainable goal. Its pragmatic reduction is a model that the data appear not to contradict, as assessed by various model-diagnostic procedures.

In theory and reality, there is a single valid model (the process); in practice, we improvise with the information we possess, and the intermediate goal of variable selection has different targets depending on the extent of the information, ignoring the fact that there is only *one* valid model. The pretense that the model we have selected is the valid model is a common logical inconsistency that does considerable harm to the integrity of the statistical practice. Attempts at addressing this problem (Draper, 1995; Chatfield, 1995; and Longford, 2007) have been largely ignored because of the complexity involved. They entail taking into account the model uncertainty, acknowledging that the model-selection process is also subject to sampling variation.

The model in (3.1) ascribes a different status to the covariates in **Z** than to the outcomes in **y**, even when there is no distinction in the way their values are collected in a survey. Both **y** and **Z** are attributes of the members of the population that are not amenable to any control, unlike a treatment assigned by randomisation in an experimental study. In particular, any causal inference is highly problematic when **Z** is observed just as passively as **y**, without exercising any influence (control) over its values. The regression in (3.1), summarised by the vector of parameters β, is a comparison of subpopulations (strata) defined by the values of **Z**, and it offers no basis for statements about manipulation — what would happen if a particular unit had a different value of **Z**. There would be an answer, in principle, if the valid data-generating model were known. In practice, such a model is not known and the recorded variables are usually a small subset of the variables that would have a role in such an ideal model.

In the modelling perspective, longitudinal analysis combines aspects of multivariate and multilevel analyses. It is multivariate, because one or several variables are observed on several occasions, and the study of the associations of these (time-specific) versions of the variable(s) is of obvious interest. It is multilevel, because the observations on a subject at the time points are naturally clustered, and the subjects may be further clustered within families, areas (locations), schools, businesses and similar organisations. The purpose of this chapter is to elaborate these links and perspectives, with an emphasis on taking advantage of their strengths in responding to the various complexities encountered in the analysis of longitudinal data.

The next section introduces the univariate longitudinal setting and the following section discusses nonresponse. Section 3.4 extends the models for multivariate outcomes. Section 3.5 discusses modelling of univariate outcomes in greater detail, studying dependence across time and variance heterogeneity. Section 3.6 deals with multivariate versions of these models. Computational issues, model fitting and graphical presentation, are addressed in Section 3.7. The chapter is concluded with a discussion.

3.2 Inferential Targets

Assuming that the values of the vector of outcomes $\mathbf{x}^{(t)}$ are well defined for any time-point $t \in (0,T)$, or beyond, we may associate each member j of $\mathscr{P}$ with a multivariate function $F_j(t)$ of time. This function, describing the growth, evolution or development, is a relevant target of inference. Inference about its behaviour in the near future amounts to extrapolation, but we can learn from its behaviour in the past, assuming some form of stationarity. The observations $\mathbf{x}_j^{(t)}$ at time points $t = 1,\ldots,T$ inform about F_j only partially. If all the subjects in the sample are observed in a regimented fashion, at time points $t = 1,\ldots,T$, then we have no information about the behaviour of F_j between any two (integer) time points. This suggests that we may learn more by implementing designs with unevenly set time points t. The vectors of outcomes may have uneven lengths, and the time points for a unit need not be distributed evenly. However, the choice of the time points t has to be noninformative for every unit, independently of the functions F_j. This is ensured when the time points are set by design. When the observational units (subjects) volunteer to provide the information, (e.g., as patients or customers), or become data donors opportunistically e. g., by being met at a railway station or a shopping centre, we have to be concerned about the good representation of the sample, as well as by the non-ignorable nature of the time-selection process.

The model in (3.1) has no straightforward adaptation for unevenly distributed time points. For each unit i we posit a model

$$y_{ih} = f_i(t_{ih}) + \varepsilon_{ih}\,,$$

where t_{ih} is the time at the observation h of unit i and ε_{ih} are a random sample from a (univariate) centred normal distribution, $\mathscr{N}(0,\sigma^2)$. We may specify a separate model for the variance σ^2, relating it to time t. The unit-specific functions f_i may involve some coefficients ξ_i, for which another model would be defined, linking the units to vectors ξ_i:

$$\xi_i = \nu + \delta_i\,, \tag{3.2}$$

where δ_i is a random sample from a multivariate distribution. Instead of ν we may have a model that relates the expectations $\mathrm{E}(\xi)$ to a (linear) function of some covariates defined for the units. The decomposition in (3.2) connects the unit-specific functions f_i and enables us to describe the population of units in terms of a typical unit given by the parameter vector ν and unit-level variation described by the distribution of δ_i. The link between f_i and ξ_i need not be linear, and so the function f that corresponds to ν is, in general, not a population average of the functions f_i.

3.3 Incompleteness

By complete data we understand a valid entry for every data item that was intended to be collected by the design (protocol). A typical protocol calls for collecting a rectangular dataset, a list of variables recorded at each of a set of time points for every unit in the sample. Incompleteness, broadly interpreted as failure to adhere to the design, is common especially when the units are human subjects for whom the interview and measurement (elicitation) process are an unwelcome distraction. A record comprising entirely of missing values (unit nonresponse) or lost in the process of transfer from the interviewer (data collector) to the (secondary) analyst through the database constructor, may be dropped from the analysis. If no trace is left after such records in the database the analyst knows nothing about their existence.

A record comprises subrecords for the time points, and any of these subrecords may be missing (time-point or *wave* nonresponse). Unless the analyst is aware, or infers from the patterns in the data, that the design called for the collection of a rectangular dataset, the dataset can be subjected to an analysis as if it were complete. Similarly, a subrecord may be empty or incomplete, involving item nonresponse. The design, however, is important. Pretending that the incomplete dataset is complete results in invalid inferences — inappropriate claims of unbiasedness and efficiency.

Even if the design did not call for a rectangular dataset, we may pose the problem of the analysis as involving missing values, values the addition of which would make the dataset rectangular and amenable to a relatively simple analysis. Of course, this approach is not practical when a large fraction of the values in the hypothetical rectangular dataset are missing (and have to be imputed) and the pattern of nonresponse is varied. When practical, this approach is relatively simple to implement because we are privy to the details of the nonresponse process.

3.4 From Univariate to Structured Multivariate Data

We develop models for multivariate longitudinal data within a more general framework of multivariate structured outcomes from univariate models and data by adding dimensions. We use the term *dimension* similarly to the term *factor* in ordinary regression (and the software `GLIM`; Francis, Green and Payne, 1993). Thus the various outcomes recorded on an occasion are a dimension, and the times of observation are another dimension. We refer to the outcomes as *components* of the vector of outcomes, although the components themselves can be multivariate; for example, a row of the timepoint-by-variable matrix of outcomes of a subject is a component.

For a univariate outcome y we consider linear regression on some covariates $\mathbf{x}$:

$$y = \mathbf{x}\beta + \varepsilon\,,$$

with the usual assumptions of independence, normality and homoscedasticity. We address departures from normality by generalized linear models, for which an alternative distributional assumption is required, together with a link function that relates the underlying linear predictors to the conditional expectations of the outcomes, $\mathrm{E}(y\,|\,\mathbf{x})$.

The structure of clustering, of sets of units having more similar values of the outcome y than the units in general, is introduced by assuming that the units within clusters are correlated. The simplest way of doing this is by the *compound symmetry* model, in which

$$\mathbf{y}_j = \mathbf{X}_j\beta + \delta_j + \varepsilon_j\,, \tag{3.3}$$

where $\mathbf{y}_j = (y_{1j},\ldots,y_{n_j j})^\top$ is the vector of outcomes in cluster j, $\mathbf{X}_j$ is the regression matrix for these units, composed of the rows $\mathbf{x}_{ij}$; δ_j, $j = 1,\ldots,n$, are a random sample from $\mathcal{N}(0,\sigma_\mathrm{B}^2)$; the $n = n_1 + \cdots + n_m$ elements of ε_j are a random sample from $\mathcal{N}(0,\sigma^2)$; and the two random samples are independent. The within-cluster correlation $\rho = \sigma^2/(\sigma^2+\sigma_\mathrm{B}^2)$ summarises the relative similarity of the units within clusters.

For observational (elementary) units within clusters we can distinguish between variables that are defined for the units (elements) and for the clusters. The latter variables are expanded for the elements so that all units within a cluster have the same value as their cluster. Variables defined for units could, in principle, have values that are constant within clusters. At the other extreme, the values could have identical means, or even identical distributions within the clusters. Such variables are called *balanced* with respect to clustering. As a convention, we include the intercept, represented by the vector of unities $\mathbf{1}$, among the balanced variables. For the vectors of covariates $\mathbf{x}_{ij}$ we have the following decomposition of the matrix of crossproducts:

$$\mathbf{X}^\top\mathbf{X} = \mathbf{B} + \mathbf{W}\,, \tag{3.4}$$

where

$$\mathbf{B} = \sum_{j=1}^{m} n_j\,(\bar{\mathbf{x}}_j - \bar{\mathbf{x}})^\top\,(\bar{\mathbf{x}}_j - \bar{\mathbf{x}})$$

$$\mathbf{W} = \sum_{j=1}^{m}\sum_{i=1}^{n_j} (\mathbf{x}_{ij} - \bar{\mathbf{x}}_j)^\top\,(\mathbf{x}_{ij} - \bar{\mathbf{x}}_j)$$

and $\bar{\mathbf{x}}_j$ is the sample mean within cluster j and $\bar{\mathbf{x}}$ the overall sample mean. Balanced variables contribute only to $\mathbf{W}$ and cluster-level variables only to $\mathbf{B}$.

We have to draw a distinction between the sample and population versions of the summaries **B** and **W**, as well as any other quantities we define later. Although the models we consider have fixed values of the covariates, **X**, in a typical sampling process applied in a (human) population the values of **X** are random. That is, in a hypothetical replication of the survey, a different matrix **X** would be realised. In the (sampling) design-based perspective, the values of **X** and **y** are fixed in the population, and the sampling process is the sole source of variation. That is, if a subject happened to be included in the sample in two replications, his or her values of **x** would be the same, and he or she would be in the same cluster.

The design-based perspective has in the past been regarded as not constructive, and the inferential effort in many areas has drifted toward model-based approaches. However, there are areas where the balance is being restored. For example, the potential outcomes framework for observational studies (Holland, 1986; Rubin, 2005) shifts the focus from the association of **X** and **y** to the analysis of the (treatment) assignment process. This analysis is model-based, but it is only an intermediary to the substantive analysis which follows, and which is simple, related to the analysis in an experimental setting, and has more in common with the design-based paradigm.

In the model-based paradigm, the similarity of the units within clusters can be interpreted in terms of differing within-cluster associations of **X** and **y**. The model in (3.3) corresponds to parallel within-cluster regressions, which have identical regression slopes, but different intercepts $\beta_0 + \delta_j$. This characterisation uncovers its relative rigidity. Much greater flexibility is attained by allowing some (or all) regression slopes to vary from cluster to cluster. The within-cluster slope for a variable that is constant within clusters is not identified. Therefore it is meaningful to consider varying slopes only with respect to variables defined for the elements. We split the covariates into the two groups, $\mathbf{X} = (\mathbf{X}^{(1)}, \mathbf{X}^{(2)})$, where $\mathbf{X}^{(1)}$ are defined for elements and $\mathbf{X}^{(2)}$ for clusters; we assume that none of the variables in $\mathbf{X}^{(1)}$ is constant within clusters. Then the compound symmetry model is

$$\mathbf{y}_j = \mathbf{X}_j^{(1)}\beta^{(1)} + \mathbf{X}_j^{(2)}\beta^{(2)} + \delta_j + \varepsilon_j .$$

Its obvious generalisation is

$$\mathbf{y}_j = \mathbf{X}_j^{(1)}\beta^{(1)} + \mathbf{X}_j^{(2)}\beta^{(2)} + \mathbf{X}_j^{(1)}\delta_j + \varepsilon , \tag{3.5}$$

where δ_j is a random sample from a centred multivariate normal distribution, $\mathcal{N}(\mathbf{0}, \Sigma_{\mathrm{B}})$. We have to extend the definition of the multivariate normal distribution to singular (degenerate) distributions for which Σ_{B} is singular. Let p_h be the number of covariates (columns) in $\mathbf{X}^{(1)}$ and $\mathbf{X}^{(2)}$. Then Σ_{B} is a $p_2 \times p_2$ variance matrix. In Σ_{B}, it is meaningful to constrain some variances to zero. This corresponds to constant within-cluster slopes with respect to the corresponding covariates. When a variance is constrained to zero, then so are all the covariances in the same row and column. We have to obey the rules of invariance with respect to linear transformations (Longford, 2007, Chapter 9), which dictate that the intercept should be associated with a variance to be estimated so long as any other covariate is. A categorical variable with K distinct labels is represented among covariates by $K-1$ indicator variables. When such a variable is defined for elements, the invariance rules imply that either all the $K-1$ variables are associated with variances to be estimated, or none are. After all, the values of the indicator variables are contingent on the choice of the reference, which in most cases is arbitrary or opportunistic.

3.5 Univariate Observations at Time Points

In longitudinal analysis, each unit j is observed at a (finite) sequence of time points

$$\mathbf{t}_j = \left(t_j^{(1)}, t_j^{(2)}, \ldots, t_j^{(n_j)} \right)^\top .$$

When all units are observed at the same set of time points, $\mathbf{t}_j \equiv \mathbf{t}$, the outcomes form a sample from a multivariate (normal) distribution, so that

$$\mathbf{y}_j \sim \mathscr{N}(\mu, \Sigma) .$$

Structures can be imposed on the vector μ and variance matrix Σ, such as linear growth and compound symmetry, but these are useful only when the number of time points, p, is large, so that a linear function, represented by two parameters, or a quadratic function, by three, provides a much more compact description for the growth (development, expansion, decay, or the like) than the components of μ. The unstructured vector μ is 'always correct', but may be ineffective, in that it restricts the inferences that can be made to the specific time points.

In contrast, a functional expression for μ is in general incorrect, but the bias it entails may be offset by the reduced sampling variance associated with its estimation. The multivariate perspective is inflexible — it cannot be adapted for inferences about other time points, by inter- or extrapolation. The functional perspective caters for such inferences by prediction, although the issues of correctness, and its scope being limited to the particular context, delimit its application, especially for extrapolation.

Without a structure imposed on μ, the design has to ensure that the time points in $\mathbf{t}$ are the ones for which inference is desired. With a structure on μ, the design has to ensure that the function underlying the expectations μ, $\mu(t)$, can be estimated with desired precision and predictions based on it have sufficient quality.

Similar comments can be made about specifying Σ. Without a structure imposed on Σ, each covariance in Σ is a unique quantity, although in most contexts we can reasonably assume that greater distance of the time points is associated with lower correlation. For the variances in Σ, a reasonable assumption may be that they are constant or increasing with the distance in time, but a function underlying them amounts to an assumption highly contingent on (the choice and coding of) the time points $\mathbf{t}$.

A parametric structure can be imposed on Σ or on its inverse, Σ^{-1}, called the *concentration matrix* or, in principle, on any transformation of Σ. Working with Σ^{-1} is particularly attractive when considering the Markov property of conditional independence of any two observations given an observation that separates them:

$$\left(y_{t_1} \,|\, y_{t_2}, y_{t_3}\right) \sim \left(y_{t_1} \,|\, y_{t_2}\right),$$

for the outcomes at any time points $t_1 < t_2 < t_3$. The corresponding matrix Σ^{-1} is tridiagonal:

$$\left\{\Sigma^{-1}\right\}_{kh} = 0$$

whenever $|k-h| \geq 2$. When the number of time points is greater, this constraint may be considered also for $|k-h| \geq 3$ or 4; that is, entries outside the diagonal strip of Σ^{-1} of the given width vanish.

The parameters in μ and Σ are indivisible in the following sense. When no structure is imposed on μ a structure should not be imposed on Σ either. Imposing a structure on μ is simpler than on Σ, because it is a unidimensional object. Therefore a structure may be imposed on μ, but not on Σ, but this is mainly a pragmatic matter reflecting our inability or lack of confidence about specifying suitable submodels.

Observations of the outcomes Y in a longitudinal analysis may be accompanied by the values of covariates. These may be defined for the subjects and for the (elementary) observations. Adjustment for the variables defined for subjects may be made by multivariate regression:

$$\mathbf{y}_j = \mathbf{x}_j \mathbf{B} + \gamma_j, \tag{3.6}$$

where $\mathbf{y}_j$ is the vector of outcomes for subject j, $\mathbf{x}_j$ the vector of values of the covariates, $\mathbf{B}$ the matrix of regression parameters, and the deviations γ_j are a random sample from a multivariate normal distribution, $\mathcal{N}(\mathbf{0}, \Gamma)$.

Covariates that are specific to time points cannot be accommodated in the model in (3.6) because the vector $\mathbf{y}_j$ is treated like a single unit. The problem does not arise with hierarchical models in which observations and subjects form separate levels of nesting:

$$y_{ij} = \mathbf{x}_j^{(1)} \beta^{(1)} + \mathbf{x}_j^{(2)} \beta^{(1)} + \mathbf{x}_j^{(1)} \delta_j + \varepsilon_{ij}, \tag{3.7}$$

with assumptions similar to those in (3.5). The variables in $\mathbf{x}_j^{(1)}$ are defined for the occasions and those in $\mathbf{x}_j^{(2)}$ for subjects. In the variance matrix $\Sigma_{\mathrm{B}} = \mathrm{var}(\delta_j)$, we can introduce constraints analogous to those in (3.5), so that an expression more accurate than (3.7) is

$$y_{ij} = \mathbf{x}_j^{(1)} \beta^{(1)} + \mathbf{x}_j^{(2)} \beta^{(2)} + \mathbf{z}_j \delta_j + \varepsilon_{ij}, \tag{3.8}$$

where $\mathbf{z}$ is a subset of the variables in $\mathbf{x}^{(1)}$. The interpretation in terms of varying regression slopes also carries over to the longitudinal setting. The within-subject regression slopes with respect to the variables in $\mathbf{z}$ vary, and with respect to their complement in $\mathbf{x}^{(1)}$ are constant.

The vector $\mathbf{x}^{(1)}$ contains the variable(s) that represent time. For linear within-subject regressions, time is represented by a single variable, but growth may follow any other pattern. To cater for the possibilities, transformations of the time have to be included in $\mathbf{X}^{(1)}$, and some of them also in $\mathbf{Z}$. Invariance with respect to linear transformations dictates that a variable included in $\mathbf{Z}$ should be included also in $\mathbf{X}^{(1)}$. Further, when a hierarchy is defined for the variables in $\mathbf{X}$, such as in polynomial regression, then this hierarchy should also be reflected in the model choice. For example, if the cubic term, t^3, is included in $\mathbf{X}^{(1)}$, then so should be the linear and quadratic terms. Similarly, if t^3 is included in $\mathbf{Z}$, then so should be the linear and quadratic terms. However, if t^3 is included in $\mathbf{X}^{(1)}$, it does not have to be included in $\mathbf{Z}$ although if t^2 is included, then so should be the linear term t.

The two-level model (Longford, 1993; Verbeke and Molenberghs, 2000; and Goldstein, 2003) can be applied when observations are made at a given (fixed) set of time points, but some limitations arise for the combination $\mathbf{z}_j\delta_j$. For r time points, the largest possible dimension of δ_j is r. The multivariate model in (3.6) corresponds to r-variate δ_j with $\mathbf{Z}$ comprising the unity (intercept) and the indicators of the categories $2, 3, \ldots, r$. Other options correspond to a reparametrisation of such a vector $\mathbf{Z}$. When the observations are not made in a regimented fashion, the number of variables in $\mathbf{Z}$ may still have to be restricted. To see this, consider a design with time points that for every subject are drawn from the same set, such as $1, \ldots, 10$, but not every subject has all the ten observations. Then $\mathbf{Z}$ should not contain more than ten covariates (columns). A direct analogy can be drawn with the models for the analysis of covariance (ANCOVA). The models in (3.8) differ from them solely by associating the subject specific deviations δ_j with randomness; in ANCOVA they are (fixed) parameters, subject only to the constraints of identifiability.

A subject-level variable $X^{(2)}$ is by definition constant within subjects, and so the within-subject regression with respect to $X^{(2)}$ is not well defined. The only reason why such a variable might be included in $\mathbf{Z}$ is to model variance heterogeneity — the dependence of the variance on the covariates. In general, for the model in (3.8), we have the identities

$$\begin{aligned} \operatorname{var}(y_{ij}) &= \sigma^2 + \mathbf{z}_{ij}\Sigma_{\mathrm{B}}\mathbf{z}_{ij}^{\top} \\ \operatorname{cov}(y_{ij}, y_{i'j}) &= \mathbf{z}_{ij}\Sigma_{\mathrm{B}}\mathbf{z}_{i'j}^{\top} \end{aligned} \tag{3.9}$$

for $i \neq i'$. Both expressions are quadratic functions of the components of $\mathbf{z}$. Therefore, exploring the properties of $\operatorname{var}(y)$ and $\operatorname{cov}(y_1, y_2)$ as functions of $\mathbf{z}$ is relatively simple, although the components of $\mathbf{z}$ may be interrelated, such as the indicator variables for a categorical variable, or the linear and quadratic terms of a polynomial. The range of the values of the time t is usually limited, so we can evaluate $\operatorname{var}(y)$ as a function of t unambiguously when $\mathbf{z}$ contains only functions of time. Otherwise we have to consider a few (typical) values of the other variables and evaluate $\operatorname{var}(y)$ for each of them.

3.5.1 Example

Figure 3.1 gives an example of a longitudinal dataset with observations at time points 1, 2, ..., 12 for 40 subjects. The outcomes are generated according to the model in (3.8) with no covariates, except for the time and its transformations. For the regression $\mathbf{x}^{(1)}\beta$ we use a cubic polynomial in t, and for the variation $\mathbf{z}\delta_j$ a quadratic polynomial in t. Thus, the values $\mathbf{x}^{(1)}\beta + \mathbf{z}\delta_j$, $j = 1,\ldots,40$, are cubic polynomials with the same cubic coefficient but different quadratic (and linear and absolute) coefficients. The data are generated with

$$\beta = (1,\, 0.3,\, 0.024,\, 0.0011)^\top ,$$

$\sigma^2 = 0.25$ and

$$\Sigma = \begin{pmatrix} 0.60000 & 0.04000 & 0.00030 \\ 0.04000 & 0.02000 & 0.00015 \\ 0.00030 & 0.00015 & 0.00009 \end{pmatrix} .$$

The curves (trajectories) $\mathbf{x}^{(1)}\beta + \mathbf{z}\delta_j$ are plotted in panel A. We refer to them as smooth or underlying trajectories, because they are devoid of the inexplicable contribution ε. This random term is commonly referred to as an error. In most contexts, this label is inappropriate and misleading. It would be appropriate if the model we specify were correct (as it is in a simulation) and if all subjects behaved according to this model with σ^2, and the elementary-level deviations ε arose as a result of an imperfect measurement process. In most cases, the model is incorrect, and a particular positive value of σ^2 is appropriate because subjects do not behave according to any conceivable formula, but there are some equations (models) that approximate the behaviour reasonably well. The approximation is in error, not the behaviour.

Panel B presents the trajectories as they would be observed, made coarse by the elementary-level deviations ε. It is difficult to infer the patterns of the trajectories, smooth or coarse, from the parameter values in β, Σ and σ^2, except perhaps for the extent of the average curvature (from β_3) and the extent of inexplicability (from σ^2); using a simulated sample from the fitted model is much more reliable. Such a sample, replicated several times, also has an important diagnostic value, as discussed in Section 3.5.3.

The variance of an observation, as a function of time t,

$$\operatorname{var}(y\,|\,t) = \sigma^2 + (1,\, t,\, t^2)\,\Sigma\,(1,\, t,\, t^2)^\top ,$$

is drawn in panel C, together with the indication of σ^2 as its 'constant' contributor (drawn by dashes). There is no profound reason why the elementary-level variance should be constant; it is merely a convenient assumption. Without it, we would have to posit a particular form for how σ^2 depends on t. Alternatives plausible in some settings are that the correlation of two outcomes of the same subject is constant, or the ratio of the within- and between-subject variances is constant.

There is a trade-off between a within-subject variance σ_t^2 and the subject-level matrix Σ. That is, up to a point, a change in one or several values of σ_t^2 can be

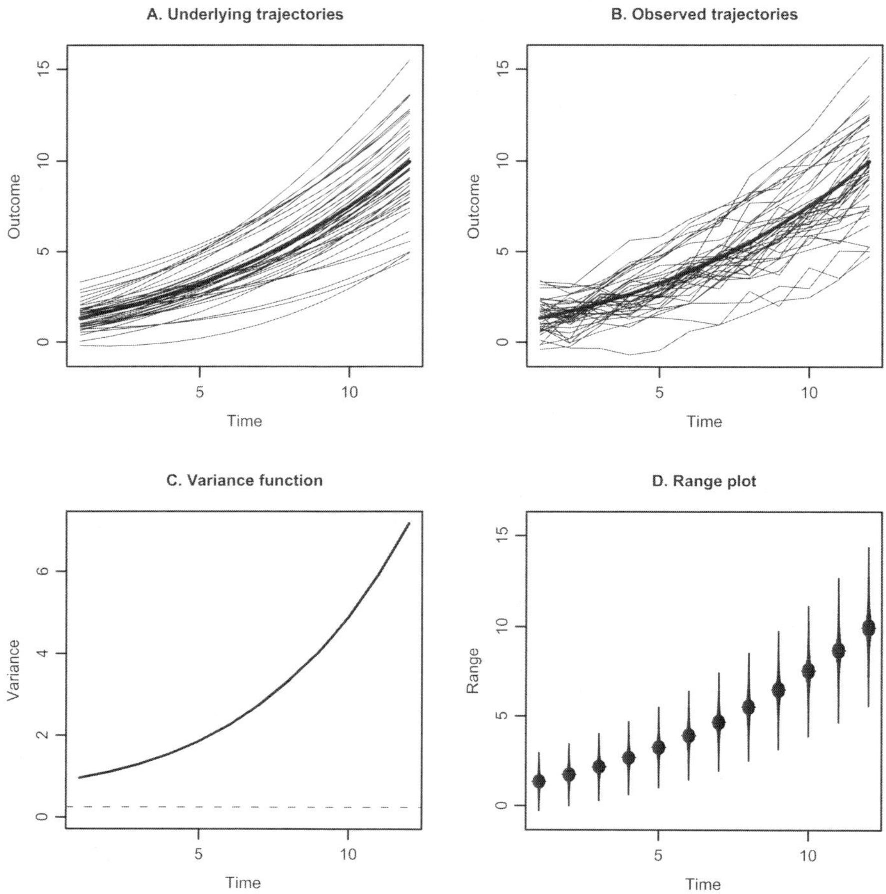

Fig. 3.1 Graphical representation of balanced longitudinal data. A simulated example.

compensated by changes in Σ, so long as the 'new' Σ remains non-negative definite. In principle, σ^2 is identifiable from the data, because it represents the *independent* contribution to the variance var(y). The variance matrix Σ characterises the covariance structure of the observations of a subject. However, large samples are required to separate the two components of variance, σ_t^2 and Σ, with any meaningful reliability.

Panel D of Figure 3.1 illustrates the distributions of the outcomes within the time points. It does not contain all the information about the underlying process, because it gives no indication of the covariance structure of the outcomes. In this respect, there is no replacement for the simulated trajectories in panels A and B.

Nonlinear transformations can alter the pattern of the trajectories substantially, from convex to approximately linear to concave. With a nonlinear transformation, we manipulate the underlying distributions (normal for ε and δ), the covariance structure, and the pattern of the variances σ_t^2. In theory, only one family (class of equivalence) of transformations yields normally distributed outcomes, so arranging for all the distributional assumptions to hold is next to impossible. In practice, there is a considerable leeway in the choice of a transformation to make the assumptions of normality palatable. In fact, in many settings we can focus on transformations that bring about variance homogeneity (independence of the variances var(y) and σ_t^2 on time t) as well.

3.5.2 The Time-Selection Process

In many longitudinal studies, the values of the time points t are not set by design, prior to data collection. For example, a study may rely on subscribing individuals turning up at a given location for a particular service, such as health care, advice with jobs search, a form of entertainment, and similar. In such settings, the realised values of t may be *informative*, and the process that generates its values *nonignorable*. The observed data are not a good reflection of the process we set out to study.

This problem does not have a solution, in that there is no straightforward way of adjusting the analysis so that it would be suitable for inferences about the entire evolution of the outcome variables, or about the values of the outcome variables at time points selected by design, with the subjects exercising no choice in the matter.

3.5.3 Simulation-Based Diagnostics

Established methods for model diagnostics are difficult to adapt for longitudinal analysis because of a combination of concerns about normality, appropriate covariance structure and heteroscedasticity. The following generic procedure, introduced by Rubin (1984), can be applied. We define a data summary called *feature*; this can be a single quantity, a vector, a table, a diagram, or their combination (a *multifeature*). We evaluate (or apply) this feature to the realised dataset, thus obtaining the *realised* feature. Next, we simulate datasets from the model fit using the same design (sample sizes and values of the covariates) as the realised data, and evaluate the feature on each replicate dataset. We shuffle the one realised and the several simulated features, and ask a third party (a colleague) to identify one of them as being exceptional. If he or she points to the realised feature (without knowing that it is based on the real dataset and the others are not), we conclude that the model is not appropriate, because if it were, as it is with the simulated data, then the features would not look (or be) different. It is advantageous to generate 19, 49 or 99 replicate datasets, so that we would have 20, 50 or 100 datasets and could relate the probability of identifying the realised dataset by chance to the size of a test in hypothesis testing. The price for greater accuracy is having to generate a greater number of replicates (a serious problem only with very large datasets), and presenting a more cumbersome task for the colleague. See Longford (2001) for an example.

3.6 Multivariate Observations at Time Points

Suppose the observations of a set of variables at a time point t are well described by a multivariate normal distribution $\mathcal{N}(\mu_t, \Sigma_t)$, specific to the time point t. We are concerned about the evolution of these distributions across the time points. This entails specifying models for the vectors of expectations μ_t and variance matrices Σ_t, but also for the correlation structure of vectors of observations at distinct (consecutive) time points. This is necessary even in the stationary case, when the matrices Σ_t are identical. For example, the assumption that vectors of outcomes $\mathbf{y}_t$ and $\mathbf{y}_{t'}$ are independent for distinct time points $t \neq t'$ is in most settings untenable, and so is the assumption of perfect correlation, $\mathbf{C}_{tt'} = \mathrm{cov}(\mathbf{y}_t, \mathbf{y}_{t'}) = \Sigma_t$.

Multivariate longitudinal outcomes are represented by a matrix of variables $\mathbf{Y}$, comprising vectors of variables $\mathbf{y}_t$ at a time point as its rows and the time series of univariate longitudinal outcomes $\mathbf{y}^{(k)} = \left(y_1^{(k)}, y_2^{(k)}, \dots y_T^{(k)}\right)^\top$ as its columns. An ideal solution for the correlation structure across the time points would allow an (arbitrary) univariate longitudinal model for each component $\mathbf{y}^{(k)}$ and a rich variety of dependence structures implied by the covariance matrices $\mathbf{C}_{tt'}$. Of course, imposing constraints such as non-negative covariances in $\mathbf{C}_{tt'}$ and higher correlations for pairs of time points t and t' in greater proximity, is reasonable in most contexts. We seek models mainly for short time series (small T), so we are not concerned about stationarity and other properties that are related to large T.

3.6.1 *Autoregression*

The univariate autoregression has an obvious multivariate analogue,

$$\mathbf{y}_{t+1} = \mathbf{a}_t + \mathbf{B}_t \mathbf{y}_t + \varepsilon_t\,, \qquad (3.10)$$

where $\mathbf{a}_t$ is a vector and $\mathbf{B}_t$ a matrix of coefficients and ε_t a centred random vector independent of $\mathbf{y}_1, \dots, \mathbf{y}_t$. To maintain multivariate normality, we assume that $\varepsilon_t \sim \mathcal{N}(\mathbf{0}, \Xi)$ and $\mathbf{y}_1 \sim \mathcal{N}(\mu_1, \Sigma_1)$. Then

$$\begin{aligned} \mathbf{C}_{t,t} &= \Sigma_t = \mathbf{B}_t \Sigma_{t-1} \mathbf{B}_t^\top + \Xi \\ \mathbf{C}_{t,t+1} &= \Sigma_t \mathbf{B}_t^\top . \end{aligned}$$

An important special case arises when $\mathbf{B}_t$ is diagonal. This does not correspond to independent autoregressions, because dependence is still injected by the covariance structure of ε_t, as well as the initial covariance matrix Σ_1. When $\mathbf{y}_t$ comprises closely related variables, the components of ε_t are correlated.

3.6.2 Moving Average

The univariate moving average model has a similar extension for multivariate outcomes. Each time point t is associated with an independent random vector ε_t with centred multivariate normal distribution $\mathcal{N}(\mathbf{0}, \Xi)$, and the vector of outcomes is assumed to be generated according to the model

$$\mathbf{y}_t = \mu_t + \mathbf{A}_0\varepsilon_t + \mathbf{A}_1\varepsilon_{t-1} \tag{3.11}$$

for some matrices of constants $\mathbf{A}_0$ and $\mathbf{A}_1$. A joint distribution has to be specified for the start of the series, $(\mathbf{y}_1, \mathbf{y}_2)$. The essential multivariateness of such a moving average arises as a result of the covariance structure of Ξ *combined* with the non-zero off-diagonal elements of $\mathbf{A}_0$ and $\mathbf{A}_1$. The models in (3.10) and (3.11) are for the respective first-order autoregressive and moving-average models. Their generalisation to higher-order models is straightforward. However, such models are of limited use with short time series typically encountered in longitudinal analysis. Autoregression and moving average yield distinct sets of models, so that, at least in principle, the issue of distinguishing between them, e.g., by hypothesis testing or information criteria, may arise. In practice, such tests have limited power even in the univariate case, so the data-based choice between them is unlikely to be feasible in a multivariate setting. The two kinds of models can be combined, in analogy with the univariate case.

3.6.3 Two-Level Models

The multivariate version of the compound symmetry model in (3.3) is

$$\mathbf{Y}_j = \mathbf{X}_j\mathbf{B} + \mathbf{1}\delta_j^\top + \mathbf{E}_j,$$

where $\mathbf{Y}_j$ is the $T \times K$ (times-by-variables) matrix of outcomes for subject j, $\mathbf{X}_j$ the corresponding matrix of covariates, $\mathbf{B}$ is a matrix of regression parameters, δ_j a random sample from a multivariate normal distribution (one vector per subject), $\mathbf{t}_j$ is the vector of time points, and $\mathbf{E}_j$ a matrix; the rows $\mathbf{e}_{ij}$ of $\mathbf{E}_j$ are mutually independent random vectors (a multivariate random sample), both within a matrix $\mathbf{E}_j$ and across them, from another multivariate normal distribution. The two random samples, δ_j (for subjects) and $\mathbf{e}_{ij}$ (for occasions within subjects), are mutually independent.

A column of the matrix $\mathbf{X}_j$ is time and some others are its transformations. These can be associated with subject-level variation by the model

$$\mathbf{Y}_j = \mathbf{X}_j \mathbf{B} + \mathbf{1}\delta_j^{(0)\top} + \mathbf{t}_j \delta_j^{(1)\top} + \mathbf{E}_j, \tag{3.12}$$

where the matrix $\Delta_j = (\delta_j^{(0)}, \delta_j^{(1)})$ is a random sample from a multivariate normal distribution. In this model, the subjects have different associations with time (varying coefficients $\delta_j^{(1)}$). The model can be supplemented with transformations of time, injecting more flexibility in how the values of the variables within subjects evolve. Covariances in $\mathrm{var}(\Delta_j)$ are essential, because the evolution of the variables is unlikely to be independent (unrelated). Of course, $\mathbf{E}_j$ induces some dependence among the rows of $\mathbf{Y}_j$, but we can regard $\mathrm{E}(\mathbf{Y}_j \,|\, \mathbf{E}_j) = \mathbf{X}_j \mathbf{B} + \mathbf{1}\delta_j^{(0)\top} + \mathbf{t}_j \delta_j^{(1)\top}$ as an underlying trend, and study its dependence structure.

We distinguish among variables defined for subjects, which are represented in each $\mathbf{X}_j$ by a column of constants, and variables defined for occasions. The time, represented in $\mathbf{X}_j$ by a column $\mathbf{t}_j$, is one such variable. Variables that are not functions of time, but are recorded on every occasion (observation), may also be included in $\mathbf{X}_j$. Such variables are usually called *time-varying*. They can be associated with subject-level variation to model the varying within-subject regressions of the outcomes on them. Associations of variables with the outcomes have to be interpreted with care when the values of these variables are recorded passively, without (experimental) control over them. In the framework of causal analysis, they may be 'intermediate' variables, affected by the earlier outcomes, and so their associations with the outcomes differ from the causal effects of these variables.

3.7 Maximum Likelihood Estimation

Maximisation of the likelihood with the normality assumptions is conceptually simple and is relatively easy to implement because the likelihood for all the models we consider has an analytical form. Some difficulties are caused by the large number of parameters some of which are connected by the assumed structures. The constraints of nonnegative definiteness are difficult to enforce. Other difficulties arise in the model specification, because there is no obvious way of defining a sequence of nested models that would represent gradual increase in model complexity. Connection of the substantive information with such constraints is particularly difficult to establish. In principle, we could define the joint distribution of all the outcomes directly. In such a definition, it is difficult to reflect the structure of observations within time points.

Likelihood maximisation involves iterative procedures, and these require a (good) initial solution. Initial solutions are frequently the fits of some very simple submodels which are obtained by a simple algorithm. A practical initial solution for fitting the model in (3.12) or its generalisations is the set of univariate multilevel model fits. These themselves require iterations, but they are much simpler than a 'multivariate' iteration. The univariate model fits are useful also for exploring informally the choice of models for the marginals, the components of $\mathbf{X}_j \mathbf{B}$.

Large variance matrices (of model parameters) are estimated by an algorithm that does not internally respects the nonnegative definiteness of the estimated variance matrices. In a large (estimated) matrix $\hat{\Omega}$, the presence of a negative eigenvalue is not obvious, so the problem might be ignored, until we come across a negative value of a quadratic form $\mathbf{c}^\top \hat{\Omega} \mathbf{c}$ for a vector of constants $\mathbf{c}$ or wish to draw a random sample from the fitted distribution. The constraints of nonnegative definiteness are difficult to implement in a full-proof fashion, because they involve a trade-off between slowing down the convergence rate and ensuring that the solution moves smoothly from one iteration to the next along (or close to) the boundary of the parameter space defined by nonnegative definiteness.

Alternative solutions estimate decompositions of the variance matrices, such as the Cholesky or single-value, but the structures we want to impose on the variance matrices are very difficult to convert to the constraints on these decompositions.

There is no comprehensive software for multivariate random coefficient models, but software for univariate models can be adapted for the purpose. `MLwin` (Rasbash *et al*, 2005) and the software `nlme` described in Pinheiro and Bates (2000) are well suited for this purpose. For methods, examples and general background, we recommend Diggle *et al* (2002). Laird and Ware (1982) is a paper of historical importance, outlining the application of random coefficient models for longitudinal analysis. There is extensive Bayesian literature on longitudinal analysis, much of it centred around or using the `WinBugs` software (`wwww.mrc-bsu.ca.ac.uk/bugs`).

3.7.1 Graphics – Initial Data Exploration

The first step in an initial exploration of the data is to plot the trajectories (evolutions) for each variable separately. The next step entails representing the dependence of the observations across the variables. Plotting the trajectories of the distinct variables side-by-side, with the subject marked for each trajectory is effective only for a few subjects (e.g., a random sample drawn from the data), so that the trajectories of a subject could be easily identified in the adjacent panels. In multivariate models with random slopes, the variances and correlations of the observations are time-specific, and so we can study their evolution by plotting them as functions of time. This can be effectively implemented by a matrix plot (function `pairs` in `R`), with the variances plotted in the diagonal panels and the correlations plotted in the off-diagonal panels. More information is displayed when the correlations are plotted under the diagonal and the covariances above it.

Figure 3.2 presents a bivariate longitudinal dataset. The relatively smooth lines in the top panels are for the underlying trends, devoid of the within-subject variation. The average trend (the regressions) are drawn by thick lines in the top panels. They enable, however crudely, to gain an impression of the correlation of the two outcomes (components). Comparisons within columns help us to assess the impact of the within-subject variation, commonly interpreted as noise or error, although an attribution of ε to a replication-specific random variable (due to the subject's

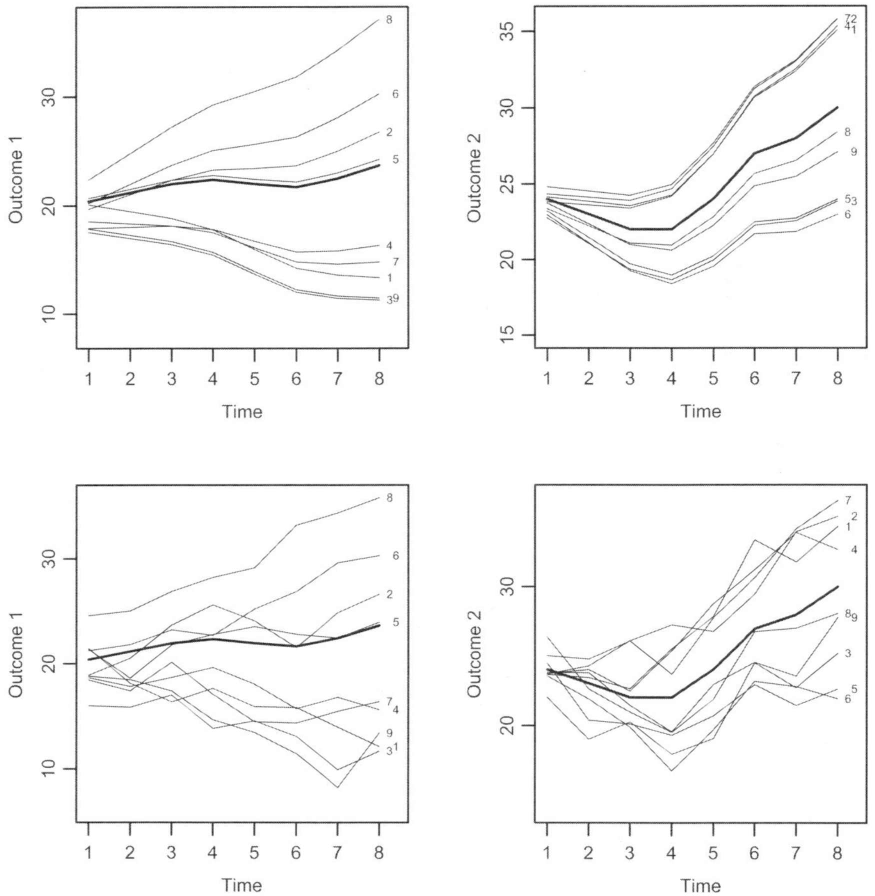

Fig. 3.2 A bivariate longitudinal dataset. A simulated example. The top panels display the underlying trends and the bottom panels the values of the observations. The thick solid line indicates the marginal (population) mean.

inconsistency in the response or imperfection of the measurement/recording process) is not always warranted.

In this example, the subject-level variance matrix was specified as

$$\Omega = \begin{pmatrix} 1.042 & -0.112 & 0.376 & -0.028 \\ -0.112 & 1.602 & 0.104 & 0.268 \\ & & & \\ 0.376 & 0.104 & 0.928 & 0.026 \\ -0.028 & 0.268 & 0.026 & 0.337 \end{pmatrix},$$

constructed from an eigenvalue decomposition to ensure nonnegative definiteness. The additional space in the display separates the rows and columns that correspond to the outcomes, and within each 2×2 matrix, the first component corresponds to the intercept and the second to (linear) time. The within-subject variance matrix is

$$\Sigma = \begin{pmatrix} 1.8 & 1.0 \\ 1.0 & 1.4 \end{pmatrix},$$

and the vectors of the population means for the two components are

$$\mu_1 = (20.4, 21.2, 22.0, 22.4, 22.0, 21.7, 22.5, 23.7)^\top$$
$$\mu_2 = (24.0, 23.0, 22.0, 22.0, 24.0, 27.0, 28.0, 30.0)^\top.$$

Figure 3.3 summarizes the marginal distributions graphically, highlighting the increasing variation with time.

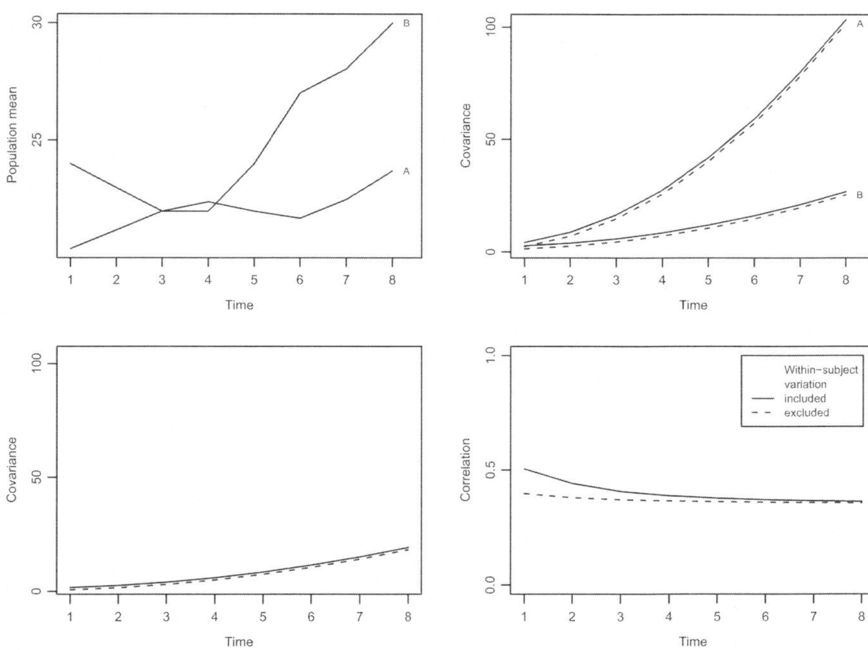

Fig. 3.3 The (marginal) summaries of a bivariate longitudinal series: trend (expectations), variances, covariances and correlations. Simulated data, with the parameters given in the text.

3.8 Discussion

Longitudinal analysis refers to analysis that involves a time dimension. In this respect it is multivariate, although it may involve other aspects of multivariateness, such as several outcomes being observed at each time point. Longitudinal data comprise repeated observations on subjects, so that their change (growth, decay or development) can be studied. The temporal dependence can be accounted for by regression or correlation structures, or their combinations, and the subject-to-subject variation by random coefficients. In the model construction, for estimation and prediction, we can draw on models for time series (autoregression and moving average) and for random coefficients. These are most conveniently specified with the assumptions of normality and linearity, for which estimation procedures are relatively simple, based on maximum likelihood. Transformations and the generalized linear modelling framework cater for departures from normality.

Designing longitudinal studies and dealing with nonresponse, and designing studies which anticipate nonresponse, are challenging problems that do not have a universal solution because of the intricate interplay of the correlation structure of the outcome variables with the quality of the estimation. Survey expenses are an important consideration, especially in studies that take place over a long period of time (several years) and in populations that, in general, do not have a stake in the survey and regard responding as a distraction from their everyday affairs. Methods for dealing with nonresponse and with data that do not fit into neat rectangular data structures have an important role in the analysis of such surveys.

References

Chatfield, C. (1995). Model uncertainty, data mining and statistical inference. *Journal of the Royal Statistical Society, Ser. A, 158,* 419-466.

Diggle, P., Heagerty, P., Liang, K.-Y., & Zeger, S. L. (2002). *Analysis of longitudinal data* (2nd ed.). Oxford: Oxford University Press.

Draper, D. (1995). Assessment and propagation of model uncertainty. *Journal of the Royal Statistical Society, Ser. B, 57,* 45-97.

Francis, B., Green, M., & Payne, C. (1993). *The GLIM system. Release 4 manual.* Oxford: Oxford University Press.

Goldstein, H. (2003). *Multilevel statistical models* (3rd ed.). London: E. Arnold.

Holland, P. W. (1986). Statistics and causal inference. *Journal of the American Statistical Association, 81,* 945-970.

Laird, N. M., & Ware, J. H. (1982). Random-effects models for longitudinal data. *Biometrics, 38,* 963-974.

Little, R. J. A., & Rubin, D. B. (2002). *Statistical analysis with missing data* (2nd ed.). New York: Wiley.

Longford, N. T. (1993). *Random coefficient models.* Oxford: Oxford University Press.

Longford, N. T. (2001). Simulation-based diagnostics in random-coefficient models. *Journal of the Royal Statistical Society, Ser. A, 64,* 259-273.

Longford, N. T. (2007). *Studying human populations. An advanced course in statistics.* New York: Springer-Verlag.

Pinheiro, J. C., & Bates, D. M. (2000). *Mixed-effects models in* `S` *and* `Splus`. New York: Springer-Verlag.

Rasbash, J., Charlton, C., Browne, W. J., Healy, M., & Cameron, B. (2005). `MLwin` version 2.02. Centre for Multilevel Modelling, University of Bristol.

Rubin, D. B. (1984). Bayesianly justifiable and relevant frequency calculations for the applied statistician. *Annals of Statistics, 12,* 1151-1172.

Rubin, D. B. (2005). Causal inference using potential outcomes: design, modelling, decisions. 2004 Fisher Lecture. *Journal of the American Statistical Association, 100,* 32-331.

Verbeke, G., & Molenberghs, G. (2000). *Linear mixed models for longitudinal data.* New York: Springer-Verlag.

Chapter 4
Longitudinal Research Using Mixture Models

Jeroen K. Vermunt

Abstract This chapter provides a state-of-the-art overview of the use of mixture and latent class models for the analysis of longitudinal data. It first describes the three basic types mixture models for longitudinal data: the mixture growth, mixture Markov, and latent Markov model. Subsequently, it presents an integrating framework merging various recent developments in software and algorithms, yielding mixture models for longitudinal data that can (1) not only be used with categorical, but also with continuous response variables (as well as combinations of these), (2) be used with very long time series, (3) include covariates (which can be numeric or categorical, as well as time-constant or time-varying), (4) include parameter restrictions yielding interesting measurement models, and (5) deal with missing values (which is very important in longitudinal research). Moreover, it discusses other advanced models, such as latent Markov models with dependent classification errors across time points, mixture growth and latent Markov models with random effects, and latent Markov models for multilevel data and multiple processes. The appendix shows how the presented models can be defined using the Latent GOLD syntax system (Vermunt and Magidson, 2005, 2008).

4.1 Introduction

The aim of this chapter is to provide a state-of-the-art overview of the use of mixture and latent class models for the analysis of longitudinal data. While in the more formal statistical literature the term "latent class model" is typically reserved for a specific type of mixture model (Everitt and Hand, 1981; McLachlan and Peel, 2000), namely for the mixture model for categorical responses described by Lazarsfeld and Henry (1968) and Goodman (1974), in applied fields these terms are used

Jeroen K. Vermunt
Department of Methodology and Statistics, Tilburg University, The Netherlands
e-mail: `j.k.vermunt@uvt.nl`

K. van Montfort et al., *Longitudinal Research with Latent Variables*,
DOI 10.1007/978-3-642-11760-2_4,

interchangeably. This is also what I will do in this chapter; that is, I will use the terms mixture model and latent class model to denote latent variable models containing one or more discrete latent variables.

In the context of longitudinal research, a mixture model is a latent variable model containing a single or multiple time-constant or time-varying discrete latent variables. The best-known examples are the latent class or mixture growth model (Muthén 2004; Nagin, 1999; Vermunt, 2007), the mixture Markov model (Goodman, 1961; Poulsen, 1990; van de Pol and Langeheine, 1990; Vermunt, 1997a), and the latent or hidden Markov model (Baum, Petrie, Soules, and Weiss, 1970; Bartolucci, Pennoni, and Francis, 2007; Collins and Wugalter, 1992; Mooijaart and van Montfort, 2007; Poulsen, 1990; van de Pol and de Leeuw, 1986; Vermunt, Langeheine, and Böckenholt, 1999; Wiggins, 1973).

Diggle, Liang, and Zeger (1994) distinguished three main approaches for analyzing longitudinal data: (1) marginal or population-average models, (2) random-effects, subject-specific, or growth models, and (3) conditional or transitional models. Marginal models focus on the change in univariate distributions, growth models study individual-level change over time, and transitional models describe changes between consecutive time points. These three approaches do not only differ with regard to the questions they address, but also in how they deal with the dependencies between the repeated measures. Because of their structure, transitional models take the bivariate dependencies between observations at consecutive occasions into account. Growth models capture the dependencies using latent variables (random effects). In marginal models, dependencies are not explicitly modeled, but dealt with as found in the data and in general are taken into account in a more ad hoc way in the estimation procedure. Variants of transitional, growth, and marginal models have been developed for both continuous and categorical response variables.

Discrete latent variables may be introduced in longitudinal data models for various purposes, the most important of which are dealing with unobserved heterogeneity, dealing with measurement error, and clustering. Or more specific, in context of the three approaches described above, latent classes can be introduced in growth models for clustering and dealing with unobserved heterogeneity (yielding mixture growth models), and in transitional models for dealing with measurement error, static or dynamic clustering, and dealing with unobserved heterogeneity (yielding mixture and latent Markov models). Hagenaars (1990) and Bergsma, Croon, and Hagenaars (2009) used a latent class marginal model for dealing with measurement error in categorical responses.

Starting point of this chapter are the simplest variants of the three basic mixture models for longitudinal data: the mixture growth, mixture Markov, and latent Markov model. Recent developments in software and algorithms have resulted in many extensions of these basic models; that is, mixture models for longitudinal data can nowadays (1) not only be used with categorical, but also continuous response variables (as well as combinations of these), (2) be used with very long time series, (3) include covariates (which can be numeric or categorical, as well as time-constant or time-varying), (4) include parameter restrictions yielding interesting measurement models, and (5) deal with missing values (which is very important in longitudinal research). I will present an integrating framework including all these extended features. Moreover, I will discuss other more advanced features, such as latent Markov models with dependent classification errors across time points, mixture growth and latent Markov models with random effects, and latent Markov models for multilevel data and multiple processes.

There is some overlap between the current chapter and Hagenaars' chapter in this volume, which deals with longitudinal categorical data analysis using the log-linear SEM approach implemented in the LEM software (Vermunt 1997b). On the one hand, this SEM framework is more general than the framework discussed here because it allows defining any type of categorical data model. On the other hand, it is more restricted since it deals with categorical data (responses and covariates) only, and, because it is not tailored for longitudinal data analysis, it can, for example, not be used with long time series.

In the remaining of this chapter, I will first describe the three basic mixture models for longitudinal data analysis, including some of their extensions. Then a general framework is presented containing each of these as special cases, and allowing various interesting combinations. Though several other recent developments could be fit into an even more general framework, these will be discussed as separate extensions in a next section. The last section presents two applications, and the Appendix illustrates how the models concerned can be defined using the Latent GOLD syntax system (Vermunt and Magidson, 2005, 2008).

4.2 The three basic models

Before describing the three basic types of mixture models for longitudinal data, I will introduce the relevant notation. Longitudinal data sets analyzed with the models described in this chapter will typically contain information on multiple response variables from multiple subjects at multiple time points. Let y_{itj} denote the response of subject i on response variable j at occasion t, where $1 \leq i \leq N$, $1 \leq j \leq J$, and $0 \leq t \leq T_i$. Here, N is the number subjects, J the number of response variables, and $T_i + 1$ is the number of measurement occasions for subject i. Note that we use the index i in T_i to be able to deal with the rather common situation in which the number of measurement occasions differ across individuals. The vector collecting the responses of subject i at occasion t is denoted as $\mathbf{y}_{it}$ and the vector collecting all responses of subject i as $\mathbf{y}_i$.

Three remarks have to made about the response variables. First, response variables may also be referred to as output variables, dependent variables, indicators, items, manifest variables, etc. Second, response variables cannot only be categorical variables – in which case $1 \leq y_{itj} = m_j \leq M_j$, with M_j being the number of categories and m_j a particular category of response variable j – but also continuous variables or counts. As we will see below, the scale type of y_{itj} affects its conditional distribution, as well as the type of regression model one may specify to restrict its expected value. Third, often only one response variable is available, in which case the index j can be dropped, yielding the simpler notation y_{it}.

Longitudinal data models may not only contain response variables, but also predictors, also referred to as input variables, independent variables, covariates, concomitant variables, etc. The vectors of time-constant predictors and time-varying predictors at occasion t are denoted by $\mathbf{z}_i$ and $\mathbf{z}_{it}$, respectively. Note that predictors cannot only be numeric but also categorical variables, which will typically be included in the model using a series of dummies or effects. Note also that time and functions of time can be included in the vector of time-varying predictors.

What makes a statistical model a latent class or mixture model is that it contains either a time-constant or a time-varying (or dynamic) discrete latent variable. These two types of latent variables are denoted by w_i and x_{it}, respectively, their number of categories by L and K, and one of their categories by ℓ and k_t. That is, $1 \leq w_i = \ell \leq L$ and $1 \leq x_{it} = k_t \leq K$. To clearly distinguish the two types of latent variables, I will refer to w_i as a latent class and to x_{it} as a latent state.

4.2.1 Mixture growth model

A latent class or mixture growth model is a model for a single response variable y_{it} measured at $T_i + 1$ occasions (Nagin, 1999; Muthén, 2004; Vermunt 2007). In fact, a regression model is specified for y_{it} in which time serves as the only explanatory variable. The aim of growth models is to determine whether individuals differ with respect to the parameters of the growth model, where differences are usually modeled using random effects under the assumption that these come from a multivariate normal distribution.

There are two possible reasons for introducing latent classes in a growth model. First, one may wish to identify (interpretable) clusters of individuals with similar growth parameters. This is similar to the aim of a standard latent class model, with the difference that the observed variables used to find the clusters are repeated measurements of a single response variable rather than multiple items or indicators. A second reason for using a mixture growth model is more technical; that is, one may wish to specify a model with random effects without making strong distributional assumptions about the random effects. This yields what is referred to as a non-parametric maximum likelihood (NPML) approach to random effects modeling, which cannot only be used in the context of longitudinal data analysis but in any type of two-level regression model (Aitkin, 1999; Skrondal and Rabe-Hesketh, 2004; Vermunt 2004; Vermunt and van Dijk, 2001).

A mixture growth model is a statistical model for $f(\mathbf{y}_i|\mathbf{z}_i)$, the probability density of the $T_i + 1$ responses of subject i collected in the vector $\mathbf{y}_i$ conditional on a set of time variables collected in the vector $\mathbf{z}_i$. It can be formulated using the following three equations:

$$f(\mathbf{y}_i|\mathbf{z}_i) = \sum_{\ell=1}^{L} P(w_i = \ell) f(\mathbf{y}_i|w_i = \ell, \mathbf{z}_i) \tag{4.1}$$

$$f(\mathbf{y}_i|w_i = \ell, \mathbf{z}_i) = \prod_{t=0}^{T_i} f(y_{it}|w_i = \ell, \mathbf{z}_{it}), \tag{4.2}$$

$$g[E(y_{it}|w_i = \ell, \mathbf{z}_{it})] = \beta_{0\ell} + \sum_{p=1}^{P} \beta_{p\ell} z_{itp}, \tag{4.3}$$

The first of these three equations indicates that the density $f(\mathbf{y}_i|\mathbf{z}_i)$ is a weighted average of class-specific densities $f(\mathbf{y}_i|w_i = \ell, \mathbf{z}_i)$, where the class proportions $P(w_i = \ell)$ serve as weights. More intuitively, the likelihood of the set of responses $\mathbf{y}_i$ depends on the class membership of person i (on w_i). But because the class membership is unknown, the likelihood is obtained by averaging over the L classes. Note that this kind of reasoning applies to any type of mixture or latent class model.

The second equation states that the joint distribution of $\mathbf{y_i}$ given w_i and $\mathbf{z}_i$ (appearing in the equation 4.1) can be obtained as a product of the $T_i + 1$ univariate marginal distributions $f(y_{it}|w_i = \ell, \mathbf{z}_{it})$. This expresses that the responses are assumed to be independent across time points given a person's class membership, which in the latent class analysis literature is usually referred to as the local independence assumption. The specific form chosen for $f(y_{it}|w_i = \ell, \mathbf{z}_{it})$ depends on the scale type of y_{it}. For example, with binary responses on will often use a binomial distribution, with continuous responses a normal distribution, and with counts a Poisson distribution.

The third equation shows that the responses are related to the time variables using a regression model from the generalized linear modeling (GLM) family (Agresti, 2002). After applying an appropriate transformation $g(\cdot)$, which in GLM terminology is referred to as a link function, the expected value of y_{it} is modeled as a linear function of a set of P time variables. For example, with $P = 2$, $z_{it1} = t$, and $z_{it2} = t^2$, the expected value of y_{it} would be a quadratic function of time. A key feature is that the regression parameters capturing the time dependence of the responses are assumed to differ across latent classes; that is, each class has its own pattern of change. Note that by defining a regression model for y_{it} one restricts the density $f(y_{it}|w_i = \ell, \mathbf{z}_{it})$ which appears in equation 4.2. In fact, we have a latent class model with restrictions on the class-specific response probabilities/densities which are specified by assuming that the class-specific means are functions of time.

The basic model described in equations (4.1)–(4.3) can be extended in various ways. One important extension is the inclusion of covariates in the model for w_i. Similarly to the model proposed by Dayton and Macready (1988) and van der Heijden, Dessens, and Böckenholt (1996) in the context of standard latent class analysis, this involves replacing $P(w_i = \ell)$ in equation (4.2) by $P(w_i = \ell|\mathbf{z}_i)$ and defining a multinomial logit model for w_i; that is,

$$P(w_i = \ell|\mathbf{z}_i) = \frac{\exp(\gamma_{0\ell} + \sum_{q=1}^{Q} \gamma_{q\ell} z_{iq})}{\sum_{\ell'=1}^{L} \exp(\gamma_{0\ell'} + \sum_{q=1}^{Q} \gamma_{q\ell'} z_{iq})}, \tag{4.4}$$

where for identification we may for example set $\gamma_{0L} = \gamma_{qL} = 0$, yielding what is usually referred to as a baseline category logit model (Agresti, 2002).

Another extension is the inclusion of other predictors than time in the model for y_{it} (in equation 4.3). These could serve as control variables when one is interested determining class-specific change patterns after accounting for the fact that other variables may partially explain the observed change. But other predictors may also be the ones of main interest, in which case the aim of the analysis changes somewhat and the mixture variable will mainly be used to capture unobserved heterogeneity using the NPML approach mentioned above.

4.2.2 Mixture Markov model

As mentioned in the introduction, rather than using a growth model, longitudinal data may also be modeled using a transitional or conditional model. The best-known model from this family is the (first-order) Markov model, which assumes that y_{it} depends on y_{it-1} but not on values at earlier occasions. Similarly to mixture growth models, in mixture Markov models, one will typically have a single response variable. The main reason for using a mixture variant of a Markov model is to deal with unobserved heterogeneity; that is, to account for the fact that transition probabilities/densities are not homogeneous, but instead may differ across (unobserved) subgroups. A more substantive reason may be to find meaningful clusters of individuals with different change patterns. An example of the latter is the application by Dias and Vermunt (2007) in which market segments were identified based on website users' search patterns.

The mixture Markov can be formulated as follows:

$$f(\mathbf{y}_i) = \sum_{\ell=1}^{L} P(w_i = \ell) f(y_{i0}|w_i = \ell) \prod_{t=1}^{T_i} f(y_{it}|y_{it-1}, w_i = \ell). \tag{4.5}$$

As can be seen, the L latent classes are assumed to differ with respect to the initial-state and transition densities. Variants of this model for continuous response variables – referred to as mixture dynamic regression and mixture autoregressive models – were proposed by Kaplan (2005) and Wong and Li (2000). However, most applications of the mixture Markov model concern categorical response variables (Dias and Vermunt, 2007; Poulsen, 1990), in which case the model may also be written as

$$P(\mathbf{y}_i) = \sum_{\ell=1}^{L} P(w_i = \ell) P(y_{i0} = m_0|w_i = \ell) \left[\prod_{t=1}^{T_i} P(y_{it} = m_t|y_{it-1} = m_{t-1}, w_i = \ell) \right]; \tag{4.6}$$

that is, in terms of initial-state and transition probabilities.

Various special cases of the mixture Markov model can be obtained by restricting the transition probabilities. A well-documented special case is the mover-stayer model (Goodman, 1961), which is a two-class model ($L = 2$) where one class (say the second) contains respondents who have a zero probability of making a transition: $P(y_{it} = m_t | y_{it-1} = m_{t-1}, w_i = 2) = 0$ for $m_t = m_{t-1}$. Another special case is a Markov model with a random responder class for which the measurements are independent across time points: $P(y_{it} = m_t | y_{it-1} = m_{t-1}, w_i = 2) = P(y_{it} = m_t | w_i = 2)$.

Various extensions of the simple models described in equations (4.5) and (4.6) are possible, the most important of which is the introduction of predictors affecting the class membership, the initial state, and the transitions. The first extension was discussed above in the context of mixture growth models (see equation 4.4). Covariates can be allowed to affect the initial state and the transitions by defining regression models for y_{i0} and y_{it}, which in the case of a categorical response will be logistic regression models. With Q predictors in the model for y_{i0} and P time-varying predictors in the model for y_{it} conditional on y_{it-1}, we get

$$P(y_{i0} = m | w_i = \ell, \mathbf{z}_{i0}) = \frac{\exp(\beta^0_{\ell m} + \sum_{q=1}^{Q} \beta^0_{q+L,m} z_{i0q})}{\sum_{m'=1}^{M} \exp(\beta^0_{\ell m'} + \sum_{q=1}^{Q} \beta^0_{q+L,m'} z_{i0q})}, \tag{4.7}$$

$$P(y_{it} = m | y_{it-1} = n, w_i = \ell, \mathbf{z}_{it}) = \frac{\exp(\beta_{\ell nm} + \sum_{p=1}^{P} \beta_{p+L,nm} z_{itp})}{\sum_{m'=1}^{M} \exp(\beta_{\ell nm'} + \sum_{p=1}^{P} \beta_{p+L,nm'} z_{itp})}. \tag{4.8}$$

As in a standard multinomial logit model, identifying restrictions on $\beta^0_{\ell m}$ and $\beta^0_{q+L,m}$ are required, for example, they may be fixed to 0 for $m = M$. The same applies to the $\beta_{\ell nm}$ and $\beta_{p+L,nm}$ parameters for which one constraint is needed for each origin state n. A coding referred to as transition coding by Vermunt and Magidson (2008) involves setting $\beta_{\ell nn} = \beta_{p+L,nn} = 0$; that is, the coefficients are fixed to 0 for $m = n$, which implies that the free coefficients can be interpreted as effects on the logit of a transition from n to m.

4.2.3 Latent Markov model

Whereas mixture growth and mixture Markov models contain a static categorical latent variable (w_i), a latent Markov model is a mixture model with a dynamic categorical latent variable – denoted by x_{it}. One of the key elements of this model is that latent-state transitions occurring over time are modeled using a first-order Markov structure. The second key element is that the latent states are connected to one or more observed response variables via a latent class structure with conditional densities $f(y_{itj} | x_{it} = k_t)$. The latent Markov model – which is also referred to as hidden Markov model (Baum et al., 1970; McDonald and Zucchini, 1997), Markov switching or regime switching model (Goldfeld and Quandt, 1973), and latent transition model (Collins and Wugalter, 1992) – can be defined as follows (Poulsen, 1990; van de Pol and de Leeuw, 1986, Wiggins, 1973):

$$f(\mathbf{y}_i) = \sum_{k_0=1}^{K} \sum_{k_1=1}^{K} \dots \sum_{k_{T_i}=1}^{K} P(x_{i0} = k_0) \left[\prod_{t=1}^{T_i} P(x_{it} = k_t | x_{it-1} = k_{t-1}) \right] \left[\prod_{t=0}^{T_i} \prod_{j=1}^{J} f(y_{itj} | x_{it} = k_t) \right]. \quad (4.9)$$

Besides the Markov assumption for the latent states and the local independence assumption for the responses within occasions, the latent Markov model assumes that responses are independent across occasions conditional on the latent states. The latter implies that the observed associations across time points are assumed to be explained by the autocorrelation structure for the latent states.

The typical applications of this model concern either a single continuous response variable (Schmittmann, Dolan, van der Maas, and Neale, 2005; Dias, Vermunt, and Ramos, 2009), a single categorical response variable (Magidson, Vermunt, and Tran, 2009; Poulsen, 1990; van de Pol and de Leeuw, 1986; Wiggins, 1973), or multiple categorical responses (Bartolucci, Pennoni, and Francis, 2007; Collins and Wugalter, 1992; Paas, Vermunt, and Bijmolt, 2007). With a single continuous response, the model may either be used for clustering or for dealing with unobserved heterogeneity, where contrary to the mixture models described above respondents may switch across clusters or mixture components over time. When applied with a single categorical response variable, one will typically assume that the number of latent states equals the number or categories of the response variable: $K = M$. Moreover, model restrictions are required to obtain an identified model, the most common of which are time-homogeneous transition probabilities or time-homogeneous misclassification probabilities. The aim is to split observed changes in the response into a true change component and a measurement error component. When used with multiple indicators, the model is a longitudinal data extension of the standard latent class model (Hagenaars, 1990). The time-specific latent states can be seen as clusters or types which differ in their responses on the J indicators, and the Markovian transition structure is used to describe and predict changes that may occur across adjacent measurement occasions.

The most straightforward extension of the latent Markov model presented in equation (4.9) involves the inclusion of explanatory variables affecting the initial state and the transition probabilities. Special cases are the multiple-group latent Markov model proposed by van de Pol and Langeheine (1990), the latent Markov model with covariates proposed by Vermunt, Langeheine and Böckenholt (1999), and the input-output model described by Mooijaart and van Montfort (2007). Models with predictors can be defined using similar logistic equations as we used for the mixture Markov model (see equations 4.7 and 4.8), but now for x_{i0} and x_{it} instead of y_{i0} and y_{it} and without conditioning on w_i; that is,

$$P(x_{i0} = k|\mathbf{z}_{i0}) = \frac{\exp(\alpha^0_{0k} + \sum_{q=1}^{Q} \alpha^0_{qk} z_{i0q})}{\sum_{k'=1}^{K} \exp(\alpha^0_{0k'} + \sum_{q=1}^{Q} \alpha^0_{qk'} z_{i0q})},$$

$$P(x_{it} = k|x_{it-1} = n, \mathbf{z}_{it}) = \frac{\exp(\alpha_{0nk} + \sum_{p=1}^{P} \alpha_{pnk} z_{itp})}{\sum_{k'=1}^{K} \exp(\alpha_{0nk'} + \sum_{p=1}^{P} \alpha_{pnk'} z_{itp})}.$$

Again, identifying restrictions are needed on the α^0_{0k}, α^0_{qk}, α_{0nk}, and α_{pnk} parameters, where for the latter two one may again use transition coding.

Other extensions include models with predictors affecting the responses, mixture variants with a time-constant latent variable w_i, models with restrictions on the transition probabilities $P(x_{it} = k_t|x_{it-1} = k_{t-1})$ or the response densities $f(y_{it}|x_{it} = k_t)$, models that relax the assumption that measurement errors are independent across occasions, and models with multiple dynamic latent variables. These and other extensions will be discussed below.

4.3 The mixture latent Markov model

4.3.1 The general model

In the previous section, we described three types of mixture models for longitudinal data analysis. These models contained either a time-constant (w_i) or time-varying (x_{it}) discrete latent variables. In this section, I present the mixture latent Markov with covariates, which can be seen as the encompassing model which contains the three models discussed above as special cases, as well as which allows various interesting extensions and combinations of these. The presented mixture latent Markov model is an expanded version of the mixed Markov latent class model proposed by van de Pol and Langeheine (1990) in the sense that it cannot only be used with categorical but also with continuous responses, it may contain time-constant and time-varying covariates, and it can be used when the number of time points is large. For simplicity of exposition, here, I will restrict myself to models with a single time-constant and a single time-varying latent variable, but in the next section I will present extensions for multiple time-constant and multiple time-varying latent variables.

The general model of interest is the following mixture latent Markov model:

$$f(\mathbf{y}_i|\mathbf{z}_i) = \sum_{\ell=1}^{L}\sum_{k_0=1}^{K}\sum_{k_1=1}^{K}\dots\sum_{k_{T_i}=1}^{K} P(w_i=\ell,\mathbf{x}_i=\mathbf{k}|\mathbf{z}_i)f(\mathbf{y}_i|w_i=\ell,\mathbf{x}_i=\mathbf{k},\mathbf{z}_i) \tag{4.10}$$

$$= \sum_{\ell=1}^{L}\sum_{k_0=1}^{K}\sum_{k_1=1}^{K}\dots\sum_{k_{T_i}=1}^{K} P(w_i=\ell|\mathbf{z}_i)P(x_{i0}=k_0|w_i=\ell,\mathbf{z}_{i0}) \left[\prod_{t=1}^{T_i} P(x_{it}=k_t|x_{it-1}=k_{t-1},w_i=\ell,\mathbf{z}_{it})\right] \left[\prod_{t=0}^{T_i}\prod_{j=1}^{J} f(y_{itj}|x_{it}=k_t,w_i=\ell,\mathbf{z}_{it})\right]. \tag{4.11}$$

As many statistical models, the model in equations (4.10) and (4.11) is a model for $f(\mathbf{y}_i|\mathbf{z}_i)$, the (probability) density associated with the responses of subject i conditional on his/her observed covariate values. The right-hand side of equation (4.10) shows that we are dealing with a mixture model containing a time constant latent variable (w_i) and $T+1$ realizations of a time-varying latent variable (collected in the vector $\mathbf{x}_i$). The total number of mixture components (or latent classes) for individual i equals $L\cdot K^{T_i+1}$, which is the product of the number of categories of w_i and x_{it} for $t=0,1,2,...,T_i$. Equation (4.10) shows that, as in any mixture model, $f(\mathbf{y}_i|\mathbf{z}_i)$ is obtained as a weighted average of class-specific probability densities – here $f(\mathbf{y}_i|w_i=\ell,\mathbf{x}_i=\mathbf{k},\mathbf{z}_i)$ – where the (prior) class membership probabilities or mixture proportions – here $P(w_i=\ell,\mathbf{x}_i=\mathbf{k}|\mathbf{z}_i)$ – serve as weights (Everitt and Hand, 1981; McLachlan and Peel, 2000).

Equation (4.11) shows the specific structure assumed for the mixture proportions and the class-specific densities. The assumption for $P(w_i=\ell,\mathbf{x}_i=\mathbf{k}|\mathbf{z}_i)$ is that conditional on w_i and $\mathbf{z}_i$, x_{it} is associated only with $x_{i,t-1}$ and $x_{i,t+1}$ and thus not with the states occupied at the other time points – the well-know first-order Markov assumption. For $f(\mathbf{y}_i|w_i=\ell,\mathbf{x}_i=\mathbf{k},\mathbf{z}_i)$ two assumptions are made: (1) conditionally on w_i, x_{it}, and $\mathbf{z}_{it}$, the J responses at occasion t are independent of the latent states and the responses at other time points, and (2) conditionally on w_i, x_{it}, and $\mathbf{z}_{it}$, the J responses at occasion t are mutually independent, which is referred to as the local independence assumption in latent class analysis (Goodman, 1974).

As can be seen from equation (4.11), the models of interest contain four different kinds of model probabilities/densities:

- $P(w_i=\ell|\mathbf{z}_i)$ is the probability of belonging to a particular latent class conditional on a person's covariate values,
- $P(x_{i0}=k_0|w_i=\ell,\mathbf{z}_{i0})$ is an initial-state probability; that is, the probability of having a particular latent initial state conditional on an individual's class membership and covariate values at $t=0$,
- $P(x_{it}=k_t|x_{it-1}=k_{t-1},w_i=\ell,\mathbf{z}_{it})$ is a latent transition probability; that is, the probability of being in a particular latent state at time point t conditional on the latent state state at time point $t-1$, class membership, and time-varying covariate values,

- $f(y_{itj}|x_{it} = k_t, w_i = \ell, \mathbf{z}_{it})$ is a response density, that is, the density corresponding to the observed value for person i of response variable j at time point t conditional on the latent state occupied at time point t, class membership w_i, and time-varying covariate values.

Typically, these four sets of probabilities/densities will be parameterized and restricted by means of regression models from the generalized linear modeling family. As shown in various examples in the previous section, this is especially useful when a model contains covariates, where time itself may be one of the time-varying covariates of main interest.

The three key elements of the mixture latent Markov model described in equation (4.11) are that it can take into account (1) unobserved heterogeneity, (2) autocorrelation, and (3) measurement error. Unobserved heterogeneity is captured by the time-constant latent variable w_i, autocorrelations are captured by the first-order Markov transition process in which the state at time point t may depend on the state at time point $t-1$, and measurement error or misclassification is accounted for by allowing an imperfect relationship between the time-specific latent states x_{it} and the observed responses y_{itj}. Note that these are three of the main elements that should be taken into account in the analysis of longitudinal data; that is, the inter-individual variability in patterns of change, the tendency to stay in the same state between consecutive occasions, and the spurious change resulting from measurement error in observed responses.

4.3.2 Estimation, missing data, and time-unit setting

Parameters of the mixture latent Markov model can be estimated by means of maximum likelihood (ML). For that purpose, it advisable to use a special variant of the expectation maximization (EM) algorithm that is usually referred to as the forward-backward or Baum-Welch algorithm (Baum et al., 1970; McDonald and Zucchini, 1997). This is an EM algorithm in which the E step, which involves computing the relevant set posterior distributions given the current parameter estimates and the observed data, is implemented in a way that is tailored to the models we are dealing with. More specifically, this special algorithm is needed because our model contains a potentially huge number of entries in the joint posterior latent distribution $P(w_i = \ell, \mathbf{x}_i = \mathbf{k}|\mathbf{y}_i, \mathbf{z}_i)$, except for cases where T, L and K are all small. For example, in a fairly moderate sized situation where $T_i = 10$, $L = 2$ and $K = 3$, the number of entries in the joint posterior distribution already equals $2 \cdot 3^{11} = 354294$, a number which is impossible to process and store for all N subjects as has to be done within standard EM. The Baum-Welch algorithm circumvents the computation of this joint posterior distribution making use of the conditional independencies implied by the model; that is, rather than computing the joint distribution and subsequently obtaining the relevant marginals, it computes the relevant marginals directly. For more details, we refer to Vermunt, Tran, and Magidson (2008) who also provided the generalized version of the Baum-Welch algorithm which is required for the estimation of the mixture latent Markov model presented in equation (4.11) and which is implemented in the Latent GOLD 4.5 program (Vermunt and Magidson, 2008). Rather than using ML estimation, it is also possible to estimate these models using Bayesian estimation procedures, an excellent overview of which is provided by Frühwirth-Schnatter (2006).

A common phenomenon in the analysis of longitudinal data is the occurrence of missing data. Subjects may have missing values either because they refused to participate at some occasions or because it is part of the research design. A nice feature of the approach described here is that it can easily accommodate missing data in the ML estimation of the unknown model parameter. Let δ_{it} be an indicator variable taking on the value 1 if subject i provides information for occasion t and 0 if this information is missing. The only required change with missing data is the following modification of the model for the response density $f(\mathbf{y}_i|w_i=\ell,\mathbf{x}_i=\mathbf{k},\mathbf{z}_i)$:

$$f(\mathbf{y}_i|w_i=\ell,\mathbf{x}_i=\mathbf{k},\mathbf{z}_i)=\prod_{t=0}^{T_i}[P(\mathbf{y}_{it}|x_{it}=k_t,w_i=\ell,\mathbf{z}_{it})]^{\delta_{it}}.$$

For $\delta_{it}=1$, nothing changes compared to what we had before. However, for $\delta_{it}=0$, the time-specific conditional density becomes 1, which means that the responses of a time point with missing values are skipped. Actually, for each pattern of missing data, we have a mixture latent Markov for a different set of occasions. Two limitations of the ML estimation procedure with missing values should be mentioned: (1) it can deal with missing values on response variables, but not with missing values on covariates, and (2) it assumes that the missing data are missing at random (MAR). The first limitation may be problematic when there are time-varying covariates for which the values are also missing. However, in various special cases discussed below – the ones that do not use a transition structure – it is not a problem if time-varying covariates are missing for the time points in which the responses are missing. The second limitation concerns the assumed missing data mechanism: MAR is the least restrictive mechanism under which ML estimation can be used without the need of specifying the exact mechanism causing the missing data; that is, under which the missing data mechanism is ignorable for likelihood-based inference (Little and Rubin, 1987; Schafer, 1997). It is possible to relax the MAR assumption by explicitly defining a not missing at random (NMAR) mechanism as a part of the model to be estimated (Fay, 1986; Vermunt 1997a).

An issue strongly related to missing data is the one of unequally spaced measurement occasions. As long as the model parameters defining the transition probabilities are assumed to be occasion specific, no special arrangements are needed. If this is not the case, unequally spaced measurements can be handled by defining a grid of equally spaced time points containing all measurement occasions. Using this technique, the information on the extraneous occasions can be treated as missing data for all subjects. An alternative is to use a continuous-time rather than a discrete-time framework (Böckenholt, 2005), which can be seen as the limiting case in which the elapsed time between consecutive time points in the grid approaches zero.

	Model name	Transition structure	Unobserved heterogeneity	Measurement error
I.	Mixture latent Markov	yes	yes	yes
II.	Mixture Markov	yes	yes	no
III.	Latent Markov	yes	no	yes
IV.	Standard Markov*	yes	no	no
V.	Mixture latent growth	no	yes	yes
VI.	Mixture growth	no	yes	no
VII.	Standard latent class	no	no	yes
VIII.	Independence*	no	no	no

*: This model is not a latent class model.

Another issue related to missing data is the choice of the time variable and the corresponding starting point of the process. The most common approach is to use calender time as the time variable and to define the first measurement occasion to be $t = 0$. However, one may, for example, also use age as the relevant time variable, as I do in the second empirical example. Although children's ages at the first measurement vary between 11 and 17, I use age 11 as $t = 0$. This implies that for a child that is 12 years of age information at $t = 0$ is treated as missing, for a child that is 13 years of age information a $t = 0$ and $t = 1$ is treated as missing, etc.

4.3.3 The most important special cases

Table 4.3.3 lists the various special cases that can be derived from the mixture latent Markov model defined in equation in (4.11) by assuming that one or more of its three elements – transition structure, measurement error, and unobserved heterogeneity – is not present or needs to be ignored because the data is not informative enough to deal with it. Models I-III and V-VII are latent class models, but IV and VIII are not. Model VII differs from models I-VI in that it is a model for repeated cross-sectional data rather than a model for panel data. Below we describe the various special cases in more detail.

4.3.3.1 Mixture latent Markov

First of all, it is possible to define simpler versions of the mixture latent Markov model itself. Actually, the mixed Markov latent class model proposed by van de Pol and Langeheine (1990) which served as an inspiration for our model is the special case in which responses are categorical and in which no covariates are present. van de Pol and Langeheine (1990) proposed a variant in which the four types of model probabilities could differ across categories of a grouping variable (see also Langeheine and van de Pol, 2002). A similar model is obtained by replacing the $\mathbf{z}_i$ and $\mathbf{z}_{it}$ by a single categorical covariate coded using a set of dummy predictors.

4.3.3.2 Mixture Markov

The mixture Markov model for a categorical response variable (Poulsen, 1990; Dias and Vermunt, 2007) is the special case of the model presented in equation (4.11) when there is a single response variable ($J = 1$) that is assumed to be measured without error, which is specified by $K = M$ and $P(y_{it} = m_t | x_{it} = k_t) = 1$ if $m_t = k_t$ and 0 otherwise. Note that y_{it} is assumed not to depend on w and $\mathbf{z}_{it}$ but only on x_t. The mover-stayer model (Goodman, 1961) can be obtained by setting $L = 2$ and fixing the transition probabilities to 0 for the second class: $P(x_{it} = k_t | x_{tt-1} = k_{t-1}, w_i = 2, \mathbf{z}_{it}) = 0$ if $k_t = k_{t-1}$ and 0 otherwise. Note that the mover-stayer constraint cannot only be imposed in the mixture Markov but also in the mixture latent Markov, in which case transitions across imperfectly measured stated are assumed not to occur in the stayer class.

4.3.3.3 Latent Markov model

The latent Markov, latent transition, or hidden Markov model (Baum et al., 1970; Collins and Wugalter, 1992; Mooijaart and van Montfort, 2007; van de Pol and de Leeuw, 1996; Vermunt, Langeheine, and Böckenholt, 1999, Wiggins, 1973) is the special case of the mixture latent Markov that is obtained by eliminating the time-constant latent variable w_i from the model; that is, by assuming that there is no unobserved heterogeneity or that it can be ignored. The latent Markov model can be obtained without modifying the formulae, but by simply assuming that $L = 1$; that is, that all subject belong to the same latent class.

4.3.3.4 Markov model

By assuming both perfect measurement as in the mixture Markov model and absence of unobserved heterogeneity as in the latent Markov model, one obtains a standard Markov model, which is no longer a mixture model. This model can further serve as a simple starting point for longitudinal applications with a single response variable, where one wishes to assume a Markov structure. It provides a baseline for comparison to the three more extended models discussed above. Use of these more extended models makes sense only if they provide a significantly better description of the data than the simple Markov model.

4.3.3.5 Mixture latent growth model

Now we turn to latent class models for longitudinal research that are not transition or Markov models. These mixture growth models assume that dependencies between measurement occasions can be captured by the time-constant latent variable w_i. The most extended variant is the mixture latent growth model, which is obtained from the mixture latent Markov model by imposing the constraint $P(x_{it} = k_t | x_{i,t-1} = k_{t-1}, w_i = \ell, \mathbf{z}_{it}) = P(x_{it} = k_t | w_i = \ell, \mathbf{z}_{it})$. This model is a variant for longitudinal data of the multilevel latent class model proposed by Vermunt (2003): subjects are the higher-level units and time points the lower-level units. It should be noted that application of this very interesting model with categorical responses requires that there be at least two response variables ($J \geq 2$).

In mixture growth models one will typically pay a lot of attention to the modeling of the time dependence of the state occupied at the different time points. The latent class or mixture approach allows identifying subgroups (categories of the time-constant latent variable w_i) with different change patterns (Nagin, 1999). The extension provided by the mixture latent growth model is that the dynamic dependent variable is itself a (discrete) latent variable which is measured by multiple indicators.

4.3.3.6 Mixture growth model

The mixture or latent class growth model (Nagin, 1999, Muthén, 2004; Vermunt, 2007) for a categorical response variable can be seen as a restricted variant of the mixture latent growth model; i.e., as a model for a single indicator measured without error. The extra constraint is the same as the one used in the mixture Markov model: $K = M$ and $P(y_{it} = m_t | x_{it} = k_t) = 1$ if $m_t = k_t$ and 0 otherwise. A more natural way to define the mixture growth model is by omitting the time-varying latent variable x_{it} from the model specification as was done in equations (4.1) and (4.2).

4.3.3.7 Standard latent class model

When we eliminate both w_i and the transition structure, we obtain a latent class model that assumes observations are independent across occasions. This is a realistic model only for the analysis of data from repeated cross-sections; that is, to deal with the situation in which observations from different occasions are independent because each subject provides information for only one time point.

4.4 Other extensions

The previous section presented a general mixture model for longitudinal analysis, which contained the three basic models and various of their extensions as special cases. This section describes several other interesting extensions, which could be fit into an even more general mixture model for the longitudinal analysis.

4.4.1 Ordered states

The first extension concerns a latent Markov model for dichotomous or ordered polytomous responses in which the latent states can be interpreted as ordered categories. Examples are developmental stages of children, disease stages of patients, and states representing degrees of agreement in attitude measurement. It may of course turn out that the estimation of an unrestricted latent Markov model yields the hypothesized ordering of the latent states. However, it is also possible to force the latent states to be ordered by imposing constraints on the model parameters.

One class of restrictions concerns the relationship between latent states and responses. Bartolucci, Pennoni, and Francis (2007) and Vermunt and Georg (2002) presented various of such models, which can be seen as longitudinal data variants of the discretized item response theory models described by Heinen (1996) and Vermunt (2001). Two possible restrictions for multicategory items are

$$\log \frac{P(y_{itj} = m | x_{it} = k)}{P(y_{itj} = m - 1 | x_{it} = k)} = \beta_{0jm} + \beta_{1j}\, v_k,$$

and

$$\log \frac{P(y_{itj} \geq m | x_{it} = k)}{P(y_{itj} < m | x_{it} = k)} = \beta_{0jm} + \beta_{1j}\, v_k,$$

where the former defines an adjacent category ordinal logit model for y_{itj} and the latter a cumulative logit model (Agresti, 2002). Note that v_k represents the location of latent state k, which can either be fixed a priori or treated as a free parameter to be estimated. Vermunt and Hagenaars (2004) gave an extended overview of longitudinal models for ordinal responses, which also included various types of mixture models.

Another way to obtain latent states that can be interpreted as ordered categories is via restrictions on the transition probabilities. An example in which latent states represent (ordered) developmental stages was provided by Collins and Wugalter (1992). According to the underlying developmental psychology theory, children may make a transition to a next stage but will never return to a previous stage. In terms of the latent Markov model parameters, this means that $P(x_{it} = k_t | x_{it-1} = k_{t-1}) = 0$ for $k_t < k_{t-1}$.

4.4.2 Continuous latent variables

The mixture models discussed so far contained only discrete latent variables. However, in many applications, it may be useful to include also continuous latent variables in the model, which can play the role of latent factors in a measurement model or the role of random effects in a regression model. Below, I describe several situations in which time-constant or time-varying continuous latent variables may be used in the model for the transitions or in the model for the responses. I will denote continuous latent variables by F.

4.4.2.1 Time-constant affecting transitions

As a way to account for unobserved heterogeneity, it may be useful to expand the latent Markov model with random effects in the regression models for the initial state and transition probabilities. An example was provided by Pavlopoulos, Muffels, and Vermunt (2009) in an application of wage mobility. Their model contains two continuous latent variables, one affecting the initial state and the other the transitions.

Note that not only continuous random effects can be used to model unobserved heterogeneity, but also the mixture variable w_i can be used for this purpose. The choice between the two approaches depends on the assumptions one wishes to make about the nature of the unobserved heterogeneity; that is, whether it can be assumed to be continuous and normally distributed or whether a discrete specification – for example, using a mover-stayer structure – is more appropriate.

4.4.2.2 Time-constant affecting responses

Not only the transitions, but also the responses can be affected by time-constant continuous latent variables. In latent Markov models this would be a way to model dependencies between responses across occasions using an approach which is similar to the random-effects latent class models proposed in the biomedical field (Hadgu and Qu, 1998). Such a model is obtained by replacing $f(y_{itj}|x_{it} = k_t)$ with $f(y_{itj}|x_{it} = k_t, F_i)$ and defining a regression model for y_{itj} where F_i enters as one of the predictors.

In mixture growth modeling, it is very common to use a combination of discrete and continuous latent variables, where the continuous latent variables capture the unobserved heterogeneity within latent classes (Muthén, 2004; Vermunt, 2007). This involves replacing $f(y_{it}|w_i = \ell, \mathbf{z}_{it})$ by $f(y_{it}|w_i = \ell, \mathbf{z}_{it}, \mathbf{F}_i)$ or, equivalently, by allowing $\beta_{0\ell}$ and $\beta_{p\ell}$ (see equation 4.3) to be random effects.

4.4.2.3 Time-varying affecting responses

Rather than using time-constant continuous latent variables, it is also possible to work with time-varying continuous latent variables. One possible application is in a latent Markov model for multiple responses which cannot be assumed to be locally independent within time points. The time-varying continuous latent variables would capture unobserved time-specific factors which vary across individuals and which are independent across occasions. Such a model can be obtained by replacing $f(y_{itj}|x_{it}=k_t)$ by $f(y_{itj}|x_{it}=k_t, F_{it})$ and defining a regression model for y_{itj} where F_{it} enters as a predictor.

Another, very different, type of use of time-varying continuous variables in latent Markov models is as common factors in a factor analytic model for the response variables. In other words, the continuous latent variables define a factor analytic measurement model for the responses. Changes in the factor mean(s) could be modeled using either a mixture growth or a latent Markov model, which defines two longitudinal data variants of the mixture factor analysis model proposed by Yung (1997).

For the situation that there is one common factor, the variant using a latent Markov structure to model the change in the factor means may have the following form:

$$f(\mathbf{y}_i) = \sum_{k_0=1}^{K} \sum_{k_1=1}^{K} \cdots \sum_{k_{T_i}=1}^{K} P(x_{i0}=k_0) \left[\prod_{t=1}^{T_i} P(x_{it}=k_t|x_{it-1}=k_{t-1}) \right]$$
$$\left\{ \prod_{t=0}^{T_i} \int \left[f(F_{it}|x_{it}=k_t) \prod_{j=1}^{J} f(y_{itj}|F_{it}) \right] dF_{it} \right\},$$

where the last part shows that the distribution of the latent factor F_{it} depends on x_{it} and that F_{it} affects the responses. Regression models for F_{it} and y_{itj} complete the model specification.

4.4.3 Multiple, multilevel, and higher-order processes

This subsection presents extensions of latent Markov models for multiple, multilevel, and higher-order processes. These have in common that they require including an additional time-constant or time-varying discrete latent variable in the model. I will use a number as a subscript to denote the latent variable number (e.g., x^1_{it} and x^2_{it}), and an asterisk to refer to a latent variable at a higher level of a nested structure (e.g., x^*_{it}).

4.4.3.1 Parallel processes

The latent Markov models described so far assume that there is a single Markov process of interest, which is possibly affected by time-constant and time-varying predictors. Suppose one has a categorical time-varying predictor which cannot be assumed to be measured without error. As suggested by Vermunt, Langeheine, and Böckenholt (1999) as a possible extension of their model, a latent Markov structure could also be defined for such a time-varying predictor. This yields a latent Markov model with two latent variables x_{it}^1 and x_{it}^2, where x_{it}^1 is related to the first set of J^1 response variables and x_{it}^2 to the other set of J^2 responses. Assuming that there are no (other) covariates, such a model has the following form:

$$f(\mathbf{y}_i) = \sum_{k_0^1=1}^{K^1} \sum_{k_1^1=1}^{K^1} \dots \sum_{k_{T_i}^1=1}^{K^1} \sum_{k_0^2=1}^{K^2} \sum_{k_1^2=1}^{K^2} \dots \sum_{k_{T_i}^2=1}^{K^2} P(x_{i0}^1 = k_0^1, x_{i0}^2 = k_0^2)$$
$$\left[\prod_{t=1}^{T_i} P(x_{it}^1 = k_t^1, x_{it}^2 = k_t^2 | x_{it-1}^1 = k_{t-1}^1, x_{it-1}^2 = k_{t-1}^2)\right]$$
$$\left\{\prod_{t=0}^{T_i} \left[\prod_{j=1}^{J^1} f(y_{itj} | x_{it}^1 = k_t^1)\right] \left[\prod_{j=J^1+1}^{J^1+J^2} f(y_{itj} | x_{it}^2 = k_t^2)\right]\right\}.$$

Additional attention is required with respect to the joint probability of x_{it}^1 and x_{it}^2 given x_{it-1}^1 and x_{it-1}^2, which may be decomposed in a specific way and/or modeled using a logistic regression equation. A meaningful specification is, for example, a model in which x_{it}^1 and x_{it}^2 are both affected by x_{it-1}^1 and x_{it-1}^2 but are not associated with one another, yielding what is sometimes referred to as a cross-lagged panel model. This involve decomposing the joint transition probability of x_{it}^1 and x_{it}^2 by

$$P(x_{it}^1 = k_t^1 | x_{it-1}^1 = k_{t-1}^1, x_{it-1}^2 = k_{t-1}^2) P(x_{it}^2 = k_t^2 | x_{it-1}^1 = k_{t-1}^1, x_{it-1}^2 = k_{t-1}^2).$$

Another possibility is that the causal effect goes in one direction; that is, x_{it}^2 affects x_{it}^1 but x_{it}^1 is not affected by x_{it}^2 or x_{it-1}^2. This can be specified as follows:

$$P(x_{it}^1 = k_t^1 | x_{it-1}^1 = k_{t-1}^1, x_{it}^2 = k_t^2) P(x_{it}^2 = k_t^2 | x_{it-1}^2 = k_{t-1}^2). \tag{4.12}$$

A specification for correlated processes that are not causally related is obtained by allowing x_{it1} and x_{it2} to be associated and omitting the cross-lagged direct effects from the logistic model for x_{it1} and x_{it2}.

4.4.3.2 State-trait models

Eid and Langeheine (1999) proposed a discrete latent variable variant of the state-trait model. This model is obtained by expanding the latent Markov model with a J time-constant discrete latent variables, each of which affects one of the J responses.

The time-varying latent variable (representing the state) is assumed to be independent of the J time-constant latent variables (representing the traits). A state-trait model can be defined as follows:

$$f(\mathbf{y}_i) = \sum_{\ell^1=1}^{L^1} \sum_{\ell^2=1}^{L^2} \cdots \sum_{\ell^J=1}^{L^J} \sum_{k_0=1}^{K} \sum_{k_1=1}^{K} \cdots \sum_{k_{T_i}=1}^{K} P(w_i^1 = \ell^1, w_i^2 = \ell^2, ..., w_i^J = \ell^J)$$
$$P(x_{i0} = k_0) \left[\prod_{t=1}^{T_i} P(x_{it} = k_t | x_{it-1} = k_{t-1}) \right]$$
$$\left[\prod_{t=0}^{T_i} \prod_{j=1}^{J} f(y_{itj} | x_{it} = k_t, w_i^j = \ell^j) \right].$$

A more restricted variant of this model is obtained by assuming that the states are independent across occasions: $P(x_{it} = k_t | x_{it-1} = k_{t-1}) = P(x_{it} = k_t)$.

Eid and Langeheine (1999) worked with categorical y_{itj} variables for which they defined logistic models. These contained main effects of the state at time point t and the trait for response j but no interaction term; that is,

$$\log \frac{P(y_{itj} = m | x_{it} = k, w_i^j = \ell)}{P(y_{itj} = M | x_{it} = k, w_i^j = \ell)} = \beta_{0jm} + \beta_{1jkm} + \beta_{2j\ell m}.$$

4.4.3.3 Second-order model

As indicated earlier, one of the key assumptions of the latent Markov model is that the latent state transitions can be described with a first-order Markov structure. This assumption can be relaxed, for example, by allowing x_{it} to be affected not only by x_{it-1}, but also by x_{it-2}, which involves replacing $P(x_{it} = k_t | x_{it-1} = k_{t-1})$ by $P(x_{it} = k_t | x_{it-1} = k_{t-1}, x_{it-2} = k_{t-2})$ for $t \geq 2$. Though most software for latent Markov modeling does not allow defining such a second-order process, it can be defined with a trick which involves using a second time-varying latent variable x_{it}^2. The cross-lagged effect of x_{it}^1 (the variable of interest) on x_{it}^2 is restricted in such a way that $P(x_{it}^2 = k_t | x_{it-1}^1 = k_{t-1}) = 0$ for $k_t \neq k_{t-1}$, which implies that the lag one of the second latent variable (x_{it-1}^2) is in fact the lag two of the first latent variable (x_{it-2}^1). The second-order latent Markov model can now be obtained by allowing the transition probability for x_{it}^1 to depend on the lag of the second latent variable, which yields a model of the form

$$f(\mathbf{y}_i) = \sum_{k_0^1=1}^{K^1} \sum_{k_1^1=1}^{K^1} \dots \sum_{k_{T_i}^1=1}^{K^1} \sum_{k_0^2=1}^{K^2} \sum_{k_1^2=1}^{K^2} \dots \sum_{k_{T_i}^2=1}^{K^2} P(x_{i0}^1 = k_0^1) P(x_{i0}^2 = k_0^2)$$

$$P(x_{i1}^1 = k_1^1 | x_{i0}^1 = k_0^1) \left[\prod_{t=2}^{T_i} P(x_{it}^1 = k_t^1 | x_{it-1}^1 = k_{t-1}^1, x_{it-1}^2 = k_{t-1}^2) \right]$$

$$\left[\prod_{t=1}^{T_i} P(x_{it}^2 = k_t^2 | x_{it-1}^1 = k_{t-1}^1) \right]$$

$$\left[\prod_{t=0}^{T_i} \prod_{j=1}^{J} f(y_{itj} | x_{it}^1 = k_t^1) \right].$$

4.4.3.4 Processes for nested time units

Another interesting extension of the simple latent Markov model was recently presented by Rijmen et. al (2008). In their application there were two nested time units: the higher-level concerned changes occurring between days and the lower-level changes occurring between (non-sleeping) hours within days. The proposed model consists of two nested latent Markov models, one for between-day transitions and one for within-day transitions. A slight expansion of our notation is needed to write down the relevant model formulae. Let h, i, and t be the indices for a person, a day, and an hour, respectively. For the rest, notation is kept as much as possible as above, with the exception that quantities referring to the higher-level process get an asterisk as a superscript. The higher-level (between-day) model for person h can now be defined as

$$f(\mathbf{y}_h) = \sum_{k_0^*=1}^{K^*} \sum_{k_1^*=1}^{K^*} \dots \sum_{k_{T_h^*}^*=1}^{K^*} P(x_{h0}^* = k_0^*) \left[\prod_{i=1}^{T_h^*} P(x_{hi}^* = k_i^* | x_{hi-1}^* = k_{i-1}^*) \right]$$

$$\left[\prod_{i=0}^{T_h^*} f(\mathbf{y}_{hi} | x_{hi}^* = k_i^*) \right],$$

which has the structure of a standard latent Markov model. The lower-level (within-day) model describing the hourly changes specifies a latent Markov model for $f(\mathbf{y}_{hi} | x_{hi}^* = k_i^*)$,

$$f(\mathbf{y}_{hi}|x^*_{hi}=k^*_i) = \sum_{k_0=1}^{K}\sum_{k_1=1}^{K}\cdots\sum_{k_{T_i}=1}^{K} P(x_{hi0}=k_0|x^*_{hi}=k^*_i)$$
$$\left[\prod_{t=1}^{T_{hi}} P(x_{hit}=k_t|x_{hit-1}=k_{t-1},x^*_{hi}=k^*_i)\right]$$
$$\left[\prod_{t=0}^{T_{hi}}\prod_{j=1}^{J} f(y_{hitj}|x_{hit}=k_t,x^*_{hi}=k^*_i)\right].$$

Note that this is, in fact, a mixture latent Markov model in which the initial-state and transition probabilities and possibly also the response densities depend on the higher-level latent state occupied by person h at day i (x^*_{hi}).

4.4.3.5 Multilevel data

Vermunt (2003, 2004) proposed multilevel extensions of various types of mixture models that may also be useful in longitudinal data analysis. That is, when the observations for which we have longitudinal data are nested within higher-level units. Examples are longitudinal data on children which are nested within school, repeated measures data on patients nested within hospitals, and panel data from respondents nested within regions.

Palardy and Vermunt (in press) presented a multilevel mixture growth model for such data sets and illustrated the model with an application in which higher-level units (schools) are clustered based on the learning rates of children. Vermunt (2004) presented an application using a similar, but slightly simpler, multilevel mixture growth model. Denoting a higher-level unit by h, the higher-level part of this model is

$$f(\mathbf{y}_h|\mathbf{z}_h) = \sum_{\ell^*=1}^{L^*} P(w^*_h=\ell^*)\prod_{h=0}^{I^*_h} f(\mathbf{y}_{hi}|w^*_h=\ell^*,\mathbf{z}_{hi}),$$

where I_h is the number of persons belonging to higher-level unit or group h. The lower-level part is

$$f(\mathbf{y}_{hi}|w^*_h=\ell^*,\mathbf{z}_{hi}) = \sum_{\ell=1}^{L} P(w_{hi}=\ell)\prod_{t=0}^{T_{hi}} f(y_{hit}|w_{hi}=\ell,w^*_h=\ell^*,\mathbf{z}_{hit}).$$

As in the mixture growth model described in equations (4.1) and (4.3), the regression model for y_{hit} specifies how the higher- and lower-level latent classes differ in term of the growth parameters.

Yu and Vermunt (in progress) developed a multilevel extension of the latent Markov model. The structure of this model is similar to that of a mixture latent Markov model, with the important difference that the mixture is at the group level and thus not at the individual level. The model can be formulated as follows:

$$f(\mathbf{y}_i) = \sum_{\ell^*=1}^{L^*} P(w_h^* = \ell^*) \prod_{i=1}^{I_h^*} f(\mathbf{y}_{hi}|w_h^* = \ell^*).$$

The lower-level part defines the structure for $f(\mathbf{y}_{hi}|w_h^* = \ell^*)$ which is the same as the lower-level part of the multilevel process model described above, except for that the conditioning is on $w_h^* = \ell^*$ instead of $x_{hi}^* = k_i^*$; that is,

$$f(\mathbf{y}_{hi}|w_h^* = \ell^*) = \sum_{k_0=1}^{K} \sum_{k_1=1}^{K} \dots \sum_{k_{T_i}=1}^{K} P(x_{hi0} = k_0|w_h^* = \ell^*) \left[\prod_{t=1}^{T_{hi}} P(x_{hit} = k_t|x_{hit-1} = k_{t-1}, w_h^* = \ell^*)\right] \left[\prod_{t=0}^{T_{hi}} \prod_{j=1}^{J} f(y_{hitj}|x_{hit} = k_t, w_h^* = \ell^*)\right].$$

4.4.3.6 Dependent classification errors

One of the assumptions of the latent Markov model is that responses are independent across time points conditional on the latent states, an assumption that may be unrealistic in certain applications. However, it is sometimes possible to relax this assumption, which is sometimes referred to as ICE (independent classification errors).

Above, we already discussed a non-ICE model; that is, a latent Markov model with a time-constant continuous latent variable affecting the responses at the different time points. In this model, it is assumed that an unobserved individual factor is causing correlations between measurement errors. This is a good non-ICE model when these correlations are (almost) equally strong between each pair of occasions.

However, typically, correlations between errors are much stronger between adjacent time points. Possible mechanisms leading to such correlated errors are that making an error at one occasion increases the likelihood of making an error at the next occasion (Manzoni et al., in progress), or that experiencing a transition increases the likelihood of making an error (see also Hagenaars, 1988). Bassi et al. (2000) proposed a non-ICE latent Markov model for employment status measurements obtained using a very specific retrospective data collection design (see also Hagenaars' chapter in this volume).

Here, I would like to discuss the non-ICE specification proposed by Manzoni et al. (in progress). Their application concerned a latent Markov model with two measures of a person's monthly employment status (employed, self employed, unemployed, and not employed) for a period of about a year. The first measure is a retrospective report of the last year and the second is a retrospective report on the same period collected ten years later. The aim of the analysis was to determine the quality latter report. Because respondents are likely to misplace or forget unemployment spells when these occurred a long time ago, it is clearly incorrect to assume that errors in the second measure are uncorrelated across occasions. Manzoni et al. proposed a correlated measurement error model which involves replacing the response probability $P(y_{it2} = m_{t2}|x_{it} = k_t)$ by $P(y_{it2} = m_{t2}|x_{it} = k_t, y_{it-1,2} = m_{t-1,2}, x_{it-1} = k_{t-1})$; that is, a model in which $y_{it,2}$ is not only affected by x_{it}, but also by $y_{it-1,2}$ and x_{it-1}. Moreover, restrictions were imposed on the way the lagged observed and latent states affect the measurement error. One restriction yielded a specification in which respondents making an error at $t-1$ have a different (higher) probability of making an error at t. So, in fact, two sets of error probabilities were estimated, one for respondents reporting correctly at $t-1$ ($m_{t-1,2} = k_{t-1}$) and another for respondents reporting incorrectly ($m_{t-1,2} \neq k_{t-1}$). Various alternative specifications were also investigated.

4.5 Applications

This section presents two applications of the mixture models for longitudinal data described in this chapter. The first application concerns a repeated measures experimental study and is used to illustrate the mixture growth model, including the more advanced model with continuous random effects. The second application concerns a longitudinal survey and is used to illustrate the latent Markov and mixture latent Markov model, as well as the latent Markov model for parallel processes. For parameter estimation, I used version 4.5 of the Latent GOLD program (Vermunt and Magidson, 2005, 2008). Examples of syntax files can be found in the Appendix.

4.5.1 A mixture growth model

The empirical example I will use to illustrate mixture growth modeling is taken from Hedeker and Gibbon's (1996) MIXOR program. It concerns a dichotomous outcome variable "severity of schizophrenia" measured at 7 occasions (consecutive weeks). This binary outcome was obtained by collapsing a severity score ranging from 1 to 7 into two categories, where a 1 indicates that the severity score was at least 3.5 (severe), and 0 that is was smaller than 3.5 (non severe). In total, there is information on 437 cases. However, for none of the cases there is complete information. For 42 cases, we have observations at 2, for 66 at 3, for 324 at 4, and for 5 at 5 time points. There are 434, 426, 14, 374, 11, 9, and 335 observations at the 7 time points. Besides the repeated measures for the response variable, there is one time-constant predictor, treatment (0=control group; 1=treatment group). The treatment is a new drug that is expected to decrease the symptoms related to schizophrenia. The main research question to be answered with this data set is whether the treatment reduces the symptoms related to schizophrenia.

Table 4.1 Test results for the mixture growth models estimated with the schizophrenia data

Model	Log-likelihood	BIC	# Parameters
A1: 1-class growth	-704	1421	2
A2: 2-class growth	-625	1286	6
A3: 3-class growth	-608	1277	10
A4: 4-class growth	-601	1287	14
B2: 2-class growth with squared time for class 2	-620	1282	7
B3: 3-class growth with squared time for class 3	-597	1261	11
C2: B2 with random intercept	-601	1250	8
C3: B3 with random intercept	-595	1263	12

Whereas Vermunt (2007) used the same data set for a more extended comparison of various types of growth models, here the focus will be on the mixture growth models described in this chapter. More specifically, it will be shown that two growth classes can be identified – one class with decreasing severity and one class without – and that patients receiving the treatment are much more likely to be belong to the decreasing severity class than the control group. Moreover, it will be shown that using random effects may yield a simpler solution with a smaller number of latent classes.

In the analysis of this data set, I followed Hedeker and Gibbon's (1996) suggestion to set $P = 1$, with $z_{it1} = \sqrt{t}$, and to use a binary logit model. This yields a model in which the logit of severity is a function of the square root of time. Though there is no strong theoretical motivation for using this functional form for the time dependence, there is a good empirical motivation: in a simple model without latent classes nor random effects, this model fits the time-specific response probabilities much better than a linear or a quadratic model, and almost as well as a model with an unrestricted time dependence.

Table 4.1 reports the log-likelihood value, the number of parameters, and the BIC value obtained by applying various of the models described in the previous two sections to the schizophrenia data set. Models A1-A4 are 1 to 4-class mixture growth models using the $\sqrt{t}$ time dependence and containing treatment as a covariate affecting the class membership. Based on the BIC value, one would select the 3-class model as the best one. Models B2 and B3 modify models A2 and A3 in the sense that one latent class (the last one) has a different (quadratic) time dependence. This is specified by defining $z_{it2} = t$ and $z_{it3} = t^2$, and setting the parameters corresponding to these two terms to 0 in all but class K and the parameter corresponding to z_{it1} to 0 in class K. As can be seen from the BIC values, Models B2 and B3 fit better than Models A2 and A3, which indicates that it makes sense to assume another type of time dependence for one of the classes. It can also be seen that the 3-class model is still preferred to the 2-class model. Models C2 and C3 are variants of Models B2 and B3 containing a random intercept to allow for within class heterogeneity. As can be seen, these models have lower BIC values than Models B2 and B3. Moreover,

Table 4.2 Parameter estimates obtained with Model C2

Model for Responses	Class 1			Class 2		
	β or λ	s.e.	z-value	β or λ	s.e.	z-value
Intercept	9.16	1.22	7.49	6.95	0.92	7.56
StdDev Random Intercept	3.50	0.55	6.42	3.50	0.55	6.42
TIME				- 3.82	1.01	-3.77
SQ-TIME				1.13	0.30	3.77
SQRT-TIME	-4.98	0.65	-7.66			
Model for Latent Classes	**Class 1**					
	γ	s.e.	z-value			
Intercept	-0.64	0.31	-2.08			
Treatment	1.80	0.36	4.99			

under this specification, the simpler 2-class model (C2) performs better than the 3-class model (C3).

Table 4.2 reports the parameter estimates obtained with Model C2. For each latent class, we have a set of parameters describing the time dependence of the logit of the probability of being in the severely schizophrenic state – Intercept and SQRT-TIME in class 1 and Intercept, TIME, and SQ-TIME in class 2 – as well the standard deviation of the random effect indicating how much the intercept varies within classes. The size of latter parameter, which is assumed to be equal across latent classes, indicates that there is quite some variation within classes. Figure 4.5.1 depicts the estimated growth curves for the two latent classes, which are obtained by marginalizing over (integration out) the continuous random effects. Class 1 contains the patients for which the probability of severe symptoms of schizophrenia decreases during the study. It can now also be seen why the quadratic curve was needed for class 2: after a small drop in weeks 1 and 2, the probability of a severe form of schizophrenia increased again, a pattern that cannot be described by a monotonic function.

Out of the total sample, 66% is estimated to belong to latent class 1 and 34% to latent class 2. These numbers are 76% and 24% for the treatment group and 35% and 65% for the control group. The treatment effect on class member is given in terms of a logistic regression coefficient and its asymptotic standard error in the lower part of Table 4.2 – the odds of begin in class 1 instead of 2 is exp(1.80) higher for the treatment than for the control group. The encountered treatment effect shows, on the one hand, that there is a rather strong relation between treatment and class membership, but, on the other hand, that this relationship is far from perfect.

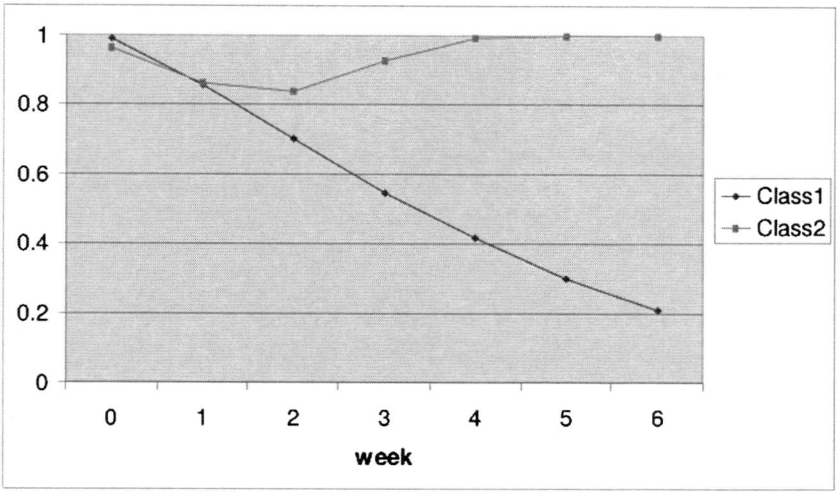

Fig. 4.1 Class-specific trajectories obtained with Model C2.

4.5.2 A mixture Latent Markov model

The latent Markov models described above will be illustrated with the nine-wave National Youth Survey (Elliott, Huizinga, and Menard, 1989) for which data were collected annually from 1976 to 1980 and at three year intervals after 1980. At the first measurement occasion, the ages of the 1725 children varied between 11 and 17. To account for the unequal spacing across panel waves and to use age as the time scale, we define a model for 23 time points ($T+1=23$), where $t=0$ corresponds to age 11 and the last time point to age 33. For each subject, we have observed data for at most 9 time points (the average is 7.93) which means that the other time points are treated as missing values.

We study the change in a dichotomous response variable "drugs" indicating whether young persons used hard drugs during the past year (1=no; 2=yes). It should be noted that among the 11 years of age nobody in the sample reported to have used hard drugs, which is something that will be taken into account in our model specification. Time-varying predictors are age and age squared, and time-constant predictors are gender and ethnicity. In the second step of the analysis, I will introduce alcohol use during the past year as a time-varying covariate containing measurement error.

A preliminary analysis showed that there is a clear age-dependence in the reported hard-drugs use which can well be described by a quadratic function: usage first increases with age and subsequently decreases. That is why we used this type of time dependence in all reported models. Age and age-squared are used as

Table 4.3 Test results for the Markov models estimated with the drugs use data

Model	Log-likelihood	BIC	# Parameters
A1. Markov	-4143	8330	6
A2. Latent Markov with K=2	-4009	8078	8
A3. Mover-stayer latent Markov with K=2	-4000	8068	9
A4. Mixture latent Markov with L=2 and K=2	-3992	8066	11
A5. A4 with Gender & Ethnicity effects on W_i	-3975	8061	15
B1. A5 with Markov model for Alcohol	-9328	18789	18
B2. B1 with Alcohol affecting X_{it}	-9185	18520	20
B3. B2 with Alcohol measured with error	-8912	17989	22

time-dependent covariates in the regression model for the latent transition probabilities (see also equation 4.8); that is,

$$\log \frac{P(x_{it} = k' | x_{it-1} = k, w_i = \ell, \text{age}_{it})}{P(x_{it} = k | x_{it-1} = k, w_i = \ell, \text{age}_{it})} = \alpha_{0kk'} + \alpha_{\ell kk'} + \alpha_{L+1,kk'} \cdot \text{age}_{it} + \alpha_{L+2,kk'} \cdot (\text{age}_{it})^2,$$

where the α coefficients are fixed to 0 for $k' = k$ and for $\ell = 1$. For the initial-state, we do not have a model with free parameters but we simply assume that all children start in the no-drugs state at age 11.

Table 4.3 reports the fit measures for the estimated models, where Models A1 to A4 do not contain covariates gender and ethnicity. Among these models, the most general model – the mixture latent Markov model – performs best. By removing measurement error, simplifying the mixture into a mover-stayer structure, or eliminating the mixture structure, the fit deteriorates significantly. Model A5 is a mixture latent markov model in which we introduced covariates in the model for the mixture proportions: sex and/or ethnicity seem to be significantly related to the mixture component someone belongs to.

As a final step, we investigated whether alcohol use affects hard drugs use. We specified three additional models: Model B1 in which alcohol does not affect drugs use, Model B2 in which alcohol use at age t affects the transitions in the model for drugs, and Model B3 in which alcohol use is treated as a time-varying covariate measured with error. The latter model is a latent Markov model for two parallel processes. We used a specification in which alcohol use affects the drugs-use transitions but in which the reversed effect is absent (see equation 4.12). In Models B1 and B2, we specified a Markov model without measurement error for alcohol use in order to be able to compare the BIC values across these three models. Note that as far as the modeling of drugs use is concerned, Model B1 is, in fact, equivalent to Model A5, but their log-likelihood values cannot be compared because alcohol is introduced as an additional response variable in Model B1. Comparison of the fits measures for Models B1 and B2 shows that alcohol use has a significant effect on the drugs use transitions, and comparison of the fit measures for Models B2 and B3 shows that there is evidence that alcohol use is measured with error.

One set of parameters of the final model (B3) are the probabilities of the measurement models for drugs and alcohol. These show that the latent states are rather strongly connected to the two observed states: $P(y_{it1} = 1|x_{it}^1 = 1) = 0.99$ and $P(y_{it1} = 2|x_{it}^1 = 2) = 0.83$ for drugs use; $P(y_{it2} = 1|x_{it}^2 = 1) = 0.87$ and $P(y_{it1} = 2|x_{it}^2 = 2) = 0.99$ for alcohol use.

The most relevant coefficients in the model for the drugs use transitions are the effects of alcohol (x_{it}^2) and of w_i. The former show that being in the latent alcohol use state increases the probability of moving into the drugs use state ($\alpha = 4.61; S.E = 1.33$) and decreases the probability of exciting the drugs use state ($\alpha = -1.86; S.E = 0.52$). The parameters for w_i show that class 1 is the low-risk class having a lower probability than class 2 of entering into the use state ($\alpha = -1.19; S.E = 0.36$) and a much higher probability of leaving the non-use state ($\alpha = 4.16; S.E. = 0.63$). This means that class 1 contains young people that quit the drug-use state quickly when they get into this state.

The parameters in the logistic regression model for w_i shows that males are less likely to be in the low-risk class than females ($\gamma = -0.67; S.E. = 0.20$). Moreover, blacks are more likely ($\gamma = 0.41; S.E = 0.26$), hispanics less likely ($\gamma = -0.75; S.E. = 0.52$), and other ethnic groups less likely ($\gamma = -0.09; S.E = 0.70$) to be in the low-risk class than whites, but these ethnicity effects are non significant.

Appendix: Examples of Latent GOLD syntax files

The Latent GOLD 4.5 software package (Vermunt and Magidson, 2008) implements the mixture models described in this article. In this appendix, I provide examples of syntax files used for the empirical applications.

The data should be in the format of a person-period file, where for Markov type models it is important to include also periods with missing values in the file since each next record for the same subject is assumed to be the next time point. The definition of a model contains three main sections: "`options`", "`variables`" and "`equations`".

The mixture growth models A1 to A4 from Table 4.1 can be defined as follows:

```
options
   output parameters standarderrors estimatedvalues;
 variables
   caseid id;
   dependent severity binomial;
   independent sqrttime, treatment;
   latent W nominal 2;
 equations
   W        <- 1 + treatment;
   severity <- 1 | W + sqrttime | W;
```

In the above `options` section, only the commands related to the output options are shown. It is indicated that we wish to output parameters and standard errors of the parameters, as well as the estimates for the model probabilities.

In the `variables` section we define the `caseid` variable connecting the multiple records of a person, the `latent`, `dependent`, and `independent` variables to be used in the analysis, as well as various attributes of these variables, such as their scale types and, for categorical latent variables, also their number of categories. Note that the model above is a two-class mixture model since we specified "`latent W nominal 2;`".

The `equation` section contains 2 equations: one for the mixture variable (`W`) and another for the response variable. The logit model for `W` contains an intercept (the term "`1`") and the effect of treatment. The model for the response variable `severity` contains an intercept and an effect of square root time. Both parameters are assumed to vary across latent classes, which is achieved by the conditioning "`| W`".

The more complex final two-class model C2 – containing a continuous random effect and a different time dependence for classes 1 and 2 – is defined as follows:

```
options
  output parameters standarderrors estimatedvalues;
variables
  caseid id;
  dependent severity binomial;
  independent sqrttime, time, sqtime, treatment;
  latent W nominal 2, F continuous;
equations
  W        <- 1 + treatment;
  severity <- 1 | W + (b1) sqrttime | W + (b2) time | W
               + (b3) sqtime | W + F;
  b1[2]=0; b2[1]=0; b3[1]=0;
```

As can be seen, the model contains two additional predictors (`time` and `sqtime`) and a continuous latent variable (`F`). These are all used as predictors in the regression model for the response variable. It can also be seen that three of the regression coefficient get labels, which is needed to be able to define the three constraints at the bottom. These restrictions indicate that `sqrttime` has no effect in class 2, and that `time` and `sqtime` have no effect in class 1.

The syntax for Markov models is somewhat more complicated than for growth models. As an example, this is the setup for model A5 appearing in Table 4.3, a mixture latent Markov model with two covariates affecting the mixture distribution and with a quadratic time dependence of the transition logits:

```
options
  missing includeall;
  output parameters=first standarderrors estimatedvalues;
variables
  caseid id;
  dependent drugs nominal;
  independent gender nominal, ethnicity nominal, age, age2;
```

```
  latent W nominal 2, X nominal dynamic 2;
equations
  W      <- 1 + gender + ethnicity;
  X[=0] <- (-100) 1;
  X      <- (~tra) 1 | X[-1] + (a~tra) W | X[-1]
            + (~tra) age | X[-1] + (~tra) age2 | X[-1];
  drugs <- (b~err) 1 | X;
```

Compared to the specification above, the `options` section contains the statement "`missing=includeall`" indicating that records with missing values should be retained in the analysis and the output option "`parameters=first`" requesting dummy coding with the first category as the reference category for nominal variables. A new element in the `variables` section is the keyword "`dynamic`" which indicates that the nominal latent variable `X` may change its value over time (in this case, it is a two-state time-varying latent variable).

The `equations` section contains 4 equations: one for the mixture variable (`W`), one for the initial state (`X[=0]`), one for the state at time point t (`X`) conditional on the state at $t-1$ (`X[-1]`), and one for the response variable at time point t (`drugs`). The logit model for `W` contains an intercept as well as effects of gender and ethnicity. The model for `X[=0]` contains an intercept that is fixed to -100, which indicates that everyone starts in latent state 1. The model for `X` is parameterized in such a way that the intercept and the effects of `W`, `age`, and `age2` can be interpreted as effects on the logit of a transition (as in the equation provided in the text). This is achieved by the conditioning "`| X[-1]`" combined with "~tra" in the parameter label, which yields a coding for the logit coefficients in which the no transition category serves as the reference category. The model for the response variable `drugs` contains an intercept which varies across latent states, with the same type of coding as used for the transition (for the dependent variable called error coding). Note that removing "~tra" and "~err" does not change the model but only the identifying constraints that are imposed in the parameter set concerned. As can be seen, two parameter sets get labels (`a` and `b`), which will be used below to define models with parameter restrictions.

The 2-class mixture can be changed into a mover-stayer structure with the additional line "`a = -100;`" which fixes the transition probabilities to 0 for the second class. A latent Markov model is obtained either by removing `W` from the `variables` and `equations` sections or by setting its number of categories to 1. A standard Markov is obtained with the extra line "`b = -100;`". This fixes the logit parameters in the model for the response variable to -100, which because of the special error coding (induced with "~err") yields a perfect relationship between `X` and `drugs`.

The model in which alcohol is used as a time-varying covariate measured with error (Model B3 of Table Table 4.3 is obtained by including `alcohol` as a second dependent variable and defining a second dynamic latent variable `X2`. The `equations` section of this more advanced model contains also equations for the initial state and the transitions of `X2`, includes `X2` in the equation for `X`, and defines a measurement equation for `alcohol`; that is,

```
equations
  W       <- 1 + gender + ethnicity;
  X[=0]   <- (-100) 1;
  X2[=0]  <- 1;
  X       <- (~tra) 1 | X[-1] + (~tra) W | X[-1]
             + (~tra) age | X[-1] + (~tra) age2 | X[-1]
             + (~tra) X2 | X[-1];
  X2      <- (~tra) 1 | X2[-1];
  drugs   <- 1 | X;
  alcohol <- 1 | X2;
```

References

Agresti, A. (2002). *Categorical data analysis*. New York: Wiley.

Aitkin, M. (1999). A general maximum likelihood analysis of variance components in generalized linear models. *Biometrics, 55,* 218-234.

Baum, L. E., Petrie, T., Soules, G., & Weiss, N. (1970). A maximization technique occurring in the statistical analysis of probabilistic functions of Markov chains. *Annals of Mathematical Statistics, 41,* 164-171.

Bartolucci, F., Pennoni, F., & Francis, B. (2007). A latent Markov model for detecting patterns of criminal activity. *Journal of the Royal Statistical Society, Ser. A, 170,* 115-132.

Bassi, F., Hagenaars, J. A., Croon, M. A., & Vermunt, J. K. (2000). Estimating true changes when categorical panel data are affected by uncorrelated and correlated errors. *Sociological Methods and Research, 29,* 230-268.

Bergma, W., Croon, M. A., & Hagenaars, J. A. (2009). *Marginal models for dependent, clustered and longitudinal categorical data*. Dordrecht, NL: Springer.

Böckenholt, U. (2005). A latent Markov model for the analysis of longitudinal data collected in continuous time: States, durations, and transitions. *Psychological Methods, 10,* 65-82.

Collins, L. M., & Wugalter, S. E. (1992). Latent class models for stage-sequential dynamic latent variables. *Multivariate Behavioral Research, 27,* 131-157.

Dayton C.M., & Macready, G. B. (1988). Concomitant-variable latent class models. *Journal of the American Statistical Association, 83,* 173-178.

Dias, J. G., & Vermunt, J. K. (2007). Latent class modeling of website users' search patterns: Implications for online market segmentation. *Journal of Retailing and Consumer Services, 14,* 359-368.

Dias, J. G., Vermunt, J. K., & Ramos, S. (2009). Mixture hidden Markov models in finance research. In A. Fink, L. Berthold, W. Seidel, & A. Ultsch (Eds.), *Advances in data analysis, data handling and business intelligence*. Berlin-Heidelberg: Springer

Diggle, P. J., Liang, K. Y., & Zeger (1994), S. L. *Analysis of longitudinal data*. Oxford: Clarendon Press.

Eid, M., & Langeheine, R. (1999). Measuring consistency and occasion specificity with latent class models: A new model and its application to the measurement of affect. *Psychological Methods, 4,* 100-116.

Elliott, D. S., Huizinga, D., & Menard, S. (1989). *Multiple problem youth: Delinquency, substance use, and mental health problems*. New York: Springer-Verlag.

Everitt, B. S., & Hand, D. J. (1981). *Finite mixture distributions*. London: Chapman & Hall.

Fay, R. E. (1986). Causal models for patterns of nonresponse. *Journal of the American Statistical Association, 81,* 354-365.

Frühwirth-Schnatter, S. (2006). *Finite mixture and Markov switching models.* New York, NY: Springer.

Goldfeld, S., & Quandt, R. (1973). A Markov model for switching regressions. *Journal of Econometrics, 1,* 3-16.

Goodman, L. A. (1961). Statistical methods for the mover-stayer model. *Journal of the American Statistical Association, 56,* 841-868.

Goodman, L. A. (1974). The analysis of systems of qualitative variables when some of the variables are unobservable: Part I - A modified latent structure approach. *American Journal of Sociology, 79,* 1179-1259.

Hagenaars, J. A. (1988). Latent structure models with direct effects between indicators: Local dependence models. *Sociological Methods and Research, 16,* 379-405.

Hagenaars, J. A. (1990). *Categorical longitudinal data: Loglinear analysis of panel, trend and cohort data.* Newbury Park: Sage.

Hadgu, A., & Qu, Y. (1998). A biomedical application of latent class models with random effects. *Applied Statistics, 47,* 603-616.

Hedeker, D., & Gibbons, R. D. (1996). MIXOR: A computer program for mixed-effects ordinal regression analysis. *Computer Methods and Programs in Biomedicine, 49,* 157-176.

Heinen, T. (1996). *Latent class and discrete latent trait models: Similarities and differences.* Thousand Oakes, CA: Sage Publications.

Kaplan, D. (2005). Finite mixture dynamic regression modeling of panel data with implications for dynamic response analysis. *Journal of Educational and Behavioral Statistics*, 30, 169-187.

Langeheine, R., & van de Pol, F. (2002). Latent Markov chains. In J. A. Hagenaars & A. L. McCutcheon (Eds.), *Applied latent class analysis.* (pp. 304-341). Cambridge, UK: Cambridge University Press.

Lazarsfeld, P. F., & Henry, N. W. (1968). *Latent structure analysis.* Boston, MA: Houghton Mifflin.

Little, R.J., & Rubin, D. B. (1987). *Statistical analysis with missing data.* New York: Wiley.

Magidson, J., Vermunt, J. K., & Tran, B. (2009). Using a mixture latent Markov model to analyze longitudinal U.S. employment data involving measurement error. In K. Shigemasu, A. Okada, T. Imaizumi, & T. Hoshino (Eds.), *New Trends in Psychometrics* (pp. 235-242). Tokyo: Universal Academy Press.

Manzoni, A., Vermunt, J.K., Luijkx, R., & Muffels, R. (in preparation). Memory bias in retrospectively collected employment careers: A model based approach to correct for measurement error.

McDonald, I. L., & Zucchini, W. (1997), *Hidden Markov and other models for discrete valued time series.* London: Chapman and Hall.

McLachlan, G. J., & Peel, D. (2000). *Finite mixture models.* New York: Wiley.

Mooijaart, A., & van Montfort, K. (2007). Latent Markov models for catagorical variables and time-dependent covariates. In K. van Montfort, J. Oud, & A. Satorra (Eds.), *Longitudinal models in the behavioral and related sciences.* (pp. 1-18). Mahwah, NJ: Lawrence Erlbaum.

Muthén, B. (2004). Latent variable analysis: Growth mixture modeling and related techniques for longitudinal data. In D. Kaplan (Ed.), *The Sage handbook of quantitative methodology for the social sciences* (pp. 345-368). Thousand Oakes, CA: Sage.

Nagin, D. S. (1999). Analyzing developmental trajectories: A semiparametric group-based approach. *Psychological Methods, 4,* 139-157.

Paas, L. J., Vermunt, J. K., & Bijmolt, T. H, (2007). Discrete-time discrete-state latent Markov modelling for assessing and predicting household acquisitions of financial products. *Journal of the Royal Statistical Society, Ser. A (Statistics in Society), 170,* 955-974.

Palardy, G., & Vermunt, J. K. (in press). Multilevel growth mixture models for classifying group-level observations. *Journal of Educational and Behavioral Statistics.*

Pavlopoulos, D., Muffels, R., & Vermunt, J. K. (2009). Training and low-pay mobility: The case of the UK, the Netherlands and Germany. *Labour, 21,* 37-59.

Poulsen, C. S. (1990). Mixed Markov and latent Markov modelling applied to brand choice behaviour. *International Journal of Research in Marketing, 7,* 519.

Rijmen, F., Vansteelandt, K., & de Boeck, P. (2008). Latent class models for diary method data: Parameter estimation by local computations. *Psychometrika, 73,* 167-182.

Schafer, J. L. (1997). *Statistical analysis with incomplete data.* London: Chapman & Hall.

Schmittmann, V. D., Dolan, C.V., van der Maas, H. L. J., & Neale, M. C. (2005). Discrete latent Markov models for normally distributed response data. *Multivariate Behavioral Research, 40,* 461-488.

Skrondal, A., & Rabe-Hesketh, S. (2004). *Generalized latent variable modeling: Multilevel, longitudinal, and structural equation models*. London: Chapman & Hall/CRC.

van de Pol, F., & de Leeuw, J. (1986). A latent Markov model to correct for measurement error. *Sociological Methods and Research, 15,* 118-141.

van de Pol, F., & Langeheine, R. (1990). Mixed Markov latent class models. *Sociological Methodology*, 213-247.

van der Heijden, P. G. M., Dessens, J., & Böckenholt, U. (1996). Estimating the concomitant variable latent class model with the EM algorithm. *Journal of Educational and Behavioral Statistics, 21,* 215-229.

Vermunt, J. K. (1997a). *Log-linear models for event histories*. Thousand Oakes, CA: Sage.

Vermunt, J. K. (1997b). *LEM: A general program for the analysis of categorical data: User's manual*. Tilburg, The Netherlands: Tilburg University.

Vermunt, J. K. (2001). The use of latent class models for defining and testing non-parametric and parametric item response theory models. *Applied Psychological Measurement, 25,* 283-294.

Vermunt, J. K. (2003). Multilevel latent class models. *Sociological Methodology, 33,* 213-239.

Vermunt, J. K. (2004) An EM algorithm for the estimation of parametric and nonparametric hierarchical nonlinear models. *Statistica Neerlandica, 58,* 220- 233.

Vermunt, J. K. (2007). Growth models for categorical response variables: Standard, latent-class, and hybrid approaches. In K. van Montfort, J. Oud, & A. Satorra (Eds.), *Longitudinal models in the behavioral and related sciences* (pp. 139-158). Mahwah, NJ: Lawrence Erlbaum.

Vermunt, J. K., & Georg, W. (2002). Longitudinal data analysis using log-linear path models with latent variables. *Metodología de las Ciencias del Comportamiento, 4,* 37-53.

Vermunt, J. K., & Hagenaars, J. A. (2004). Ordinal longitudinal data analysis. In R. C. Hauspie, N. Cameron, & L. Molinari (Eds.), *Methods in human growth research* (pp. 374-393). Cambridge, UK: Cambridge University Press.

Vermunt, J. K., Langeheine, R., & Böckenholt, U. (1999). Latent Markov models with time-constant and time-varying covariates. *Journal of Educational and Behavioral Statistics, 24,* 178-205.

Vermunt, J. K., & Magidson, J. (2005) *Latent GOLD 4.0 user's guide*. Belmont, MA: Statistical Innovations.

Vermunt, J. K., & Magidson, J. (2008). *LG-syntax user's guide: Manual for Latent GOLD 4.5 syntax module*. Belmont, MA: Statistical Innovations.

Vermunt, J. K., Tran, B., & Magidson, J. (2008). Latent class models in longitudinal research. In S. Menard (Ed.), *Handbook of longitudinal research: Design, measurement, and analysis* (pp. 373-385). Burlington, MA: Elsevier.

Vermunt, J. K., & van Dijk, L. (2001). A nonparametric random-coefficients approach: The latent class regression model. *Multilevel Modelling Newsletter, 13,* 6-13.

Wiggins, L. M. (1973). *Panel analysis*. Amsterdam: Elsevier.

Wong, C. S, & Li, W.K. (2000). On a mixture autoregressive model. *Journal of the Royal Statistical Society, Ser. B, 62,* 95-115.

Yu, H. T., & Vermunt, J. K. (in preparation). Multilevel latent Markov model for nested longitudinal discrete data.

Yung, Y. F. (1997). Finite mixtures in confirmatory factor analysis models. *Psychometrika, 62,* 297-330.

Chapter 5
An Overview of the Autoregressive Latent Trajectory (ALT) Model

Kenneth A. Bollen and Catherine Zimmer

Abstract Autoregressive cross-lagged models and latent growth curve models are frequently applied to longitudinal or panel data. Though often presented as distinct and sometimes competing methods, the Autoregressive Latent Trajectory (ALT) model (Bollen and Curran, 2004) combines the primary features of each into a single model. This chapter: (1) presents the ALT model, (2) describes the situations when this model is appropriate, (3) provides an empirical example of the ALT model, and (4) gives the reader the input and output from an ALT model run on the empirical example. It concludes with a discussion of the limitations and extensions of the ALT model. Our focus is on repeated measures of continuous variables.

5.1 Introduction

There are two intuitive ways to approach the modeling of longitudinal data. The first relies on the idea and common observation that one of the best determinants of the current value of a variable is its value in the preceding period. So a student's reading performance in 2008 is well-determined by her reading performance in 2007, and this is true for all students in the population. This perspective can be formalized into what is known as an autoregressive model where the current value of a variable is determined by its past value. A second intuitively appealing method is to treat each subject as having a separate trajectory of change over time. Some cases might have a generally upward trend, others a downward trend, and still others might be

Kenneth A. Bollen
Odum Institute for Research in Social Science and Department of Sociology, University of North Carolina, Chapel Hill, USA
e-mail: bollen@unc.edu

Catherine Zimmer
Odum Institute for Research in Social Science, University of North Carolina, Chapel Hill, USA
e-mail: cathy_zimmer@unc.edu

K. van Montfort et al., *Longitudinal Research with Latent Variables*,
DOI 10.1007/978-3-642-11760-2_5, © Springer-Verlag Berlin Heidelberg 2010

relatively stable with regard to the outcome of interest. Here individual variability in change is permitted and each case can have different parameter values where these values describe the nature of the trajectory. This second approach we refer to as a latent (growth)[1] *curve* or *latent trajectory* model.

The autoregressive and latent curve models have long but largely independent histories. In the social and behavioral sciences autoregressive models were and are of substantial interest to economists who commonly use autoregressive time-series models to study economic indicators and lagged endogenous variables in panel data. The autoregressive models spread throughout the social and behavioral sciences beyond just economic applications. Anderson (1960), Humphreys (1960), Heise (1969), Wiley and Wiley (1970), Jöreskog (1970), and Werts, Jöreskog, and Linn (1971) provide just a few examples of publications that examined autoregressive models of a single outcome. Campbell (1963), Bohrnstedt (1969), Duncan (1969), Heise (1969), and Jöreskog (1979) are some of the earlier social science examples of authors who looked at autoregressive and cross-lagged models for two or more outcome variables in panel data. Kessler and Greenberg (1981) provided a book length treatment of these autoregressive and cross-lagged models. These have been and continue to be popular modeling approaches for longitudinal data.

The growth curve models of biostatistics have a long history (Bollen and Curran, 2006, pp. 9-14). The merger of the growth curve models with the factor analysis of longitudinal data resulted in the contemporary latent curve models and the resulting latent curve models date back to the 1950s (Bollen, 2007). Rao (1958) and Tucker (1958) were key works linking growth curve and exploratory factor analysis models. Meredith and Tisak (1984) was a seminal paper connecting confirmatory factor analysis to growth curve models leading to the latent curve model tradition that is influential today. In contrast to the autoregressive models, the repeated measures are reflective of an underlying pattern of change or trajectory. The trajectory is described by a set of parameters (e.g., random intercept and random slope) and these parameters can differ by individuals permitting a rich variety of trends across the cases in a sample.

Popularity of the autoregressive models preceded that of the growth curve models in the social and behavioral sciences. Early proponents of the growth curve model in these disciplines argued that the autoregressive and growth curve models were in direct competition (e.g., Bast and Reitsma, 1997; Kenny and Campbell, 1989; Rogosa and Willett, 1985) and some advocates argued that growth curve models were inherently superior to the autoregressive models (e.g., Rogosa, Brandt, and Zimowski, 1982, p. 744).

More recently the autoregressive and latent curve model have been combined into what is called the Autoregressive Latent Trajectory (ALT) model (Bollen and Curran, 2004; Curran and Bollen, 2001). The ALT model incorporates features of both the autoregressive and the latent curve model in a single framework. It is developed in recognition of the usefulness and appeal of each model and it permits modeling

[1] "Growth" suggests that the outcome variable is always increasing in magnitude and is misleading in those cases where the outcome decreases or is stable. For this reason, we sometimes omit this and refer to the models as latent curve or latent trajectory models.

data that has features of both models. Furthermore, it permits tests that provide information on whether the data more closely conform to the autoregressive or to the latent curve model. So if one or the other models is best, the ALT model will help to reveal that, whereas if both processes are operating both can be accommodated by the ALT model.

It also is important to distinguish the ALT model from a more established one that is a latent curve model with an autoregressive disturbance. For example, Chi and Reinsel (1989), Browne and du Toit (1991), Diggle, Liang and Zeger (1994), and Goldstein, Healy and Rasbash (1994) discuss modifications of the standard growth curve model to permit an autoregressive disturbance. In these types of models the autoregression of the disturbance is a type of nuisance association that is relegated to the disturbance and it is given little substantive explanation. In the ALT model the autoregressive relation is between the repeated measures, not the disturbances.[2] Furthermore, the lagged effect of the earlier value on the current value should be substantively meaningful when using the ALT model.

The purposes of this chapter are: (1) to present the ALT model, (2) to describe the situations when this model is appropriate, (3) to provide an empirical example of the ALT model, and (4) to give the reader the input and output from an ALT model run on the empirical example. Much of the technical presentation of the ALT model is based on Bollen and Curran (2004; 2006). Applications of the ALT are in many fields, such as psychology to study developmental psychopathology (Curran and Willoughby, 2003), daily anxiety and panic expectancy (Rodebaugh, Curran, and Chambliss, 2002), job performance over time (Zyphur, Chaturvedi and Arvey, 2008), and changes in eating behavior among first-year undergraduate women (Boyd, 2007). Addiction researchers have found the ALT model useful for studying how adolescent and peer substance use changes over time and affects each other (Simons-Morton and Chen, 2006). Wan, Zhang and Unruh (2006) used the ALT model to investigate outcome improvement in residents of nursing homes.

The next several sections present single variable and two variable ALT models, a general equation for all models, the implied moment matrices, and a section on the estimation and testing of these models. After these we present an empirical example. A conclusion summarizes the ALT model and its use.

[2] Hamaker (2005) has an interesting paper where she shows that an ALT model that has an equal autoregressive coefficient and is not written with the first wave outcome as predetermined is mathematically equivalent to an alternative growth curve model with autoregressive disturbances. These two forms of the model would have different substantive meanings in that the ALT model hypothesizes that the lagged repeated measure has an impact on the current repeated measure whereas the autoregressive disturbance model assumes that the prior disturbance influences the current disturbance. In the autoregressive disturbance model there is no direct effect of the repeated measures on other repeated measures and only a direct effect between disturbances. Also the equivalency does not hold if the autoregressive parameter differs across waves or if the first wave of the outcome is treated as a predetermined variable as recommended in Bollen and Curran (2004).

5.2 Autoregressive Latent Trajectory (ALT) Model

5.2.1 Single Variable Unconditional ALT Model

In this subsection we present the single variable, unconditional ALT model. By single variable we mean that there is only one outcome observed over time. By unconditional, we refer to the fact that the model has no explanatory variables or covariates that determine the random intercepts, random slopes, or the repeated measures other than the lagged value of the repeated measures. Suppose that y_{it} is the repeated measure of y for the ith observation at the tth time point. The ALT model is

$$y_{it} = \alpha_i + \Lambda_t \beta_i + \rho_{t,t-1} y_{i,t-1} + \varepsilon_{it} \tag{5.1}$$

where the i indexes the individual in the sample and the t indexes the time with $t = 2, 3, ..., T$. The α_i is the random intercept, β_i is the random slope, and Λ_t is the time trend variable that describes the pattern of growth so that for a linear growth model it would $0, 1, 2, ...$. The autoregressive parameter is $\rho_{t,t-1}$,[3] $y_{i,t-1}$ is the lagged value of y, and ε_{it} is the disturbance of the equation. We assume $E(\varepsilon_{it}) = 0$, $COV(\varepsilon_{it}, y_{i,t-1}) = 0$, $COV(\varepsilon_{it}, \beta_i) = 0$, and $COV(\varepsilon_{it}, \alpha_i) = 0$. We also assume $E(\varepsilon_{it}, \varepsilon_{jt}) = 0$ for all t and $i \neq j$, $E(\varepsilon_{it}, \varepsilon_{it}) = \sigma^2_{\varepsilon_t}$ for each t and i, and $COV(\varepsilon_{it}, \varepsilon_{i,t+k}) = 0$ for $k \neq 0$ though in some cases this latter restriction could be removed.

If we assume that $VAR(\beta_i)$, $VAR(\alpha_i)$, and $E(\beta_i)$ are all zero, then we get

$$y_{it} = \alpha + \rho_{t,t-1} y_{i,t-1} + \varepsilon_{it} \tag{5.2}$$

which is an autoregressive model with an intercept that does not change over time. If the true model corresponds to an autoregressive model, then we would expect the variances of the random intercepts and random slopes, and the mean of the slope to be zero in the ALT model.

Alternatively, suppose that $\rho_{t,t-1}$ in the ALT model is zero for all t. Now the resulting model is

$$y_{it} = \alpha_i + \Lambda_t \beta_i + \varepsilon_{it} \tag{5.3}$$

which corresponds to a latent curve model with random intercept α_i and random slope β_i.

These preceding constraints give us information on whether the autoregressive or latent curve model are sufficient to describe data or whether the full ALT model is required. The basic task is to estimate the ALT model. If the variances of the random intercepts and random slopes and the mean of the slope are essentially zero, then the

[3] In general we assume that $|\rho_{t,t-1}| < 1$ to insure that y_{it} does not grow infinitely as t goes to infinity. In the time series literature, this is a stationarity condition (e.g., Box and Jenkins, 1976). In nonstationary data, the autoregressive parameter can equal or exceed one but in our experience such nonstationary series are rare in panel data. This condition is not critical for our developments here.

autoregressive model is appropriate as long as $\rho_{t,t-1}$ is not zero. Alternatively, if the random intercepts and random slopes have nonzero variances and $\rho_{t,t-1}$ is always zero, then the latent curve model is preferred. If neither of these conditions are true, then the full ALT model should be considered.

One complication that we have not mentioned has to do with the first wave of data. Although Bollen and Curran (2004) show how to model all repeated measures as endogenous variables, they suggest that there are some useful simplifications that result when the first wave of the outcome is treated as a predetermined variable as is shown in Figure 5.1. One advantage follows in that we cannot estimate equation (5.1) for the first wave of data since by definition we do not have the lagged value of the first wave of the outcome variable. Treating this first wave as predetermined bypasses this problem. The equation for the first wave outcome variable then becomes

$$y_{i1} = \nu_1 + \varepsilon_{i1} \tag{5.4}$$

where ν_1 is the mean of y_{i1}.

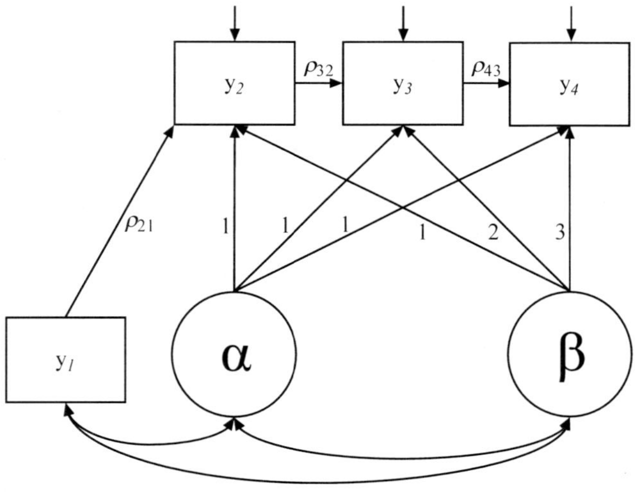

Fig. 5.1 Autoregressive Latent Trajectory (ALT) model with single variable over four waves and y_1 predetermined.

The other two equations to make the single variable ALT model complete are

$$\alpha_i = \mu_\alpha + \zeta_{\alpha i} \tag{5.5}$$

$$\beta_i = \mu_\beta + \zeta_{\beta i} \tag{5.6}$$

where μ_α and μ_β are the means of the random intercepts and random slopes, respectively, and $\zeta_{\alpha i}$ and $\zeta_{\beta i}$ are the random deviations around the respective means. The predetermined y_{i1}, α_i, and β_i are allowed to correlate.

Given the relations between the ALT model, the latent curve model, and the autoregressive model just described helps in interpreting the ALT model in equation (5.1). Consider first the latent curve model without autoregressive effects as in equation (5.3). In a latent curve model each case can have a distinct trajectory of the outcome variable. The trajectories are captured by having a random intercept and random slope that can vary by case. Once you control for the random intercepts and random slopes there is no influence of prior values of y on current values of y, that is, there is no autoregressive impact net of the trajectory parameters.

Alternatively, in the pure autoregressive model as in equation (5.2), the current y_{it} is driven by the past $y_{i,t-1}$ plus a random disturbance. Each case in the sample has the same autoregressive coefficient, $\rho_{t,t-1}$. Once the prior value of y is controlled, there are no individual trajectories for the cases in the sample.

From one perspective the ALT model is a latent curve model with random intercepts and random slopes where each individual can have a distinct trajectory. But now once we control for the random intercepts and random slopes there remains an autoregressive relationship between the ys. Taking a different perspective, the ALT model is an autoregressive model where the lagged value of a repeated measure partially determines the current value, but even taking account of the autoregressive relation each case can have a distinct trajectory. To understand the change in y we need to know the prior value of y and the individual trajectory of change for that individual. In other words both an autoregressive and growth curve model characterize the process. Neither a LCM or an autoregressive one alone is sufficient to describe the change.

5.2.2 Single Variable Conditional ALT Model

So far we have limited our description to an unconditional model where the random intercepts (α_i), random slopes (β_i), and the first wave of the repeated measures (y_{i1}) do not include covariates that determine them and they are only represented as a function of their means and deviations from their respective means (see eqs. (5.4) to (5.6)). A natural extension allows for covariates to predict α_i, β_i, and y_{i1}. To illustrate consider the incorporation of two time invariant exogenous predictors, z_{i1} and z_{i2} (though it is easy to generalize this model to any number of covariates). We modify equations (5.4) to (5.6) by adding these covariates resulting in

$$\alpha_i = \mu_\alpha + \gamma_{\alpha 1} z_{i1} + \gamma_{\alpha 2} z_{i2} + \zeta_{\alpha i} \quad (5.7)$$

$$\beta_i = \mu_\beta + \gamma_{\beta 1} z_{i1} + \gamma_{\beta 2} z_{i2} + \zeta_{\beta i} \quad (5.8)$$

$$y_{i1} = \nu_1 + \gamma_{y1} z_{i1} + \gamma_{y2} z_{i2} + \varepsilon_{i1} \quad (5.9)$$

where μ_α, μ_β, and ν_1 now represent regression intercepts rather than unconditional means. The γs represent the fixed regressions of the random intercepts (α_i), random slopes (β_i), and the predetermined y_{i1} on the two covariates. Figure 5.2 is a path diagram of the conditional ALT model for four waves of data and with two covariates. We assume that the disturbances (i.e., $\zeta_{\alpha i}$, $\zeta_{\beta i}$, ε_{i1}) have zero means and are uncorrelated with the exogenous variables (zs). Further, we permit $\zeta_{\alpha i}$, $\zeta_{\beta i}$, ε_{i1} to correlate with each other, but none of these is correlated with later values of ε_{it} where $t = 2, 3, \ldots$. Finally, we assume the exogenous variables are measured without error.

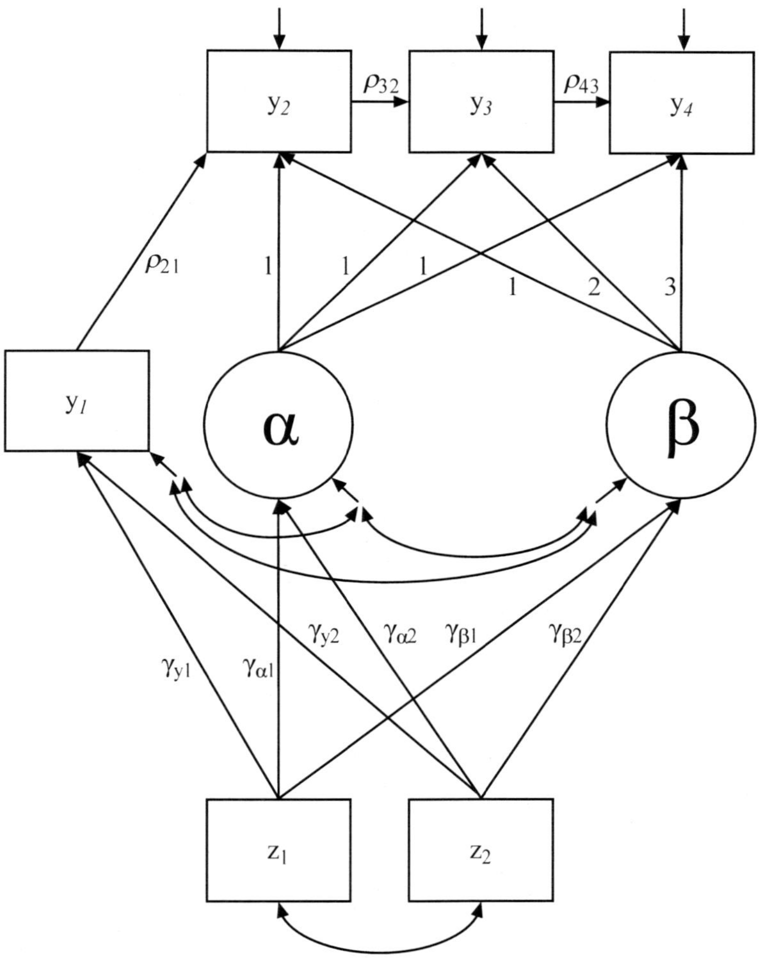

Fig. 5.2 Conditional Autoregressive Latent Trajectory (ALT) model with single variable over four waves and two covariates.

5.2.3 Bivariate Unconditional ALT Model

In the conditional univariate ALT model, we considered the influences of one or more time invariant covariates. However, there are many instances in which there might be interest in the relationship between two repeated measures, each of which is functionally related to the passage of time. We can extend the single repeated ALT model to include two or more repeated measures say, y_{it} and x_{it}. We write the bivariate ALT model for $t = 2,3,...,T$ as

$$y_{it} = \alpha_{yi} + \Lambda_{yt2}\beta_{yi} + \rho_{y_t y_{t-1}} y_{i,t-1} + \rho_{y_t x_{t-1}} x_{i,t-1} + \varepsilon_{yit} \quad (5.10)$$

$$x_{it} = \alpha_{xi} + \Lambda_{xt2}\beta_{xi} + \rho_{x_t y_{t-1}} y_{i,t-1} + \rho_{x_t x_{t-1}} x_{i,t-1} + \varepsilon_{xit} \quad (5.11)$$

We maintain similar assumptions about the disturbances (ε's) as before (means of zero, not autocorrelated, uncorrelated with the right-hand side variables and random coefficients). We permit some ε_{yit} to correlate with ε_{xit} as long as model identification is maintained. For this model we treat the $y_{i1}$and x_{i1}variables as predetermined and the random intercepts and random slopes as exogenous. Their equations are

$$y_{i1} = \nu_{y1} + \varepsilon_{yi1} \quad (5.12)$$

$$x_{i1} = \nu_{x1} + \varepsilon_{xi1} \quad (5.13)$$

$$\alpha_{yi} = \mu_{y\alpha} + \zeta_{y\alpha i} \quad (5.14)$$

$$\beta_{yi} = \mu_{y\beta} + \zeta_{y\beta i} \quad (5.15)$$

$$\alpha_{xi} = \mu_{x\alpha} + \zeta_{x\alpha i} \quad (5.16)$$

$$\beta_{xi} = \mu_{x\beta} + \zeta_{x\beta i} \quad (5.17)$$

All disturbances in these equations have means of zero. Generally, we permit ε_{yi1}, ε_{xi1}, $\zeta_{y\alpha i}$, $\zeta_{y\beta i}$, $\zeta_{x\alpha i}$, and $\zeta_{x\beta i}$ to correlate with each other, but these are assumed not to correlate with ε_{yit} and ε_{xit} for $t = 2,3,...,T$. Figure 5.3 is the path diagram of a bivariate unconditonal ALT model for four waves of data.

Each of the equations (5.10) and (5.11) are similar to the unconditional single variable ALT model except for the extra cross-lag term either $\rho_{y_t x_{t-1}} x_{i,t-1}$ in equation (5.10) or $\rho_{x_t y_{t-1}} y_{i,t-1}$ in equation (5.11). This is a noteworthy difference in that it permits the repeated measure from one series to directly impact the repeated measure of another. The bivariate ALT model not only allows the lagged dependent variable to enter the equation along with the random intercepts and random slopes, but it also permits a second repeated measure to have an impact once we control for the lagged and latent curve effects on the repeated measure. The flexibility of this model is considerable in that depending on the result of estimation the model could be an autoregressive model (when $VAR(\beta_i)$, $VAR(\alpha_i)$, $E(\beta_i)$, $\rho_{y_t x_{t-1}}$, and $\rho_{x_t y_{t-1}}$ all equal zero), a cross-lag model (when $VAR(\beta_i)$, $VAR(\alpha_i)$, and $E(\beta_i)$ all equal zero) or a latent curve model ($\rho_{y_t y_{t-1}}$, $\rho_{y_t x_{t-1}}$, $\rho_{x_t x_{t-1}}$ and $\rho_{x_t y_{t-1}}$all zero).

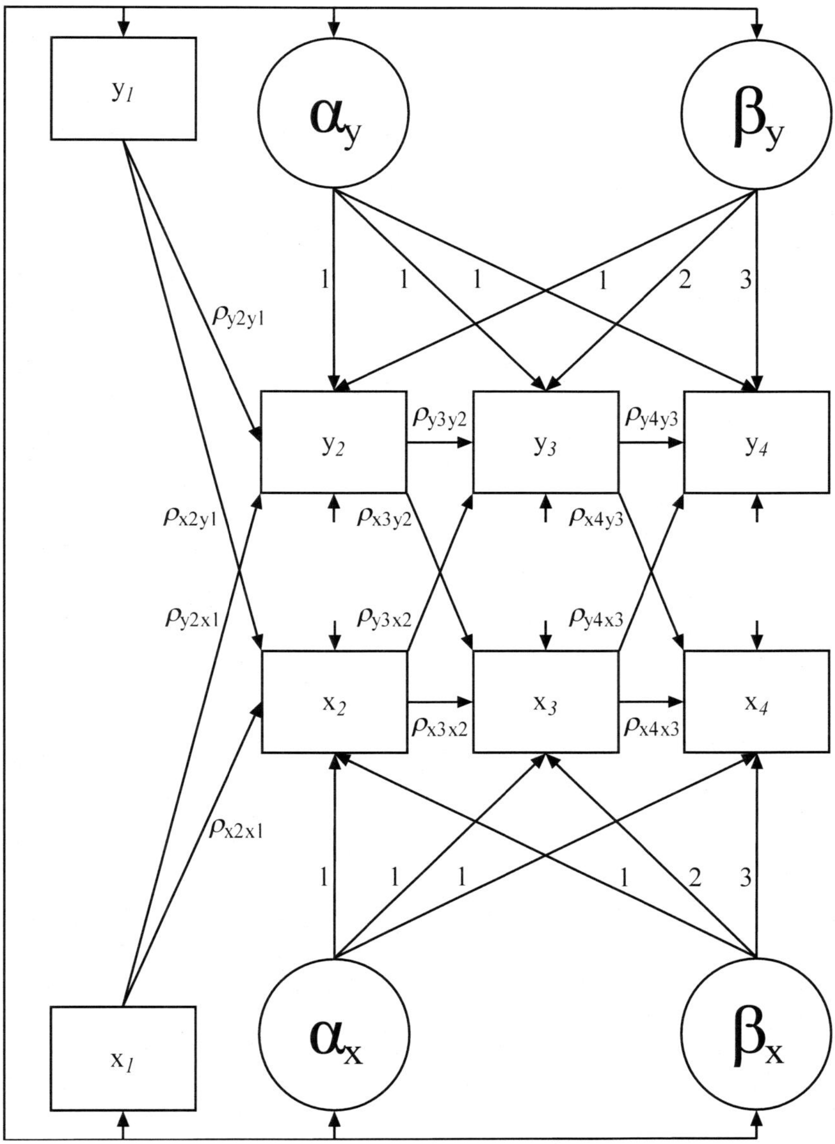

Fig. 5.3 Autoregressive Latent Trajectory (ALT) model for two variables over four waves and lagged effects between observed variables.

Furthermore additional or different lagged values of the repeated measures could enter these equations as dictated by the substantive knowledge driving the research. For instance, if the current value of a repeated measure was affected not only by say, y_{t-1}, but also by y_{t-2}, then both y_{t-1} and y_{t-2} should be included as predictors of y_t in the equations. In addition, the type of model devised for each repeated measure need not be the same. So if a latent curve model without autoregressive terms fits the x series best and an ALT model is needed for the y series, there is no reason not to include different structures for each repeated measure.

5.2.4 Bivariate Conditional ALT Model

As we described for the univariate conditional ALT model, we can incorporate one or more exogenous predictors in the bivariate ALT model as well. This is again accomplished by the extension of the equations for the random trajectories. Specifically, we modify equations (5.14) through (5.17) to include time invariant covariates z_{i1} and z_{i2} such that

$$\alpha_{yi} = \mu_{y\alpha} + \gamma_{\alpha y1} z_{i1} + \gamma_{\alpha y2} z_{i2} + \zeta_{y\alpha i} \tag{5.18}$$
$$\beta_{yi} = \mu_{y\beta} + \gamma_{\beta y1} z_{i1} + \gamma_{\beta y2} z_{i2} + \zeta_{y\beta i} \tag{5.19}$$

and

$$\alpha_{xi} = \mu_{x\alpha} + \gamma_{\alpha x1} z_{i1} + \gamma_{\alpha x2} z_{i2} + \zeta_{x\alpha i} \tag{5.20}$$
$$\beta_{xi} = \mu_{x\beta} + \gamma_{\beta x1} z_{i1} + \gamma_{\beta x2} z_{i2} + \zeta_{x\beta i} \tag{5.21}$$

As before, the set of gammas represent the fixed regressions of the random trajectory components on the two correlated exogenous variables. It is possible to have the random intercepts or random slopes as explanatory variables in equations (5.18) to (5.21). For instance, the random intercept from the y series (α_{yi}) might affect the random slope of the x series leading to $\beta_{xi} = \mu_{x\beta} + \gamma_{\beta_x \alpha_y} \alpha_{yi} + \gamma_{\beta x1} z_{i1} + \gamma_{\beta x2} z_{i2} + \zeta_{x\beta i}$ or the slope of one series could alter the slope of the other, for example, $\beta_{yi} = \mu_{y\beta} + \gamma_{\beta_y \beta_x} \beta_x + \gamma_{\beta y1} z_{i1} + \gamma_{\beta y2} z_{i2} + \zeta_{y\beta i}$.

In the bivariate unconditional ALT model, we let the initial repeated measures correlate with the random intercepts and random slopes. In the conditional bivariate ALT model, we must regress x_{i1} and y_{i1} on the set of exogenous measures. Thus, the equations for the initial measures for x_{i1} and y_{i1} are

$$y_{i1} = \nu_{y1} + \gamma_{y1} z_{i1} + \gamma_{y2} z_{i2} + \varepsilon_{yi1} \tag{5.22}$$
$$x_{i1} = \nu_{x1} + \gamma_{x1} z_{i1} + \gamma_{x2} z_{i2} + \varepsilon_{xi1} \tag{5.23}$$

The same assumptions described for the univariate conditional ALT model hold here as well.

5.3 General Equation for All Models

Up to this point we have presented unconditional and conditional ALT models for a single and two repeated measures using a scalar notation. These and variants of these models are expressable in a general matrix notation that is convenient for presenting the estimation and assessment of fit of these models. The matrix model is (Bollen and Curran, 2004):

$$\boldsymbol{\eta}_i = \boldsymbol{\mu} + \mathbf{B}\boldsymbol{\eta}_i + \boldsymbol{\zeta}_i \tag{5.24}$$

$$\mathbf{o}_i = \mathbf{P}\boldsymbol{\eta}_i \tag{5.25}$$

where the first equation provides the structural relations between variables, $\boldsymbol{\eta}_i$ is a vector that contains both the repeated measures and the random intercepts and random slopes, $\boldsymbol{\mu}$ is a vector of means or intercepts, $\mathbf{B}$ is a coefficient matrix that gives the coefficients for the relationships of $\boldsymbol{\eta}_i$s on each other, and $\boldsymbol{\zeta}_i$ is the disturbance vector for the variables in $\boldsymbol{\eta}_i$. We assume that $E(\boldsymbol{\zeta}_i) = \mathbf{0}$. The nature of the covariances of $\boldsymbol{\zeta}_i$ with $\boldsymbol{\eta}_i$ will vary depending on the model, but for identification purposes at least some of these covariances will be zero or known values. The second equation functions to pick out the observed variables, $\mathbf{o}_i$, from the latent variables of equation 5.24.

In more detail,

$$\boldsymbol{\eta}_i = \begin{bmatrix} \mathbf{y}_i \\ \mathbf{x}_i \\ \mathbf{z}_i \\ \boldsymbol{\alpha}_i \\ \boldsymbol{\beta}_i \end{bmatrix} \tag{5.26}$$

where $\mathbf{y}_i$ and $\mathbf{x}_i$ are two variables repeatedly measured for T time periods, $\mathbf{z}_i$ is a q x 1 vector of exogenous determinants of the latent trajectory parameters or of the repeated measures, α_i is the 2 x 1 vector of α_{yi} and α_{xi}, the random intercepts for the two sets of repeated measures, and β_i is the 2 x 1 vector of β_{yi} and β_{xi} the random slopes for the two repeated measures. The $\boldsymbol{\mu}$ vector is

$$\boldsymbol{\mu} = \begin{bmatrix} \boldsymbol{\mu}_y \\ \boldsymbol{\mu}_x \\ \boldsymbol{\mu}_z \\ \boldsymbol{\mu}_\alpha \\ \boldsymbol{\mu}_\beta \end{bmatrix} \tag{5.27}$$

where $\boldsymbol{\mu}_y$ and $\boldsymbol{\mu}_x$ are vectors of means/intercepts for the $\mathbf{y}_i$ and $\mathbf{x}_i$ observed repeated measures, $\boldsymbol{\mu}_z$ is the vector of means for the exogenous covariates in the model, $\boldsymbol{\mu}_\alpha$ is a vector of means/intercepts for the random intercepts, α_{yi} and α_{xi}, and $\boldsymbol{\mu}_\beta$ is a vector of the means/intercepts of β_{yi} and β_{xi}.

For the ALT model, the $\mathbf{B}$ matrix is

$$\mathbf{B} = \begin{bmatrix} \mathbf{B_{yy}} & \mathbf{B_{yx}} & \mathbf{B_{yz}} & \mathbf{B_{y\alpha}} & \mathbf{B_{y\beta}} \\ \mathbf{B_{xy}} & \mathbf{B_{xx}} & \mathbf{B_{xz}} & \mathbf{B_{x\alpha}} & \mathbf{B_{x\beta}} \\ \mathbf{0} & \mathbf{0} & \mathbf{0} & \mathbf{0} & \mathbf{0} \\ \mathbf{0} & \mathbf{0} & \mathbf{B_{\alpha z}} & \mathbf{B_{\alpha\alpha}} & \mathbf{B_{\alpha\beta}} \\ \mathbf{0} & \mathbf{0} & \mathbf{B_{\beta z}} & \mathbf{B_{\beta\alpha}} & \mathbf{B_{\beta\beta}} \end{bmatrix} \tag{5.28}$$

where the double subscript notation in the partition matrix indicates that the submatrix contains those coefficients related to effects among the subscripted variables. For instance, $\mathbf{B_{yy}}$ contains the effects of the repeated y variables on each other, and $\mathbf{B_{\beta z}}$ contains the impact of the exogenous $\mathbf{z}_i$ on the random slopes, β_{yi} and β_{xi}, for the ys and xs. The $\mathbf{z}_i$ consists of exogenous variables.

The disturbance vector for equation 5.24 is

$$\boldsymbol{\zeta}_\mathbf{i} = \begin{bmatrix} \boldsymbol{\varepsilon}_{yi} \\ \boldsymbol{\varepsilon}_{xi} \\ \boldsymbol{\varepsilon}_{zi} \\ \boldsymbol{\zeta}_{\alpha i} \\ \boldsymbol{\zeta}_{\beta i} \end{bmatrix} \tag{5.29}$$

with covariance matrix $\boldsymbol{\Sigma}_{\zeta\zeta}$. Since $\mathbf{z}_i$ is exogenous, the variance of $\boldsymbol{\varepsilon}_{zi}$ is equivalent to the variance of $\mathbf{z}_i$.

The $\mathbf{P}$ matrix is

$$\mathbf{P} = \begin{bmatrix} \mathbf{I}_T & \mathbf{0} & \mathbf{0} & \mathbf{0} & \mathbf{0} \\ \mathbf{0} & \mathbf{I}_T & \mathbf{0} & \mathbf{0} & \mathbf{0} \\ \mathbf{0} & \mathbf{0} & \mathbf{I}_q & \mathbf{0} & \mathbf{0} \end{bmatrix} \tag{5.30}$$

where $\mathbf{I}_T$ is a T x T identity matrix with dimensions that depend on the number of repeated measures and $\mathbf{I}_q$ is a q x q identity matrix with q exogenous variables. The matrix picks out the observed variables in a given model where $\mathbf{o}_i$ is

$$\mathbf{o}_i = \begin{bmatrix} \mathbf{y}_i \\ \mathbf{x}_i \\ \mathbf{z}_i \end{bmatrix} \tag{5.31}$$

Bollen and Curran (2004) demonstrate how this matrix expression enables a researcher to incorporate all of the models discussed as well as others. For instance, the standard autoregressive model for a single repeated measure has

$$\boldsymbol{\eta}_i = [\mathbf{y}_i] \tag{5.32}$$
$$\boldsymbol{\mu} = [\boldsymbol{\mu}_\mathbf{y}] \tag{5.33}$$
$$\mathbf{B} = [\mathbf{B_{yy}}] \tag{5.34}$$
$$\boldsymbol{\zeta}_\mathbf{i} = [\boldsymbol{\varepsilon}_i] \tag{5.35}$$
$$\mathbf{o}_i = \boldsymbol{\eta}_i \tag{5.36}$$

with

$$\mathbf{B_{yy}} = \begin{bmatrix} 0 & 0 & 0 & \cdots & 0 \\ \rho_{21} & 0 & 0 & \cdots & 0 \\ 0 & \rho_{32} & 0 & \cdots & 0 \\ \vdots & \vdots & \ddots & \ddots & \vdots \\ 0 & 0 & \cdots & \rho_{T,T-1} & 0 \end{bmatrix} \tag{5.37}$$

to capture a first order autoregressive relation.

The unconditional latent curve model has

$$\boldsymbol{\eta}_i = \begin{bmatrix} \mathbf{y}_i \\ \alpha_i \\ \beta_i \end{bmatrix} \tag{5.38}$$

$$\boldsymbol{\mu} = \begin{bmatrix} \mathbf{0} \\ \mu_\alpha \\ \mu_\beta \end{bmatrix} \tag{5.39}$$

where the $\mathbf{0}$ vector in $\boldsymbol{\mu}$ represents the zero fixed intercepts for the repeated measures in a latent trajectory model. The $\mathbf{B}$ matrix is

$$\mathbf{B} = \begin{bmatrix} \mathbf{0} & \mathbf{B_{y\alpha}} & \mathbf{B_{y\beta}} \\ \mathbf{0} & \mathbf{0} & \mathbf{0} \\ \mathbf{0} & \mathbf{0} & \mathbf{0} \end{bmatrix} \tag{5.40}$$

$$\mathbf{B_{y\alpha}} = \begin{bmatrix} 1 \\ 1 \\ \vdots \\ 1 \end{bmatrix} \qquad \mathbf{B_{y\beta}} = \begin{bmatrix} 0 \\ 1 \\ \vdots \\ T-1 \end{bmatrix} \tag{5.41}$$

The $\boldsymbol{\zeta_i}$ and $\mathbf{P}$ matrices are

$$\boldsymbol{\zeta_i} = \begin{bmatrix} \boldsymbol{\varepsilon}_i \\ \zeta_{\alpha i} \\ \zeta_{\beta i} \end{bmatrix} \tag{5.42}$$

$$\mathbf{P} = \begin{bmatrix} \mathbf{I}\ \mathbf{0}\ \mathbf{0} \end{bmatrix} \tag{5.43}$$

As a last example the unconditional univariate ALT model has

$$\mathbf{B} = \begin{bmatrix} \mathbf{B_{yy}} & \mathbf{B_{y\alpha}} & \mathbf{B_{y\beta}} \\ \mathbf{0} & \mathbf{0} & \mathbf{0} \\ \mathbf{0} & \mathbf{0} & \mathbf{0} \end{bmatrix} \tag{5.44}$$

where $\mathbf{B_{yy}}$ is the same as equation 5.37 and

$$\mathbf{B}_{\mathbf{y}\alpha} = \begin{bmatrix} 0 \\ 1 \\ \vdots \\ 1 \end{bmatrix} \qquad \mathbf{B}_{\mathbf{y}\beta} = \begin{bmatrix} 0 \\ 1 \\ \vdots \\ T-1 \end{bmatrix} \tag{5.45}$$

for a model where y_{1i} is predetermined. Furthermore

$$\boldsymbol{\mu} = \begin{bmatrix} \boldsymbol{\mu}_y \\ \mu_\alpha \\ \mu_\beta \end{bmatrix} \tag{5.46}$$

with

$$\boldsymbol{\mu}_y = \begin{bmatrix} \mu_{y1} \\ 0 \\ \vdots \\ 0 \end{bmatrix} \tag{5.47}$$

and

$$\boldsymbol{\zeta}_{\mathbf{i}} = \begin{bmatrix} \boldsymbol{\varepsilon}_i \\ \zeta_{\alpha i} \\ \zeta_{\beta i} \end{bmatrix} \tag{5.48}$$

The variances of ε_{1i}, $\zeta_{\alpha i}$, and $\zeta_{\beta i}$ are equal to the variances of the predetermined variables, y_{1i}, α_i, and β_i, respectively.

5.4 Implied Moment Matrices

Structural equation models (SEMs) typically involve expressing the means and covariance matrix of the observed variables as a function of the parameters ($\boldsymbol{\theta}$) in a model. These expressions of the implied mean vector ($\boldsymbol{\mu}(\boldsymbol{\theta})$) and the implied covariance matrix ($\boldsymbol{\Sigma}(\boldsymbol{\theta})$) also are referred to as the implied moment matrices and they are useful in estimation and the assessment of model fit. Bollen and Curran (2004) show that the implied mean vector is

$$\boldsymbol{\mu}(\boldsymbol{\theta}) = E(\mathbf{o}_i) = \mathbf{P}(\mathbf{I}-\mathbf{B})^{-1}\boldsymbol{\mu} \tag{5.49}$$

and the implied covariance matrix of observed variables is

$$\begin{aligned} \boldsymbol{\Sigma}(\boldsymbol{\theta}) &= [E(\mathbf{o}_i\mathbf{o}_i') - E(\mathbf{o}_i)E(\mathbf{o}_i')] \\ &= \mathbf{P}(\mathbf{I}-\mathbf{B})^{-1}\boldsymbol{\Sigma}_{\zeta\zeta}(\mathbf{I}-\mathbf{B})^{-1\prime}\mathbf{P}' \end{aligned} \tag{5.50}$$

The exact value of these implied moments depends on the value of the matrices for the particular type of ALT model, but once the matrices that correspond to the model of interest are substituted into these expressions, the implied moments are revealed.

One valuable aspect of the implied moment matrices is in determining the identification of the model parameters. A parameter is identified if it is possible to find a unique value for it. In SEMs we have

$$\boldsymbol{\mu} = \boldsymbol{\mu}(\boldsymbol{\theta}) \tag{5.51}$$

$$\boldsymbol{\Sigma} = \boldsymbol{\Sigma}(\boldsymbol{\theta}) \tag{5.52}$$

where $\boldsymbol{\mu}$ and $\boldsymbol{\Sigma}$ are the mean vector and covariance matrix of the observed variables and we have already defined their corresponding implied moments. Demonstrating that each $\boldsymbol{\theta}$ is solvable as a unique value of a function of one or more elements of $\boldsymbol{\mu}$ and $\boldsymbol{\Sigma}$ demonstrates that the parameters are identified. In general, we require four waves of data for the ALT model to be identified if the autoregressive parameter is equal over time and five waves without the equality restriction on the autoregression coefficient. If there are only three waves of data, then y_{i1} must be made endogenous and the coefficients for the paths from α_i and β_i to y_{i1} require nonlinear constraints for estimation. Bollen and Curran (2004) discuss this special case in more detail.

5.5 Estimation and Testing

SEMs are estimable with a wide variety of estimators. The most appropriate estimator depends on whether the endogenous observed variables are continuous or categorical and the distribution of these variables. In the most straightforward case of continuous endogenous variables, the Full Information Maximum Likelihood (FIML) estimator is available in all SEM software:

$$F_{ml} = \ln|\boldsymbol{\Sigma}(\boldsymbol{\theta})| + tr[\boldsymbol{\Sigma}^{-1}(\boldsymbol{\theta})\mathbf{S}] - \ln|\mathbf{S}| - p + (\bar{\mathbf{z}} - \boldsymbol{\mu}(\boldsymbol{\theta}))'\boldsymbol{\Sigma}^{-1}(\boldsymbol{\theta})(\bar{\mathbf{z}} - \boldsymbol{\mu}(\boldsymbol{\theta})) \tag{5.53}$$

where $\boldsymbol{\theta}$ is a vector that contains all of the parameters (i.e., coefficients, variances, and covariances of exogenous variables and errors) in the model that we wish to estimate, $\boldsymbol{\Sigma}(\boldsymbol{\theta})$ is the covariance matrix of the observed variables that is implied by the model structure, $\boldsymbol{\mu}(\boldsymbol{\theta})$ is the mean vector of the observed variables implied by the model, $\mathbf{S}$ is the sample covariance matrix of the observed variables, $\bar{\mathbf{z}}$ is the vector of sample means of the observed variables, and p is the number of observed variables in the model. The implied covariance matrix $[\boldsymbol{\Sigma}(\boldsymbol{\theta})]$ and the implied mean vector $[\boldsymbol{\mu}(\boldsymbol{\theta})]$ are in 5.49 and 5.50, respectively.

The classical derivation of F_{ml} begins with the assumption that the observed variables come from multivariate normal distributions (see, e.g., Bollen, 1989a, pp. 131-135). The FIML estimator of the parameters, $\widehat{\boldsymbol{\theta}}$, has several desirable properties: the estimator is consistent, asymptotically unbiased, asymptotically normally distributed, asymptotically efficient, and its covariance matrix equals the inverse of the information matrix (Lawley and Maxwell 1971). Fortunately, the FIML has desirable properties under less restrictive conditions. Browne (1984) proves that the preceding properties hold as long as the observed variables come from distributions with no excess multivariate kurtosis. There also are robustness studies that provide conditions where many of these properties hold even with excess multivariate kurtosis (see e.g., Satorra, 1990). Even when the robustness conditions fail there are corrections to the significance tests and bootstrapping procedures that permit significance tests (e.g., Satorra and Bentler, 1988; Bollen and Stine, 1990; 1993). Thus, with continuous outcome variables, estimation is possible even with excess multivariate kurtosis. Categorical dependent observed variables require procedures that take account of their categorical nature, but this is beyond the scope of our chapter. See Bollen and Curran (2006, Ch. 8) for discussion.

A first step in assessing model fit is a test of $H_0 : \boldsymbol{\Sigma} = \boldsymbol{\Sigma}(\boldsymbol{\theta})$ and $\boldsymbol{\mu} = \boldsymbol{\mu}(\boldsymbol{\theta})$ where $\boldsymbol{\Sigma}$ is the population covariance matrix of the observed variables, $\boldsymbol{\Sigma}(\boldsymbol{\theta})$ is the covariance matrix implied by the model that is a function of the parameters of the model, $\boldsymbol{\mu}$ is the population mean vector of the observed variables, and $\boldsymbol{\mu}(\boldsymbol{\theta})$ is the implied mean vector that is a function of the model parameters. These implied moment matrices were described above. If the model is true, then H_0 should be true. If the model structure is incorrect, then we should reject H_0. The test statistic of $T_{ml} = F_{ml}(N-1)$ is asymptotically distributed as a χ^2 with degrees of freedom $df = (p(p+1)/2 + p) - t$ where p is the number of observed variables and t is the number of estimated parameters. A significant chi-square test statistic is evidence against $H_0 : \boldsymbol{\Sigma} = \boldsymbol{\Sigma}(\boldsymbol{\theta})$ and $\boldsymbol{\mu} = \boldsymbol{\mu}(\boldsymbol{\theta})$ while a nonsignificant test statistic is consistent with the null hypothesis and hence the validity of the model. It is possible to compare two or more nested models where the parameters of one model are a restrictive form of the parameters of another. For instance, if we had an ALT model with no restrictions on the autoregressive parameter and a second identical to the first except that the autoregressive parameters were constrained to be equal, then the equal autoregressive ALT model would be nested in the ALT model where the autoregressive parameters were freely estimated. The difference in the chi-square test statistics for these individual ALT models would itself be asymptotically distributed as a chi-square variate with degrees of freedom equal to the difference in the degrees of freedom of the two models. The null hypothesis in this comparison of nested models is that the model with the greatest number of restrictions fits as well as the less restrictive model. A significant chi-square would be evidence in favor of the less restrictive model whereas a nonsignificant chi-square is evidence favoring the more restrictive model.

In practice, the chi-square test statistics are not the sole means of assessing model fit. Even if we use test statistics that correct for excess multivariate kurtosis, the power of the chi-square test statistic generally is large when the sample size is large. Structural misspecifications that might otherwise be judged as minor might result in a statistically significant chi-square or chi-square difference test. For this reason, researchers frequently use additional fit statistics to supplement the chi-square test statistic. There are numerous fit statistics available (Bollen and Long, 1993), but here we present several that we use in our example section: the Incremental Fit Index (*IFI*, Bollen 1989b), 1 minus the Root Mean Square Error of Approximation

$(1-RMSEA$, Steiger and Lind, 1980), and the Bayesian Information Criterion (BIC, Schwartz, 1978; Raftery, 1995):

$$IFI = \frac{T_b - T_h}{T_b - df_h} \tag{5.54}$$

$$(1-RMSEA) = 1 - \sqrt{\frac{T_h - df_h}{(N-1)df_h}} \tag{5.55}$$

$$BIC = T_h - df(\ln(N)) \tag{5.56}$$

where T_b and T_h are the likelihood ratio test statistics for a baseline and the hypothesized models, df_b and df_h are the df for the baseline and hypothesized models, N is the sample size and t is the number of free parameters in the model. The hypothesized model is simply the model that the researcher is testing and the baseline model is a highly restrictive model to which the fit of the hypothesized model is being compared. Typically the baseline model freely estimates the variances and means of the observed variables but forces their covariances to zero. A value of 1 is an ideal fit for the IFI and $(1-RMSEA)$. For the BIC, a negative value is evidence that favors the hypothesized model over the saturated model whereas a positive value favors the saturated model.[4] Although judgement is required in evaluating these fit indices, values less than .90 are typically considered to signify an inadequate fit to the data for the IFI and $(1-RMSEA)$.

5.6 Examples

5.6.1 Data

The data for these examples are repeated measures of Rosenberg's self-esteem scale from the National Longitudinal Study of Youth (NLSY). The data are organized by age of respondent rather than by wave of the survey. Using age to measure time creates missing data so we need to use the direct maximum likelihood estimator to take account of the missing values. There are 5622 respondents between the ages of 15 and 30 put in two year groupings, ages 15-16 to ages 29-30 with each assessed a minimum of once and a maximum of 6 times. The observed mean levels of self-esteem by age group are 3.058 for 15 and 16, 3.090 for 17 and 18, 3.113 for 19 and 20, 3.120 for 21 and 22, 3.125 for 23 and 24, 3.146 for 25 and 26, 3.141 for 27 and 28, and 3.127 for 29 and 30. The average mothers' education level is 11.544 years.

[4] This interpretation holds when calculating BIC as in equation (5.56), but this interpretation will not be true if different formulas are used.

5.6.2 Models

We present several models. First, we estimate the unconditional autoregressive (AR) model. Then we estimate the unconditional latent curve model (LCM). Third, we present the results from the unconditional ALT model. Fourth, and finally, we add the respondents' mothers' years of education in 1994 as an exogenous predictor to produce a conditional ALT (cALT) model. All estimation was conducted using Mplus 5.2. The programs that produced the results and the results themselves are available in Chapter 5 at the book website `http://www.econ.upf.edu/~satorra/longitudinallatent/readme.html`. Table 5.1 shows the fit statistics corresponding to the five models we estimated. The first model, AR with equal intercepts, has a statistically significant chi-square, low values of the IFI and a positive value for the BIC which suggests that the saturated model fits better than the hypothesized one. The only fit index that suggests a good fit is the $(1-RMSEA)$. We also estimated the AR model with unconstrained intercepts. The fit was very close to that of the autoregressive model we report in Table 5.1 ($T_{ML}(18) = 262.39$, $p = 0.006$; $IFI = 0.82$; $1-RMSEA = 0.95$; $BIC = 106.96$).

The second model, the Latent Curve Model (LCM), has a fit that is much better than the AR one in that the IFI and $(1-RMSEA)$ are high and the BIC is a large negative value. Combining features of both models in the ALT model we find for the first time a nonsignificant chi-square, an IFI and $(1-RMSEA)$ that are near their ideal values, and a large negative BIC. However, closer examination of the parameter estimates and their standard errors reveals that the mean, variance, and the covariances of the slope are all not significantly different from zero. This suggests that the slope factor is not needed in this model. Furthermore, the autoregressive coefficients appear near equal when their standard errors are taken into account. This led us to respecify the ALT model without the slope term and with the autoregressive parameters set equal. The fit statistics suggest that this model fits very well. This model suggests that there are stable individual differences in self-esteem and that there is an impact of past self-esteem feelings on current ones.

Table 5.1 Overall fit of Autoregressive, Latent Curve, and Autoregressive Latent Trajectory models for self-esteem, ages 15-30 (N = 5622)

Overall Fit	(1) Autoregressive Model	(2) Latent Curve Model	(3) Unconditional ALT Model	(4) No Slope Unconditional ALT Model	(5) No Slope Conditional ALT Model
T_{ML}	280.69	69.93	22.76	39.03	49.36
df	24	28	18	28	34
p-value	<0.001	<0.001	0.200	0.080	0.043
IFI	0.81	0.97	1.00	0.99	0.99
$1-RMSEA$	0.96	0.98	0.99	0.99	0.99
BIC	73.46	-171.83	-132.66	-202.73	-244.21

Which of these four models is best? The question is complicated by the fact that not all of these models are nested. However, some are. If in the unconditional ALT model (see column (3) in Table 5.1) we set the $VAR(\beta_i)$, $VAR(\alpha_i)$, $E(\beta_i)$, $COV(\beta_i, \alpha_i)$, $COV(\beta_i, y_1)$, and $COV(\alpha_i, y_1)$ to zero, then we are led to equation (5.2) which is the AR model with equal intercepts reported in column (1) of Table 5.1. A nested chi-square difference test leads to a highly significant difference ($T_{ML}(6) = 280.69 - 22.76 = 257.93$, $p < 0.001$) lending support to the ALT over the AR model. The "No Slope Unconditional ALT Model" of column (4) is nested in the "Unconditional ALT Model" of column (3) and the chi-square difference test is not significant ($T_{ML}(10) = 39.032 - 22.763 = 16.69$, $p = 0.092$) lending support to the ALT model without a slope. The "Latent Curve Model" of column (2) is not nested in the "Unconditional ALT Model" of column (3) because the ALT model treats y_1 as predetermined while the LCM model treats that variable as endogenous. Despite the nonnesting of some of these models, the other fit statistics are comparable for nonnested models. By all measures the AR model is inadequate. Considering all of the fit statistics, the "No Slope Unconditional ALT Model" appears to have the best fit among models (1) to (4).

Given that the "No Slope Unconditional ALT Model" was the best, we used it to estimate a conditional model that treats mother's education as a covariate. Though the chi-square for this model is marginally significant, the other fit statistics look excellent for this conditional model and we interpret the results of that model in detail. Table 5.2 shows the parameter estimates from the cALT model, which were taken from the Mplus 5.2 output for that model.

The first row of Table 5.2 shows the fixed relationships between the random intercepts (set at 1) and the observed repeated measures of self-esteem. The equal autoregressive effects of the self-esteem measure, the $\widehat{\rho}$ coefficients, are 0.192, showing a positive impact of past on current self-esteem. These effects are net of the random intercept effects. The residual variances ($\widehat{VAR(\varepsilon)}$) of the repeated measures are statistically significant; hence there is age-specific error in the repeated measures. They are similar in size, however, and could be constrained to be equal as another potential simplification to the model – the measurement error in the repeated measures is the same at all ages. The R-squares of all repeated measures but the first are moderate in size ranging from 0.305 to 0.369. This suggests that the random intercepts and the prior self-esteem variables explain roughly 30 to 37% of the variation in each self-esteem measure.

We turn now to the impact of mother's education on the random intercept. This is equivalent to a regression with the random intercept being the dependent variable and mother's education being the explanatory variable. There is a regression constant or fixed intercept ($\widehat{\mu}_\alpha$) and a slope ($\widehat{\gamma}_{\alpha 1}$). The slope ($\widehat{\gamma}_{\alpha 1}$) of mother's education is 0.005 so that each unit shift in education leads to an expected shift of 0.005 in the random intercept variable. The regression constant ($\widehat{\mu}_\alpha$) from this regression equation is 2.461 which is the predicted value of the random intercept when mother's education is zero, though a value of 0 for mother's education does not occur in our data. There is also significant variation of the regression residuals in the random intercepts equation of 0.025 ($= \widehat{VAR(\zeta_\alpha)}$) and an R-square

Table 5.2 ML parameter estimates and z-values in the No Slope Conditional ALT model for self-esteem, ages 15-30 ($N = 5622$)

Parameter	Model	SE 15-16	SE 17-18	SE 19-20	SE 21-22	SE 23-24	SE 25-26	SE 27-28	SE 29-30
λ_t of α	–	–	1.000	1.000	1.000	1.000	1.000	1.000	1.000
			(–)	(–)	(–)	(–)	(–)	(–)	(–)
ρ	–	–	0.192	0.192	0.192	0.192	0.192	0.192	0.192
			(7.67)	(7.67)	(7.67)	(7.67)	(7.67)	(7.67)	(7.67)
VAR (ε)	–	0.105	0.070	0.077	0.073	0.072	0.086	0.084	0.097
		(47.37)	(25.50)	(23.51)	(18.96)	(16.86)	(14.37)	(10.02)	(6.86)
μ_α	2.461	–	–	–	–	–	–	–	–
	(32.03)								
VAR(ζ_α)	0.025	–	–	–	–	–	–	–	–
	(9.66)								
ν	–	2.969	–	–	–	–	–	–	–
		(159.76)	–	–	–	–	–	–	–
COV(α, SE 15-16)	–	0.172	–	–	–	–	–	–	–
		(18.85)	–	–	–	–	–	–	–
$\gamma_{\alpha 1}$	0.005	–	–	–	–	–	–	–	–
	(4.63)	–	–	–	–	–	–	–	–
$\gamma_{SE15-16,1}$	0.008	–	–	–	–	–	–	–	–
	(5.065)	–	–	–	–	–	–	–	–
R^2	0.012 (α)	0.007	0.366	0.352	0.367	0.369	0.329	0.336	0.305

for this equation of only 0.012. Mother's education is a poor predictor of the random intercept for self-esteem. Turning to the effect of mother's education on initial self-esteem 15 to 16, we again find a low R-square (0.007), but a statistically significant intercept ($\widehat{\nu} = 2.969$) and slope ($\widehat{\gamma}_{SE15-16,1} = 0.008$). In addition, the random intercepts are significantly correlated with the self-esteem at 15 and 16 ($\widehat{COV}(\alpha, SE15-16) = 0.172$).

So what have we learned about the trajectories of self-esteem from ages 15 and 16 to ages 29 and 30? First, we found an autoregressive process with prior self-esteem having a positive effect on current self-esteem, but this is combined with a random intercept term that provides for a different constant level of self-esteem for each child. In fact, including only an autoregressive term does not lead to a good fitting model. The autoregressive and the random intercept effects explained roughly 30% to 37% of the variation in the self-esteem variables. The random slope was not needed. This implies that once we control for the random intercept and the autoregressive relationship, there is no need to add a linear trend term in self-esteem for each child. There are differences in their levels of self-esteem that tend to be constant, but that are also affected by prior self-esteem. Our conditional model revealed statistically significant positive effects of mother's education on the random intercept and on the initial self-esteem 15 and 16, but the effects were small as was the R-square.

5.7 Conclusions

This paper reviewed the ALT model which synthesizes features of the autoregressive/cross-lagged and the latent growth curve models. It permits the lagged value of a repeated measure to influence the current value while at the same time permits there to be separate over-time trajectories for individuals in the sample. As such it provides a researcher added flexibility in capturing the nature of change exhibited in panel data. Furthermore, the ALT model yields evidence relevant to whether the synthesis is required or if a researcher can get by with only the autoregressive and cross-lagged model or only the latent curve model. Obvious generalizations of the ALT model include multiple repeated measures, autoregressive models beyond lag one (e.g., AR(p) models), nonlinear trajectories, or ALT models for latent variables with separate measurement models with multiple indicators. The ALT model already includes latent variables in that the random intercept and random slope variables are latent. However, in the case of a multiple indicator model for the repeated "measure," the ALT model would allow a model of the autoregressive relation and the trajectory of the latent variables that would control for the measurement error in the indicators of the latent variables. This also would provide an estimate of the amount of measurement error in the multiple indicators. In the conditional ALT model it also would be possible to include latent exogenous variables as predictors of the random intercepts, random slopes, and the initial value of the latent repeated variable.

Despite these desirable features, several cautionary notes are in order. First, the ALT model assumes that the repeated measure has a direct impact on itself at a later point in time. A researcher should have substantive reasons to believe that this is a reasonable hypothesis and should not use the ALT model as just a way to improve model fit. A second related point is that it is possible that the autoregressive relation resides in the disturbance rather than in the repeated measures. In this situation, the disturbances should be autoregressive rather than the repeated measures since this implies a model that generally differs from the ALT.[5] Third, our presentation assumes that the researcher has the correct functional form for the latent curve trajectory in the ALT model. If, for example, we assume a linear functional form when a trajectory is nonlinear, then the autoregressive part of the ALT model might be due to the researcher using the wrong functional form (Voelkle, 2008).[6] A related point

[5] Hamaker (2005) discusses the special cases where the ALT and autoregressive disturbance model can be made statistically equivalent.

[6] We explored nonlinearity in our empirical example by using the "freed loading" model (Bollen and Curran, 2006, pp. 98-103). There was no improvement to model fit and the autoregressive parameters were still significant suggesting that the linear functional form was an appropriate starting point.

is that extrapolating trends beyond the period of observation should only be done with great caution. A linear trend might be a good approximation of a trajectory within the time period of observation, but extrapolating too far out could lead to highly inaccurate predictions if the relation is really nonlinear. Finally, throughout our presentation we assume discrete time models are good approximations to continuous time models. Many processes occur in continuous time even when the data are available only at fixed times. If the waves of data collection are too spread out relative to the timing of the relationships, then our discrete time models could be misleading. For instance, the autoregressive or ALT model might lead to inaccurate estimates of relationships if the observation interval for the discrete time model is long. Delsing and Oud (2008) present an extension of the ALT model to continuous time modeling that enables researchers to use variables observed in panel data but allow continuous rather than discrete time.

Keeping these limitations in mind, we believe that the ALT model provides a useful extension to some of the more commonly used models for panel data.

Acknowledgements Kenneth Bollen gratefully acknowledges the support from NSF SES 0617276 from NIDA 1-RO1-DA13148-01 & DA013148-05A2. We thank the editors and reviewers for valuable comments and thank Shawn Bauldry for research assistance.

References

Anderson, T. W. (1960). Some stochastic process models for intelligence test scores. In K. J. Arrow, S. Karlin, & P. Suppes (Eds.), *Mathematical methods in the social sciences.* Stanford, CA: Stanford University Press.

Bast, J. & Reitsma, P. (1997). Matthew effects in reading: A comparison of latent growth curve models and simplex models with structured means. *Multivariate Behavioral Research, 32,* 135-167.

Bohrnstedt, G. W. (1969). Observations on the measurement of change. *Sociological methodology, 1,* 113-133.

Bollen, K. A. (1989a). *Structural equation models with latent variables.* New York: Wiley.

Bollen, K. A. (1989b). A new incremental fit index for general structural equation models. *Sociological Methods & Research, 17,* 303-316.

Bollen, K. A. (2007). On the origins of latent curve models. In R. Cudeck & R. MacCallum (Eds.), *Factor analysis at 100* (pp. 79-98). Mahwah, NJ: Lawrence Erlbaum Associates.

Bollen, K. A. & Curran P. J. (2004). Autoregressive Latent Trajectory (ALT) models: A Synthesis of two traditions. *Sociological Methods & Research, 32,* 336-383.

Bollen, K. A. & Curran, P. J. (2006). *Latent curve models.* New York: Wiley.

Bollen, K. A., & Long, J. Scott. (1993). *Testing structural equation models.* Newbury Park, CA: Sage Publications.

Bollen, K. A., & Stine, R. A. (1990). Direct and indirect effects: Classical and bootstrap estimates of variability. *Sociological Methodology, 20,* 115-140.

Bollen, K. A., & Stine, R. A. (1993). Bootstrapping goodness-of-fit measures in structural equation models. In K. A. Bollen & J. Scott Long (Eds.), *Testing structural equation models* (pp. 111-135). Newbury Park, CA: Sage Publications.

Box, G. E. P., & Jenkins, G. M. (1976). *Time Series Analysis: Forecasting and Control.* San Francisco: Holden-Day.

Boyd, J. L. (2007). *Developmental and situational factors contributing to changes in eating behavior in first-year undergraduate women.* Master of Arts in Psychology. University of Waterloo, Canada.

Browne, M. W. (1984). Asymptotically distribution-free methods for the analysis of covariance structures. *British Journal of Mathematical & Statistical Psychology, 37,* 62-83.

Browne, M. W., & du Toit, S. H. C. (1991). Models for learning data. In L. Collins & J. Horn (Eds.), *Best methods for the analysis of change: Recent advances, unanswered questions, future directions* (pp. 47-68). Washington, DC: APA.

Campbell, D. T. (1963). From description to experimentation: Interpreting trends as quasi-experiments. In C. W. Harris (Ed.), *Problems in measuring change* (pp. 212-242). Madison, WI: University of Wisconsin Press.

Chi, E. M., & Reinsel, G. C. (1989). Models for longitudinal data with random effects and AR(1) errors. *Journal of the American Statistical Association, 84,* 452-59.

Curran, P. J., & Bollen, K. A. (2001). The bests of both worlds: Combining autoregressive and latent curve models. In Collins L. M. & Sayar, A.G. (Eds.), *New methods for the analysis of change* (pp. 105-136). Washington, D.C.: American Psychological Association.

Curran, P. J., & Willoughby, M. T. (2003). Implications of latent trajectory models for the study of developmental psychopathology. *Development and Psychopathology, 15,* 581-612.

Delsing, M. J. M. H., & Oud, J. H. L. (2008). Analyzing reciprocal relationships by means of the continuous-time autoregressive latent trajectory model. *Statistica Neerlandica, 62,* 58-82.

Diggle, P. J., Liang, K. Y., & Zeger, S. L. (1994). *Analysis of longitudinal data.* Oxford: Clarendon Press.

Duncan, O. D. (1969). Some linear models for two-wave, two-variable panel analysis. *Psychological Bulletin, 72,* 177-182.

Goldstein, H., Healy, M. J. R., & Rasbash, J. (1994). Multilevel time series models with applications to repeated measures data. *Statistics in Medicine, 13,* 1643-1655.

Hamaker, E. (2005). Conditions for the equivalence of the autoregressive latent trajectory model and a latent growth curve model with autoregressive disturbances. *Sociological Methods & Research, 33,* 404-416.

Heise, D. R. (1969). Separating reliability and stability in test-retest correlation. *American Sociological Review, 34,* 93-101.

Humphreys, L. G. (1960). Investigations of the simplex. *Psychometrika, 25,* 313-323.

Jöreskog, K. G. (1970). A general method for analysis of covariance structures. *Biometrika, 57,* 239-251.

Jöreskog, K. G. (1979). Statistical models and methods for analysis of longitudinal data. In K. G. Jöreskog & D. Sörbom (Eds.), *Advances in factor analysis and structural equation models.* Cambridge, Mass: Abt.

Kenny, D. A., & Campbell, D. T. (1989). On the measurement of stability in over-time data. *Journal of Personality, 57,* 445-481.

Kessler, R. C., & Greenberg, D. F. (1981). *Linear Panel Analysis.* New York: Academic Press.

Lawley, D. N., & Maxwell, A. E. (1971). *Factor analysis as a statistical method.* London: Butterworth.

Meredith, W., & Tisak, J. (1984). *Tuckerizing curves.* Paper presented at the annual meeting of the Psychometric Society, Santa Barbara, CA.

Muthén, L. K. & Muthén, B. O. (1998-2007). *Mplus User's Guide* (5th ed.). Los Angeles: Muthén & Muthén.

Rao, C. R. (1958). Some statistical methods for comparison of growth curves. *Biometrika, 51,* 83-90.

Raftery, A. E. (1995). Bayesian model selection in social research (with discussion). *Sociological Methodology, 25,* 111-163.

Rodebaugh, T. L., Curran, P. J., & Chambless, D. L. (2002). Expectancy of panic in the maintenance of daily anxiety in panic disorder with agoraphobia: A longitudinal test of competing models. *Behavior Therapy, 33,* 315-336.

Rogosa, D., & Willett, J. B. (1985). Satisfying simplex structure is simpler than it should be. *Journal of Educational Statistics, 10,* 99-107.

Rogosa, D. R., Brandt, D., & Zimowski, M. (1982). A growth curve approach to the measurement of change. *Psychological Bulletin, 92,* 726-748.

Satorra, A. (1990). Robustness issues in structural equation modeling: A review of recent developments. *Quality & Quantity, 24,* 367-386.

Satorra, A., & Bentler, P. M. (1988). Scaling corrections for chi-square statistics in covariance structure analysis. *Proceedings of the American Statistical Association,* 308-313.

Schwarz, G. (1978). Estimating the dimensions of a model. *Annals of Statistics, 6,* 461-464.

Simons-Morton, B., & Chen, R. S. (2006). Over time relationships between early adolescent and peer substance abuse. *Addictive Behaviors, 31,* 1211-1223.

Steiger, J. H., & Lind, J. M. (1980). *Statistically based tests for the number of common factors.* Paper presented at the annual meeting of the Psychometric Society, Iowa City, IA.

Tucker, L. R. (1958). Determination of parameters of a functional relation by factor analysis. *Psychometrika, 23,* 19-23.

Voelkle, M. C. (2008). Reconsidering the use of Autoregressive Latent Trajectory (ALT) models. *Multivariate Behavioral Research, 43,* 564-591.

Wan, T. T. H., Zhang, N. J., & Unruh, L. (2006). Predictors of resident outcome improvement in nursing homes. *Western Journal of Nursing Research, 28,* 974-993.

Werts, C. E., Jöreskog, K. G., & Linn, R. L. (1971). Comment on the estimation of measurement error in panel data. *American Sociological Review, 36,* 110-112.

Wiley, D. E., & Wiley, J. A. (1970). The estimation of measurement error in panel data. *American Sociological Review, 35,* 112-117.

Zyphur, M. J., Chaturvedi, S., & Arvey, R. D. (2008). Job performance over time is a function of latent trajectories and previous performance. *Journal of Applied Psychology, 93,* 217-224.

Chapter 6
State Space Methods for Latent Trajectory and Parameter Estimation by Maximum Likelihood

Jacques J. F. Commandeur, Siem Jan Koopman, and Kees van Montfort

Abstract We review Kalman filter and related smoothing methods for the latent trajectory in multivariate time series. The latent effects in the model are modelled as vector unobserved components for which we assume particular dynamic stochastic processes. The parameters in the resulting multivariate unobserved components time series models will be estimated by maximum likelihood methods. Some essential details of the state space methodology are discussed in this chapter. An application in the modelling of traffic safety data is presented to illustrate the methodology in practice.

6.1 Introduction

This chapter concerns multivariate state space analysis and discusses some particular issues of interest, see Durbin and Koopman (2001) and Commandeur and Koopman (2007).

Multivariate state space analysis is applicable to situations where two or more time series need to be analysed simultaneously. However, the material in this chapter also provides a unified treatment for univariate time series. In classical regression analysis a linear relationship is assumed between the dependent variable y_i and an independent variable x_i. The standard regression model for n realizations or

Jacques J. F. Commandeur
Department of Econometrics, VU University Amsterdam
e-mail: jcommandeur@feweb.vu.nl

Siem Jan Koopman
Department of Econometrics, VU University Amsterdam
e-mail: s.j.koopman@feweb.vu.nl

Kees van Montfort
Department of Econometrics, VU University Amsterdam
e-mail: kvmontfort@feweb.vu.nl

K. van Montfort et al., *Longitudinal Research with Latent Variables*,
DOI 10.1007/978-3-642-11760-2_6, © Springer-Verlag Berlin Heidelberg 2010

observations of y_i and covariate x_i for $i = 1, \ldots, n$ can be represented by

$$y_i = a + bx_i + \varepsilon_i,$$

where the disturbances or errors $\varepsilon_1, \ldots, \varepsilon_n$ are normally and independently distributed with mean zero and variance σ_ε^2. The coefficients a and b are unknown and fixed and are usually estimated by employing the regression method. It is implied in a classical regression analysis that the observations y_i, after the corrections for intercept and for independent variable x_i, are assumed to be independent of each other. In a time series context, it is not realistic to assume that the observations are conditionally independent because they are expected to be interrelated through time. When statistical inference is carried out when the observations are known to be subject to serial correlations (time dependencies), various problems can arise when it is based on classical regression analysis. For instance, the well-known F-test and t-test statistics do not have proper F- and t-distributions, respectively, under the null hypothesis. Time series analysis has the primary task to uncover the dynamic development of observations measured over time. By using state space methodology it is assumed that the dynamic properties cannot be observed directly from the data. The unobserved dynamic process at time t can be measured indirectly and is referred to as the state of the time series. The state of the time series may consist of several unobserved components and can be estimated by the Kalman filter.

State space methods originated in the field of control engineering, starting with the ground-breaking paper of Kalman (1960). They were initially (and still are) deployed for the purpose of accurately tracking the position and velocity of moving objects such as ships, airplanes, missiles, and rockets.

Around the eighties of the last century it was recognized by scientists involved in other fields than control engineering that these ideas could well be applied to time series analysis generally as well. Since then state space methods have been applied in a wide range of subjects, including economics, finance, political science, environmental science, the social sciences, road safety and medicine.

The outline of this chapter is as follows. In Section 6.2 we formulate the general multivariate state space model and we discuss several well-known sub models. Section 6.3 deals with the Kalman filter and the estimation of the unobserved states and the unknown model parameters. In Section 6.4 we discuss some tests to check the model assumptions such as normality, independency and homoscedasticity. Finally, we will present an empirical example.

6.2 Linear Gaussian State Space Models

A time series is a set of observations which are sequentially ordered over time. In a state space analysis the time series observations are assumed to depend linearly on a *state vector* that is unobserved and is generated by a stochastically time-varying process (a dynamic system). The observations are further assumed to be subject to measurement error that is independent of the state vector. The state vector can be estimated or identified once a sufficient set of observations becomes available. In this section we concentrate on the state space model and its special cases. In Section 6.3 we discuss methods for estimation, residual analysis and forecasting on the basis of state space models. The expositions rely mostly on the introductory textbook by Commandeur and Koopman (2007) and on the more advanced textbook by Durbin and Koopman (2001).

The general linear Gaussian state space model for the n-dimensional observation sequence $y_1, \ldots, y_n$ is given by

$$y_t = Z_t \alpha_t + \varepsilon_t, \qquad \varepsilon_t \sim \text{NID}(0, H_t), \tag{6.1}$$

$$\alpha_{t+1} = T_t \alpha_t + R_t \eta_t, \qquad \eta_t \sim \text{NID}(0, Q_t), \qquad t = 1, \ldots, n, \tag{6.2}$$

where α_t is the state vector, ε_t and η_t are disturbance vectors and the system matrices Z_t, T_t, R_t, H_t and Q_t are fixed and known but a selection of elements may depend on an unknown parameter vector. Equation (6.1) is called the *observation* or *measurement equation*, while (6.2) is called the *state* or *transition equation*. The $p \times 1$ observation vector y_t contains the p observations at time t and the $m \times 1$ state vector α_t is unobserved. The $p \times 1$ irregular vector ε_t has zero mean and $p \times p$ variance matrix H_t.

The $p \times m$ matrix Z_t links the observation vector y_t with the unobservable state vector α_t and may consist of regression variables. The $m \times m$ transition matrix T_t in (6.2) determines the dynamic evolution of the state vector. The $r \times 1$ disturbance vector η_t for the state vector update has zero mean and $r \times r$ variance matrix Q_t. The observation and state disturbances ε_t and η_t are assumed to be serially independent and independent of each other at all time points. In many standard cases, $r = m$ and matrix R_t is the identity matrix I_m. In other cases, matrix R_t is an $m \times r$ selection matrix with $r < m$. Although matrix R_t can be specified freely, it is often composed of a selection from the first r columns of the identity matrix I_m.

The initial state vector α_1 is assumed to be generated as

$$\alpha_1 \sim \text{NID}(a_1, P_1),$$

independently of the observation and state disturbances ε_t and η_t. Mean a_1 and variance P_1 can be treated as given and known in almost all stationary processes for the state vector. For nonstationary processes and regression effects in the state vector, the associated elements in the initial mean a_1 can be treated as unknown and need to be estimated. For an extensive discussion of *initialisation* in state space analysis, we refer to Durbin and Koopman (2001, Chapter 5).

6.2.1 Local Level Model and Other Unobserved Component Models

By appropriate choices of the vectors α_t, ε_t and η_t, and of the matrices Z_t, T_t, H_t, R_t and Q_t, a wide range of different time series models can be derived from (6.1) and (6.2). Here we discuss the class of *unobserved components time series models*. A number of special cases will be discussed in some detail. Special attention is given to the univariate *local level model*.

Letting

$$\alpha_t = \mu_t, \quad \eta_t = \xi_t, \quad Z_t = T_t = R_t = 1, \quad H_t = \sigma^2_\varepsilon, \quad Q_t = \sigma^2_\xi,$$

(all of order 1×1) for $t = 1, \dots, n$, model (6.1)-(6.2) reduces to the local level model as given by

$$\begin{aligned} y_t &= \mu_t + \varepsilon_t, & \varepsilon_t &\sim \text{NID}(0, \sigma^2_\varepsilon), \\ \mu_{t+1} &= \mu_t + \xi_t, & \xi_t &\sim \text{NID}(0, \sigma^2_\xi), \end{aligned} \tag{6.3}$$

for $t = 1, \dots, n$. The level component μ_t can be conceived of as the equivalent of the intercept in the classical linear regression model $y_t = \mu + \varepsilon_t$ which is obtained by setting all the level disturbances ξ_t in (6.3) equal to zero and with $\mu = \mu_1$. The key difference is that the intercept μ in a regression model is fixed whereas the level component μ_t in (6.3) is allowed to change from time point to time point.

Since the second equation in (6.3) defines a random walk, the local level model is also referred to as the random walk plus noise model (where the noise refers to the irregular component ε_t). It can be shown that the dynamic process for $x_t = y_{t+1} - y_t = \eta_t + \varepsilon_{t+1} - \varepsilon_t$, for $t = 1, \dots, n$, reduces to the moving average process $x_t = \varepsilon_t + \theta \varepsilon_{t-1}$ where θ relates to the *signal-to-noise ratio* $q = \sigma^2_\xi / \sigma^2_\varepsilon$ via a quadratic function. Furthermore, the forecasting function of observations generated by the local level model is equivalent to the *exponentially weighted moving average* scheme or *exponential smoothing*.

By defining

$$\alpha_t = \begin{pmatrix} \mu_t \\ \nu_t \end{pmatrix}, \quad \eta_t = \begin{pmatrix} \xi_t \\ \zeta_t \end{pmatrix}, \quad T_t = \begin{bmatrix} 1 & 1 \\ 0 & 1 \end{bmatrix}, \quad Z_t = \begin{pmatrix} 1 & 0 \end{pmatrix},$$
$$H_t = \sigma^2_\varepsilon, \quad Q_t = \begin{bmatrix} \sigma^2_\xi & 0 \\ 0 & \sigma^2_\zeta \end{bmatrix}, \quad \text{and} \quad R_t = \begin{bmatrix} 1 & 0 \\ 0 & 1 \end{bmatrix},$$

the scalar notation of (6.1) and (6.2) leads to

$$\begin{aligned} y_t &= \mu_t + \varepsilon_t, & \varepsilon_t &\sim \text{NID}(0, \sigma^2_\varepsilon), \\ \mu_{t+1} &= \mu_t + \nu_t + \xi_t, & \xi_t &\sim \text{NID}(0, \sigma^2_\xi), \\ \nu_{t+1} &= \nu_t + \zeta_t, & \zeta_t &\sim \text{NID}(0, \sigma^2_\zeta), \end{aligned} \tag{6.4}$$

for $t = 1,\ldots,n$, and we obtain the *local linear trend model.*

The local linear trend model requires a 2×1 state vector α_t: one element for the level component μ_t and one element for the slope component ν_t. The slope component can be conceived of as the equivalent of the regression coefficient in the classical regression model where the observed time series y_t is regressed on the independent variable time t: $y_t = \mu + \nu t + \varepsilon_t$ with $\mu = \mu_1$ and $\nu = \nu_1$. Again, the important difference is that the regression coefficient or weight ν is fixed in classical linear regression, whereas the slope ν_t in the local linear trend model is allowed to change over time.

In the situation that the observed time series consists of quarterly or monthly data, for example, the local level and the local linear trend model can be extended with a stochastic seasonal dummy component denoted here by γ_t. In the case of a quarterly time series (the seasonal length is 4), by defining

$$\alpha_t = \begin{pmatrix} \mu_t \\ \gamma_{1,t} \\ \gamma_{2,t} \\ \gamma_{3,t} \end{pmatrix}, \quad \eta_t = \begin{pmatrix} \xi_t \\ \omega_t \end{pmatrix}, \quad T_t = \begin{bmatrix} 1 & 0 & 0 & 0 \\ 0 & -1 & -1 & -1 \\ 0 & 1 & 0 & 0 \\ 0 & 0 & 1 & 0 \end{bmatrix}, \quad Z_t = \begin{pmatrix} 1 & 1 & 0 & 0 \end{pmatrix},$$

$$H_t = \sigma_\varepsilon^2, \qquad Q_t = \begin{bmatrix} \sigma_\xi^2 & 0 & 0 & 0 \\ 0 & \sigma_\omega^2 & 0 & 0 \\ 0 & 0 & 0 & 0 \\ 0 & 0 & 0 & 0 \end{bmatrix}, \qquad R_t = \begin{bmatrix} 1 & 0 \\ 0 & 1 \\ 0 & 0 \\ 0 & 0 \end{bmatrix},$$

and expanding (6.1) and (6.2) in scalar notation, we obtain

$$\begin{aligned} y_t &= \mu_t + \gamma_{1,t} + \varepsilon_t, & \varepsilon_t &\sim \mathrm{NID}(0,\sigma_\varepsilon^2), \\ \mu_{t+1} &= \mu_t + \xi_t, & \xi_t &\sim \mathrm{NID}(0,\sigma_\xi^2), \\ \gamma_{1,t+1} &= -\gamma_{1,t} - \gamma_{2,t} - \gamma_{3,t} + \omega_t, & \omega_t &\sim \mathrm{NID}(0,\sigma_\omega^2), \\ \gamma_{2,t+1} &= \gamma_{1,t}, \\ \gamma_{3,t+1} &= \gamma_{2,t}, \end{aligned} \tag{6.5}$$

for $t = 1,\ldots,n$, which is a local level and dummy seasonal model for a quarterly time series where the seasonal component is allowed to change over time. The seasonal dummy model is not the only approach to incorporate time-varying seasonal effects in unobserved components time series model. For example, the trigonometric seasonal can also be considered. For details about such alternative specifications of the seasonal we refer to Harvey (1989) and Durbin and Koopman (2001).

The textbook of Harvey (1989) was instrumental in the dissemination of state space models outside the field of control engineering. When a slope component is included in (6.5) as well, Harvey calls this model the *basic structural time series model.* A typical application of this model is for the *seasonal adjustment* of time series. A seasonally adjusted time series is defined in this context simply by $\hat{y}_t - \gamma_t$ for $t = 1,\ldots,n$.

Another extension is to include one or more cycles to any of the special models within the class of unobserved components time series models. By defining

$$\alpha_t = \begin{pmatrix} \mu_t \\ c_t \\ c_t^* \end{pmatrix}, \quad \eta_t = \begin{pmatrix} \xi_t \\ \kappa_t \\ \kappa_t^* \end{pmatrix}, \quad T_t = \begin{bmatrix} 1 & 0 & 0 \\ 0 & \rho\cos(\lambda_c) & \rho\sin(\lambda_c) \\ 0 & -\rho\sin(\lambda_c) & \rho\cos(\lambda_c) \end{bmatrix}, \quad Z_t = \begin{pmatrix} 1 & 1 & 0 \end{pmatrix},$$

$$H_t = \sigma_\varepsilon^2, \quad Q_t = \begin{bmatrix} \sigma_\xi^2 & 0 & 0 \\ 0 & \sigma_c^2(1-\rho^2) & 0 \\ 0 & 0 & \sigma_c^2(1-\rho^2) \end{bmatrix}, \quad \text{and} \quad R_t = \begin{bmatrix} 1 & 0 & 0 \\ 0 & 1 & 0 \\ 0 & 0 & 1 \end{bmatrix},$$

in (6.1) and (6.2), we obtain the following local level plus cycle model as given by

$$\begin{aligned} y_t &= \mu_t + c_t + \varepsilon_t, & \varepsilon_t &\sim \text{NID}(0, \sigma_\varepsilon^2), \\ \mu_{t+1} &= \mu_t + \xi_t, & \xi_t &\sim \text{NID}(0, \sigma_\xi^2), \\ c_{t+1} &= \rho[\cos(\lambda_c)c_t + \sin(\lambda_c)c_t^*] + \kappa_t, & \kappa_t &\sim \text{NID}(0, \sigma_c^2(1-\rho^2)), \\ c_{t+1}^* &= \rho[-\sin(\lambda_c)c_t + \cos(\lambda_c)c_t^*] + \kappa_t^*, & \kappa_t^* &\sim \text{NID}(0, \sigma_c^2(1-\rho^2)), \end{aligned} \tag{6.6}$$

for $t = 1, \ldots, n$, where $0 < \rho \leq 1$ is the *damping factor* and λ_c is the frequency of the cycle measured in radians so that $2\pi / \lambda_c$ is the *period* of the cycle. In case $\rho = 1$, the cycle reduces to a fixed sine-cosine wave but the component is still stochastic since the initial values c_1 and c_1^* are stochastic variables with mean zero and variance σ_c^2. A typical application of this model is for the signal extraction of *business cycles* from macro-economic time series.

6.2.2 Regression and Intervention Effects

Another extension of the local level and local linear trend models concerns the incorporation of fixed explanatory and intervention variables. In the case of one regression variable x_t and one intervention variable w_t, for example, we have $y_t = \mu_t + \beta x_t + \lambda w_t + \varepsilon_t$ for the local level model and a state vector of three elements is required: one for the level component μ_t, one for the regression coefficient β, and one for the regression coefficient λ. The substitution of

$$\alpha_t = \begin{pmatrix} \mu_t \\ \beta_t \\ \lambda_t \end{pmatrix}, \quad \eta_t = \xi_t, \quad T_t = \begin{bmatrix} 1 & 0 & 0 \\ 0 & 1 & 0 \\ 0 & 0 & 1 \end{bmatrix}, \quad Z_t = \begin{pmatrix} 1 & x_t & w_t \end{pmatrix},$$

$$H_t = \sigma_\varepsilon^2, \qquad Q_t = \begin{bmatrix} \sigma_\xi^2 & 0 & 0 \\ 0 & 0 & 0 \\ 0 & 0 & 0 \end{bmatrix}, \qquad R_t = \begin{bmatrix} 1 \\ 0 \\ 0 \end{bmatrix},$$

in (6.1) and (6.2) results in

$$
\begin{aligned}
y_t &= \mu_t + \beta_t x_t + \lambda_t w_t + \varepsilon_t, & \varepsilon_t &\sim \text{NID}(0, \sigma_\varepsilon^2), \\
\mu_{t+1} &= \mu_t + \xi_t, & \xi_t &\sim \text{NID}(0, \sigma_\xi^2), \\
\beta_{t+1} &= \beta_t, \\
\lambda_{t+1} &= \lambda_t,
\end{aligned}
\tag{6.7}
$$

where $\beta = \beta_1 = \beta_t$ and $\lambda = \lambda_1 = \lambda_t$ for $t = 1, \ldots, n$. This is the local level model with one continuous explanatory variable x_t and one intervention variable w_t. By adding disturbance terms to the state equation for β_t in (6.7), this regression weight is effectively subjected to a random walk, thus allowing for the estimation of time-varying regression effects.

Letting τ denote the time point at which an intervention effect occurred, variable w_t can either be coded as a pulse intervention:

$$
w_t = \begin{cases} 0, & t < \tau, \quad t > \tau \\ 1, & t = \tau \end{cases}
$$

(to model an outlier observation), or as a level intervention:

$$
w_t = \begin{cases} 0, & t < \tau, \\ 1, & t \geq \tau, \end{cases}
$$

(to model a structural break in the level of the series), or as a slope intervention:

$$
w_t = \begin{cases} 0, & t < \tau, \\ 1 + t - \tau, & t \geq \tau, \end{cases}
$$

(to model a structural break in the slope of the series). Other types of intervention effects can be modelled as well, see Box and Tiao (1975).

6.2.3 Structural Time Series Model

What emerges – and this a key advantage of state space methods – is their *structural* approach to time series analysis: the different unobserved components or building blocks responsible for the dynamics of the series such as trend, seasonal, cycle, and the effects of explanatory and intervention variables are identified separately before being put together in a state space model. It is the responsibility of the researcher to decide what components are required in a specific situation, and then to consider whether they apply to the time series under consideration. This explains why state space models are also known as *structural time series models*.

6.2.4 Multivariate Models

All this is easily extended to multivariate time series. For example, letting y_t denote a $p \times 1$ vector of observations, a multivariate local level model can be applied to the p time series simultaneously:

$$\begin{aligned} y_t &= \mu_t + \varepsilon_t, & \varepsilon_t &\sim \text{NID}(0, \Sigma_\varepsilon), \\ \mu_{t+1} &= \mu_t + \xi_t, & \xi_t &\sim \text{NID}(0, \Sigma_\xi), \end{aligned} \tag{6.8}$$

for $t = 1, \ldots, n$, where μ_t, ε_t, and ξ_t are $p \times 1$ vectors and Σ_ε and Σ_ξ are $p \times p$ variance matrices. In what is known as the *seemingly unrelated time series equations model* (6.8), the series are modelled as in the univariate situation, but the disturbances driving the level components are allowed to be instantaneously correlated across the p series. When slope, seasonal, or cycle components are involved, each of these three components also has an associated $p \times p$ variance matrix allowing for correlated disturbances across series.

If it is found that the rank r of Σ_ξ in (6.8) is smaller than p, then this indicates that the p series have r *common levels*. Such common factors may not only have a nice interpretation, but may also result in more efficient inferences and forecasts.

6.3 State Space Analysis

For given values of all system matrices – and for given initial conditions a_1 and P_1 – the state vector can be estimated in three different ways, yielding what are known as the *filtered*, the *predicted*, and the *smoothed* state vector. Depending on the types of state estimates required in the analysis, the estimates of the state vector can be obtained by performing one or two passes through the observed time series:

1. a *forward* pass, from $t = 1, \ldots, n$, using a recursive algorithm known as the *Kalman filter* enables the computation of filtered and predicted states and prediction errors;
2. a *backward* pass, from $t = n, \ldots, 1$, using output of the Kalman filter and using recursive algorithms known as *state and disturbance smoothers* enables the computation of smoothed estimates of states and disturbances.

6.3.1 Kalman Filter for Prediction, Filtering and Forecasting

The forward pass through the data with the well-known Kalman (1960) filter provides all estimates that are relevant for the filtered and the predicted state. The main purpose of the Kalman filter is to obtain optimal estimates of the state at time point t, *only* considering the observations $\{y_1, y_2, \ldots, y_{t-1}\}$. A key property of the predicted state and its related estimates is therefore that they are only based on *past values* of the observed time series. The recursive formulas for the Kalman filter are

$$\begin{aligned} v_t &= y_t - Z_t a_t, & F_t &= Z_t P_t Z_t' + H_t, \\ K_t &= T_t P_t Z_t' F_t^{-1}, & L_t &= T_t - K_t Z_t, \\ a_{t+1} &= T_t a_t + K_t v_t, & P_{t+1} &= T_t P_t L_t' + R_t Q_t R_t', \end{aligned} \quad (6.9)$$

for $t = 1, \ldots, n$. The values of a_t in (6.9) represent the predicted state, while the values of P_t quantify the estimation error variance matrix of the predicted state a_t. Under the assumption of normality, the latter variances are useful for the construction of *confidence intervals* for the predicted state, which – assuming that we are interested in their 90% confidence limits for example – can be calculated as

$$a_t \pm 1.64\sqrt{P_t},$$

for $t = 1, \ldots, n$. A modification of the Kalman filter also allows the computation of the filtered estimate of the state vectors, that is

$$a_{t|t} = a_t + P_t Z_t' F_t^{-1} v_t, \qquad P_{t|t} = P_t - P_t Z_t' F_t^{-1} Z_t P_t, \qquad t = 1, \ldots, n,$$

where $a_{t|t}$ is the optimal estimate of the state at time point t by considering the observations $\{y_1, y_2, \ldots, y_t\}$ while $P_{t|t}$ is the state filtered estimation error variance matrix. The values of v_t in (6.9) are called the *one-step ahead prediction* or *forecast* errors, since they quantify the lack of accuracy of a_t in predicting the observed value of y_t at time point t; the values of F_t are the variances of these one-step ahead prediction errors v_t.

One of the convenient features of state space methods is the ease with which they deal with two important aspects of time series analysis – forecasting and missing observations: by treating them in exactly the same way. Missing observations are handled by setting K_t and v_t in (6.9) equal to 0. Forecasts for $y_{n+1}, \ldots, y_{n+k}$ given $y_1, \ldots, y_n$ are simply obtained by applying the Kalman filter for $t = 1, \ldots, n, n+1, \ldots, n+k$ and by treating $y_{n+1}, \ldots, y_{n+k}$ as missing observations.

6.3.2 State and Disturbance Smoothing

The backward pass through the data is only required for smoothing that leads to estimates such as the smoothed states and smoothed disturbances. The main purpose of state and disturbance smoothing is to obtain estimated values of the state and disturbance vectors at time point t, considering *all* available observations $\{y_1, y_2, \ldots, y_n\}$. The recursive formulas for state smoothing are

$$r_{t-1} = Z_t' F_t^{-1} v_t + Z_t' r_t, \qquad N_{t-1} = Z_t' F_t^{-1} Z_t + L_t' N_t L_t, \quad (6.10)$$

$$\hat{\alpha}_t = a_t + P_t r_{t-1}, \qquad V_t = P_t - P_t N_{t-1} P_t, \quad (6.11)$$

for $t = n, \ldots, 1$. The recursive formulas for smoothing (6.10) are initialised with $r_n = 0$ and $N_n = 0$. The state smoothing equations (6.11) yield the smoothed state estimate $\hat{\alpha}_t$ and is defined as the optimal estimate of α_t using the full set of observations $\{y_1, y_2, \ldots, y_n\}$; the state smoothing equations also yield the corresponding smoothed state estimation error variance matrix V_t.

Analogous to the predicted state, under the assumption of normality the smoothed state estimation error variance matrix V_t is useful for the construction of confidence intervals for the smoothed state components, which – should we happen to be interested in their 90% confidence limits for example – can be calculated as

$$\hat{\alpha}_t \pm 1.64\sqrt{V_t},$$

for $t = 1, \ldots, n$.

The recursions for r_{t-1} and N_{t-1} in (6.10) also enable the computation of the smoothed estimates of the disturbances ε_t and η_t in the following way,

$$\hat{\varepsilon}_t = H_t \left(F_t^{-1} v_t - K_t' r_t \right), \qquad \mathrm{Var}(\hat{\varepsilon}_t) = H_t \left(F_t^{-1} + K_t' N_t K_t \right) H_t, \tag{6.12}$$

$$\hat{\eta}_t = Q_t R_t' r_t, \qquad \mathrm{Var}(\hat{\eta}_t) = Q_t R_t' N_t R_t Q_t, \tag{6.13}$$

for $t = n, \ldots, 1$. The equations (6.12) and (6.13) compute the smoothed observation disturbances $\hat{\varepsilon}_t$, the smoothed state disturbances $\hat{\eta}_t$, and their corresponding smoothed estimation error variance matrices $\mathrm{Var}(\hat{\varepsilon}_t)$ and $\mathrm{Var}(\hat{\eta}_t)$.

6.3.3 Diagnostic Checking

All significance tests in linear Gaussian state space models – and the construction of confidence intervals – are based on three assumptions concerning the residuals of the analysis. The residuals should satisfy independence, homoscedasticity, and normality, in this order of importance. Whether the residuals satisfy these three assumptions can be established by diagnosing what are known as the *standardised prediction errors*. They are defined as

$$\frac{v_t}{\sqrt{F_t}}, \tag{6.14}$$

for $t = 1, \ldots, n$. For the computations of the one step-ahead prediction errors v_t and their variances F_t in (6.14), we refer to the recursive formulas for the Kalman filter given in (6.9). The assumptions of independence and normality of the residuals can be diagnosed using the Box-Ljung test statistic and the Bowman and Shenton test statistic, respectively. The assumption of homoscedasticity can be checked by testing whether the variance of the standardised prediction errors in the first third part of the series is equal to the variance of the errors corresponding to the last third part of the series. For further details concerning these diagnostic tests, we refer to Harvey (1989), Durbin and Koopman (2001) and Commandeur and Koopman (2007).

A second diagnostic tool for determining the appropriateness of a model is provided by inspection of what are known as the *auxiliary residuals*. As already mentioned above, the disturbance smoothing filters applied in the backward pass through the data yield, amongst others, estimates of the smoothed observation and state disturbances, and of their variances. The auxiliary residuals are obtained by dividing the smoothed observation and state disturbances with the square root of their corresponding variances, as follows:

$$\frac{\hat{\varepsilon}_t}{\sqrt{\mathrm{Var}(\hat{\varepsilon}_t)}}, \text{ and } \frac{\hat{\eta}_t}{\sqrt{\mathrm{Var}(\hat{\eta}_t)}}, \tag{6.15}$$

for $t = 1, \ldots, n$, resulting in *standardised* smoothed disturbances. Inspection of the standardised smoothed observation disturbances (shown at the left of (6.15)) allows for the detection of possible *outlier* observations in a time series, while inspection of the standardised smoothed state disturbances (shown at the right of (6.15)) makes it possible to detect *structural breaks* in the underlying development of a time series.

Each auxiliary residuals can be considered as a t-test for the null hypothesis that there was no outlier observation (when inspecting the auxiliary residuals at the left of (6.15)) or as a t-test for the null hypothesis that there was no structural break in the corresponding unobserved component of the observed time series (when inspecting the auxiliary residuals at the right of (6.15)). Applying the usual 95% confidence limits of ± 1.96 corresponding to a two-tailed t-test, possible outlier observations or structural breaks in the unobserved components making up the state vector are thus easily detected.

6.3.4 Parameter Estimation

So far, we have presented all of the results that can be obtained with state space methods as if the disturbance variances, the fixed regression effects, the parameters ρ and λ_c associated with cycles, etcetera, are given and known. In practice, of course, these parameters are unknown, and have to be estimated.

It can be shown that the Kalman filter presented in (6.9) also provides the necessary ingredients required for evaluating the log-likelihood function, which is given by

$$\log L(y|\psi) = -\frac{np}{2}\log(2\pi) - \frac{1}{2}\sum_{t=1}^{n}\left(\log|F_t| + v_t' F_t^{-1} v_t\right), \tag{6.16}$$

where the v_t are the one-step ahead prediction errors, the F_t are their variances for $t = 1, \ldots, n$ defined in (6.9), and ψ denotes the vector of unknown parameters. The log-likelihood (6.16) is maximised with respect to ψ numerically using the score vector or the EM algorithm.

Numerical quasi-Newton methods for likelihood maximization such as the one of Broyden-Fletcher-Goldfarb-Shanno (BFGS) are generally regarded as computationally efficient in terms of convergence speed and numerical stability, see also the book of Nocedal and Wright (1999). The BFGS iterative optimization method is based on information from the gradient and terminated when some pre-chosen convergence criterion is satisfied. The convergence criterion is usually based on the gradient evaluated at the current estimate, the parameter change compared to the previous estimate or the likelihood value change compared to the previous estimate. The number of iterations required to satisfy these criteria depends on the choice of the initial parameter values, the tightness of the chosen criterion and the shape of the likelihood surface.

Several problems may arise when maximizing the likelihood function with respect to the parameter vector of a high dimension. For example, the number of required iterations may be too large for a feasible procedure, different initial parameter values and different convergence criteria may lead to different estimates. Also, flat likelihood surfaces may not allow the optimization procedure to converge.

An alternative method for computing ML estimates is the use of the EM-algorithm. The EM-algorithm is not an alternative to ML, but it is an alternative way to obtain the ML estimates. We may compare the different estimation methods in terms of required calculation time. The EM algorithm in the setting of a state space model was developed by Shumway and Stoffer (1982) and Watson and Engle (1983). The basic EM procedure works roughly as follows. Consider the joint density $p(y_1,\ldots,y_n,\alpha_1,\ldots,\alpha_n)$. The Expectation (E) step takes the expectation of the components of the joint density conditional on $y_1,\ldots,y_n$ and maximizes the resulting expression with respect to ψ. The E step mainly consists of evaluating the estimated state vector using state space smoothing algorithms. The next step is the Maximization (M) step which usually can be done analytically and is simpler than maximizing the full likelihood function directly. Given the "new" estimate from the M step, we can go back to the E step and evaluate the smoothed estimates based on the new estimate. This iterative procedure converges to the ML estimate of ψ. Under fairly weak conditions it can be proven that each iteration of the EM algorithm only increases the value of the likelihood, and that the EM estimate converges to a maximum of the likelihood. The algorithm has similar properties as a well chosen numerical ML algorithm.

6.4 An Illustration of Multivariate State Space Analysis

In this section we present the practical implications of a multivariate state space analysis. Various results of a simultaneous analysis of two time series will be discussed in some detail. In February 1983 a law was introduced in the United Kingdom (UK) obligating front seat passengers in cars (including the driver) to wear a seat belt. In Durbin and Koopman (2001) and Commandeur and Koopman (2007) the effect of this law was investigated by applying a bivariate local level with seasonal model to the log of the monthly numbers of *front* seat passengers killed or seriously injured (KSI) in cars and to the log of the monthly numbers of *rear* car seat passengers KSI, but only for the period 1969 – 1984 (thus yielding a total of $12 \times 16 = 192$ observations per series). These series were analysed previously with univariate state space models in Harvey and Durbin (1986).

In these studies the numbers of UK front seat passengers KSI in cars were treated as a *treatment* series, while the UK rear seat passengers KSI in cars were used as a *control* series, based on the assumption that the rear seat passengers KSI in cars were not affected by the introduction of this seat belt law. It was indeed found that the seat belt law resulted in a significant 28.4% to 30.5% decrease in the number of front seat passengers KSI in cars, but did not affect the number of UK rear car seat passengers KSI.

In this section we re-investigate the effect of the introduction of this law, but now applied to the same two series supplemented with monthly observations for the years 1985 – 2007, resulting in a total of $12 \times 39 = 468$ observations per series. The logs of the two series are displayed in Figure 6.1.

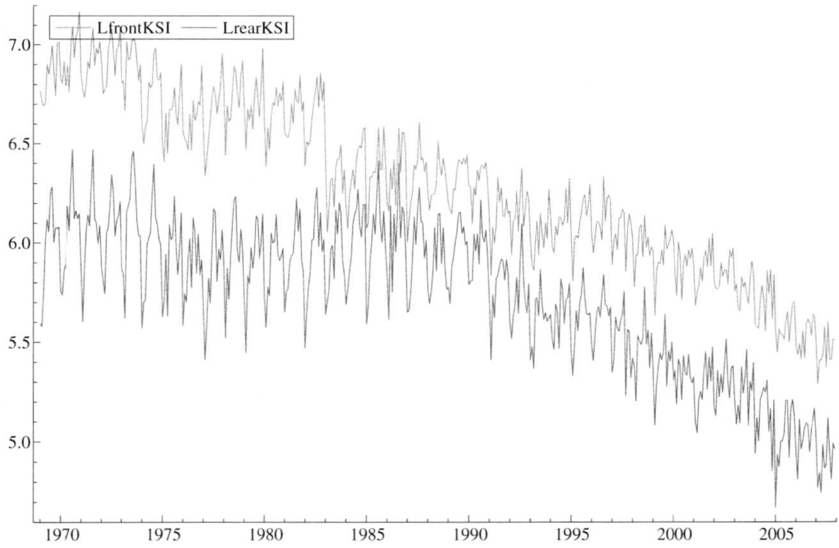

Fig. 6.1 Log of monthly numbers of front seat passengers (top) and rear seat passengers (bottom) in cars killed or seriously injured in the UK in the period 1969–2007.

These extended series not only make it possible to confirm or falsify the value and significance of the effect of the February 1983 seat belt law on front seat passengers in cars previously found in Durbin and Koopman (2001) and Commandeur and Koopman (2007) for the monthly 1969 – 1984 series, but also to investigate the effects of the introduction of two other seat belt laws in the UK: the obligation for children in the rear seat of cars to wear a seat belt in September 1989, and for adults in the rear seat of cars to wear a seat belt in July 1991. In the evaluation of the effects of the latter two laws it is typically the monthly number of rear seat passengers KSI that act as a treatment series while the monthly number of front seat passengers KSI can now be used as a control series.

All the analyses discussed in this chapter were performed in `STAMP 8` of Koopman, Harvey, Doornik, and Shephard (2007). `STAMP 8` is an easy-to-use package designed to model and forecast time series, based on uni- and multivariate structural time series models. No coding is required because all the models are simply formulated by clicking options in dialog windows. Other software packages that currently have functions for analysing time series with state space methods (but with a programmatic interface) include `SsfPack`, `R`, `Matlab`, `Eviews`, `Gauss`, `Stata`, `SAS`, `RATS`, and `Gretl`.

We start by adding three intervention variables to a bivariate local linear trend with monthly seasonal model applied to both series (in logs). These intervention variables are: the introduction of the seat belt law for car drivers and front seat car passengers in February 1983, the introduction of the seat belt law for children in the rear seat of cars in September 1989, and the introduction of the seat belt law for adults in the rear seat of cars in July 1991, all applied to both series simultaneously.

The bivariate time series analysis aims to assess the effects of the introduction of these three seat belt laws in the UK. The intervention of February 1983 is expected to affect the car drivers and front seat car passengers only, and not the rear seat car passengers. In contrast, the interventions of September 1989 and July 1991 are expected to affect the rear seat car passengers only, and not the car drivers and front seat car passengers. As we already mentioned, the car drivers and front seat car passengers series can be considered as a treatment series for the evaluation of the February 1983 intervention, while the rear seat car passengers series can be used as a control series in this case. For the evaluation of the seat belt laws implemented in September 1989 and July 1991, on the other hand, the reverse holds true: in that case it is the car drivers and front seat car passengers series that takes on the role of a control series, while the rear seat car passengers series can be used as a treatment series in these two cases.

The residual and fit diagnostics of this analysis are as follows:

```
Summary statistics
                    LfrontKSI      LrearKSI
 T                     468.00        468.00
 p                     3.0000        3.0000
 std.error           0.084885       0.10540
 Normality             1.9352        10.135
 H(150)               0.82104       0.91225
 DW                    1.9910        2.1078
 r(1)               0.0013231     -0.058876
 q                     24.000        24.000
 r(q)               -0.029281     -0.078653
 Q(q,q-p)              43.697        37.213
 Rs^2                 0.39786       0.43512
```

The Box-Ljung diagnostic tests for the independence of residuals for the front and rear seat passengers KSI series are $Q(21) = 43.697$ and $Q(21) = 37.213$, respectively. Since these should be tested against $\chi^2_{(21;0.05)} = 32.6705$, the residuals of both series are somewhat serially correlated. The tests for homoscedasticity of the residuals for the front and rear seat passengers KSI series are equal to $H(150) = 0.82104$ and $H(150) = 0.91225$, respectively. Since $F_{(150,150;0.025)} \approx 1.43$, and $1/H(150) = 1.22$ and $1/H(150) = 1.10$, the assumption of homoscedasticity is satisfied for both series. The Bowman-Shenton diagnostic tests for normality of the residuals are $N = 1.9117$ and $N = 13.679$, respectively, implying that the assumption of normality is only satisfied for the front seat passengers KSI series. This is not something to worry about very much since we are dealing with 468 observations. The values of the Akaike Information Criterion (AIC) for the two series are -4.8603 and -4.4273, respectively.

The estimates of the variance matrices (where the upper off-diagonal elements denote correlations) for this bivariate state space model are:

```
Level disturbance variance matrix:          Slope disturbance variance matrix:
             LfrontKSI     LrearKSI                      LfrontKSI     LrearKSI
LfrontKSI    0.0002752       0.8798         LfrontKSI   2.249e-008        1.000
LrearKSI     0.0002047    0.0001967         LrearKSI    3.329e-008   4.927e-008

Seasonal disturbance variance matrix:       Irregular disturbance variance matrix:
             LfrontKSI     LrearKSI                      LfrontKSI     LrearKSI
LfrontKSI   7.080e-007       0.8030         LfrontKSI     0.005460       0.5935
LrearKSI    1.186e-006   3.082e-006         LrearKSI      0.004033     0.008459
```

The t-tests for the regression weights of the three level shift intervention variables are:

```
Equation LfrontKSI: regression effects in final state at time 2007(12)

                     Coefficient         RMSE      t-value        Prob
Level break 1983(2)     -0.33634      0.05107     -6.58646 [0.00000]
Level break 1989(9)      0.04346      0.05108      0.85077 [0.39535]
Level break 1991(7)     -0.03793      0.05108     -0.74260 [0.45811]

Equation LrearKSI: regression effects in final state at time 2007(12)

                     Coefficient         RMSE      t-value        Prob
Level break 1983(2)      0.02321      0.05208      0.44564 [0.65607]
Level break 1989(9)      0.05752      0.05208      1.10445 [0.26999]
Level break 1991(7)     -0.06484      0.05206     -1.24556 [0.21357]
```

These t-tests indicate that the regression coefficient for the February 1983 level shift intervention variable applied to the front seat passengers KSI series is very significant, unlike any of the other five intervention variables. The estimated regression coefficient for the February 1983 level shift intervention variable on front seat passengers KSI is -0.33634, implying a $100 \times (\exp(-0.33634) - 1) = -28.56\%$ change in the number of front seat passengers KSI due to the introduction of this seat belt law in the UK.

Although the disturbance variances of the two slope components for both series are quite small, we decide to keep the slope components in all further multivariate analyses of these two series because the values of these components in December 2007 are found to significantly deviate from zero:

```
Equation LfrontKSI
                                Value      Prob
Slope                        -0.00395 [0.00486]

Equation LrearKSI
Slope                        -0.00476 [0.01114]
```

We now present the results of the same analysis after removing the five non-significant level shift intervention variables from the previous model. The residual and fit diagnostics are:

```
Summary statistics
                 LfrontKSI     LrearKSI
 T                  468.00       467.00
 p                  3.0000       3.0000
 std.error        0.085015      0.10567
 Normality          1.8309       9.5254
 H(151)            0.81288      0.91048
 DW                 1.9886       2.1155
 r(1)            0.0021785    -0.063392
 q                  24.000       24.000
 r(q)            -0.032486    -0.072441
 Q(q,q-p)           43.508       34.964
 Rs^2              0.39334      0.42977
```

The Box-Ljung diagnostic tests for the independence of the residuals for the front and rear seat passengers KSI series in this analysis are $Q(21) = 43.508$ and $Q(21) = 34.964$, respectively. The residuals of both series are therefore still serially correlated, although to a somewhat lesser extent than in the previous analysis. The tests for homoscedasticity of the residuals for the front and rear seat passengers KSI series for this analysis are equal to $H(151) = 0.81288$ and $H(151) = 0.91048$, respectively. Since $F_{(151,151;0.025)} \approx 1.43$, and $1/H(151) = 1.23$ and $1/H(151) = 1.10$, the assumption of homoscedasticity is still satisfied for both series. The Bowman-Shenton diagnostic tests for normality of the residuals are now $N = 1.8309$ and $N = 9.5254$, respectively, meaning that the assumption of normality is still only satisfied for the front seat passengers KSI series. Again, this is not something to worry about very much due to the large amount of observations in this data set. The AIC for the two series are now -4.8658 and -4.4351, respectively, indicating a better fit than in the previous analysis.

The estimates of the variance matrices (where the upper off-diagonal elements again denote correlations) for this analysis are:

```
Level disturbance variance matrix:        Slope disturbance variance matrix:
            LfrontKSI     LrearKSI                      LfrontKSI     LrearKSI
LfrontKSI   0.0002708       0.8734        LfrontKSI   2.408e-008        1.000
LrearKSI    0.0002110    0.0002155        LrearKSI    3.581e-008   5.325e-008

Seasonal disturbance variance matrix:     Irregular disturbance variance matrix:
            LfrontKSI     LrearKSI                      LfrontKSI     LrearKSI
LfrontKSI  7.038e-007       0.8051        LfrontKSI     0.005457       0.5927
LrearKSI   1.190e-006   3.105e-006        LrearKSI      0.004005     0.008368
```

With a value of -10.55 for the t-test, the estimated regression coefficient for the February 1983 level shift intervention variable applied to the front seat passengers KSI series is now -0.35120, implying a significant $100 \times (\exp(-0.35120) - 1) = -29.62\%$ level change in the number of front seat passengers KSI.

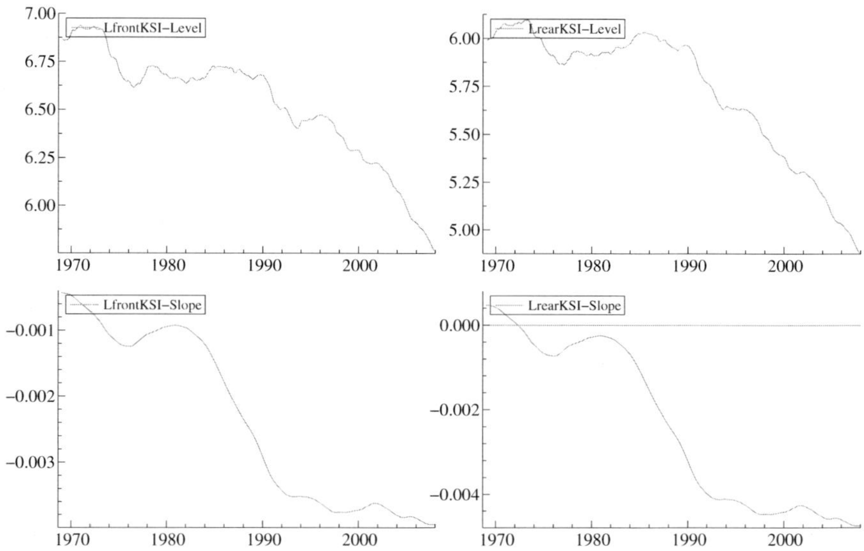

Fig. 6.2 Levels and slope components of full rank model for monthly numbers of front seat passengers (left) and rear seat passengers (right) killed or seriously injured in the UK in the period 1969–2007.

There is a perfect correlation between the slope disturbances of the two series, probably due to their very small variances. However, the just mentioned variance matrix of the level disturbances indicates that the level disturbances are also quite highly correlated. This is confirmed by the following eigenvalue decompositions of the level and slope disturbance variance matrices:

```
Analysis of variance matrices
Level disturbance variance matrix is 2 x 2 with imposed rank 2 and actual rank 2
Variance/correlation matrix
            LfrontKSI    LrearKSI
LfrontKSI   0.0002708      0.8734
LrearKSI    0.0002110   0.0002155
Cholesky decomposition LDL' with L and D
            LfrontKSI    LrearKSI
LfrontKSI       1.000      0.0000
LrearKSI       0.7791       1.000
diag(D)     0.0002708   5.112e-005
Eigenvectors and eigenvalues
              LfrontKSI    LrearKSI
LfrontKSI        0.7517      0.6596
LrearKSI         0.6596     -0.7517
eigenvalues   0.0004560  3.036e-005
percentage        93.76       6.243

Slope disturbance variance matrix is 2 x 2 with imposed rank 2 and actual rank 1
Variance/correlation matrix
            LfrontKSI    LrearKSI
LfrontKSI  2.408e-008       1.000
LrearKSI   3.581e-008  5.325e-008
```

```
Eigenvectors and eigenvalues
                  LfrontKSI    LrearKSI
LfrontKSI           -0.5580      0.8298
LrearKSI            -0.8298     -0.5580
eigenvalues     7.732e-008  4.850e-020
percentage           100.0  6.272e-011
```

The first eigenvalue of the level disturbance variance matrix explains almost 94% of the variance in this matrix. This indicates that the model for the analysis of these two series could be simplified by imposing *rank one restrictions* on both these matrices, thus treating the level and slope components as *common to both series*.

We therefore repeat the analysis only applying a level shift intervention variable in February 1983 on the front seat passengers KSI series, *and* restricting the level and slope disturbance matrices to be of rank *one*. The residual and fit diagnostics of this final model are:

```
Summary statistics
                    LfrontKSI     LrearKSI
 T                     468.00       467.00
 p                     3.0000       3.0000
 std.error           0.085009      0.10604
 Normality             1.7758       10.106
 H(151)               0.82200      0.94560
 DW                    1.9739       2.0539
 r(1)                0.010066    -0.029112
 q                     24.000       24.000
 r(q)               -0.031552    -0.075277
 Q(q,q-p)              43.087       34.490
 Rs^2                 0.39342      0.42570
```

The Box-Ljung diagnostic tests for the independence of the residuals for the front and rear seat passengers KSI series are now $Q(21) = 43.087$ and $Q(21) = 34.490$, respectively. The residuals of both series are therefore still serially correlated, although again to a somewhat lesser extent than in the previous analysis. The tests for homoscedasticity of the residuals for the front and rear seat passengers KSI series for this analysis equal $H(151) = 0.82200$ and $H(151) = 0.94560$, respectively. Since $F_{(151,151;0.025)} \approx 1.43$, and $1/H(151) = 1.22$ and $1/H(151) = 1.06$, the assumption of homoscedasticity is again satisfied for both series. The Bowman-Shenton diagnostic tests for normality of the residuals are $N = 1.7758$ and $N = 10.106$, respectively, implying that the assumption of normality is still only satisfied for the front seat passengers KSI series. The AIC for the two series are now -4.8659 and -4.428, respectively, indicating that the previous analysis results in a marginally better fit than the present one.

The estimates of the variance matrices for this last analysis are:

```
Level disturbance variance/correlation matrix:
             LfrontKSI     LrearKSI
LfrontKSI    0.0002639        1.000
LrearKSI     0.0001810    0.0001241
Level disturbance factor variance for LfrontKSI: 0.000263925
Level disturbance factor loading  for LrearKSI: 0.685819
             LfrontKSI     LrearKSI
Constant        0.0000       0.9298
```

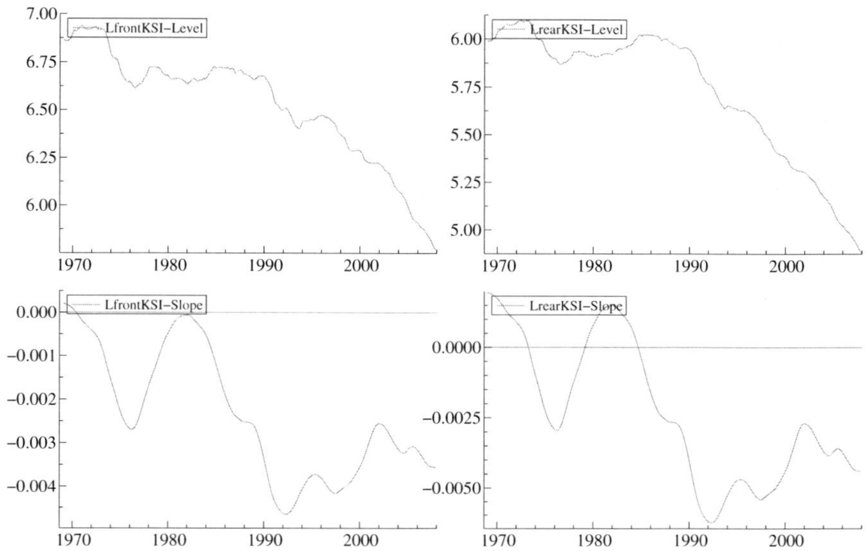

Fig. 6.3 Levels and slope components of rank one model for monthly numbers of front seat passengers (left) and rear seat passengers (right) killed or seriously injured in the UK in the period 1969–2007.

```
Slope disturbance variance/correlation matrix:
            LfrontKSI    LrearKSI
LfrontKSI  7.084e-008       1.000
LrearKSI   1.195e-007  2.014e-007
Slope disturbance factor variance for LfrontKSI: 7.08383e-008
Slope disturbance factor loading  for LrearKSI: 1.68624
            LfrontKSI    LrearKSI
Constant       0.0000    0.001602

Seasonal disturbance variance/correlation matrix:
            LfrontKSI    LrearKSI
LfrontKSI  7.048e-007      0.8072
LrearKSI   1.187e-006  3.067e-006

Irregular disturbance variance/correlation matrix:
            LfrontKSI    LrearKSI
LfrontKSI    0.005467      0.5899
LrearKSI     0.004066    0.008690
```

The t-test for the regression weight of the only level shift intervention variable is:

```
Equation LfrontKSI: regression effects in final state at time 2007(12)

                     Coefficient         RMSE     t-value        Prob
Level break 1983(2)     -0.35111      0.03124   -11.23930 [0.00000]
```

With a t-value of -11.24, the estimated regression coefficient for the February 1983 level shift intervention variable in this final analysis equals -0.35111, indicating a significant $100 \times (\exp(-0.35111) - 1) = -29.61\%$ level change in the number of front seat passengers KSI.

The most important graphical results of this final analysis are presented in Figures 6.3, 6.4, and 6.5. Figure 6.4 displays the estimated trends for the front and rear passengers series KSI series (first row in Figure 6.4), the estimated trigonometric seasonals (second row in Figure 6.4), and the corresponding irregular components (third row in Figure 6.4), while Figure 6.5 contains the correlograms of the residuals of the two series. In Figure 6.3 the common level and slope components of the two series are shown.

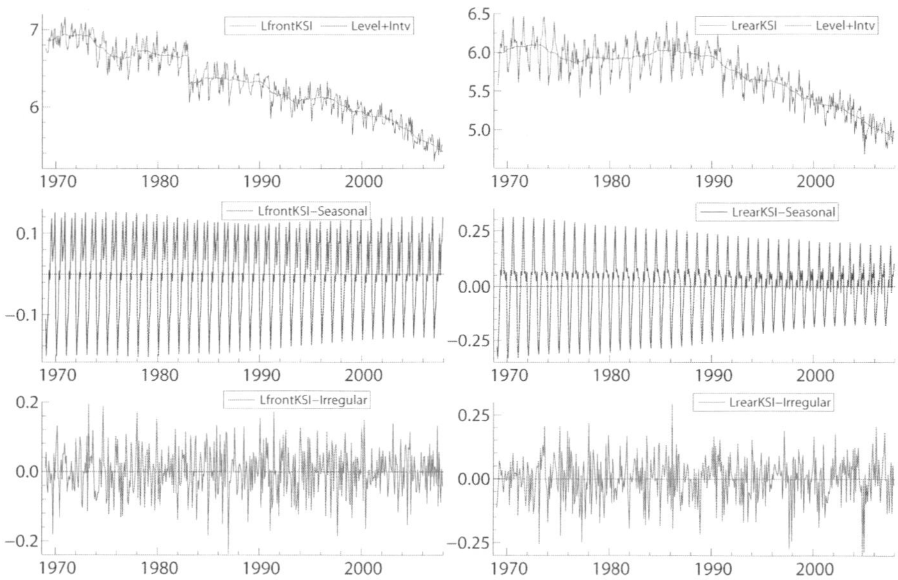

Fig. 6.4 Trends, seasonals and irregular components of rank one model for monthly numbers of front seat passengers (top) and rear seat passengers (bottom) killed or seriously injured in the UK in the period 1969–2007.

The correct implementation of the rank one restrictions is confirmed by the output of the `STAMP 8` program of Koopman, Harvey, Doornik, and Shephard (2007):

```
Level disturbance variance/correlation matrix:
              LfrontKSI     LrearKSI
LfrontKSI    0.0002639        1.000
LrearKSI     0.0001810    0.0001241
Level disturbance factor variance  for LfrontKSI: 0.000263925
Level disturbance factor loading   for LrearKSI: 0.685819

Slope disturbance variance/correlation matrix:
              LfrontKSI     LrearKSI
LfrontKSI   7.084e-008        1.000
LrearKSI    1.195e-007   2.014e-007
Slope disturbance factor variance  for LfrontKSI: 7.08383e-008
Slope disturbance factor loading   for LrearKSI: 1.68624
```

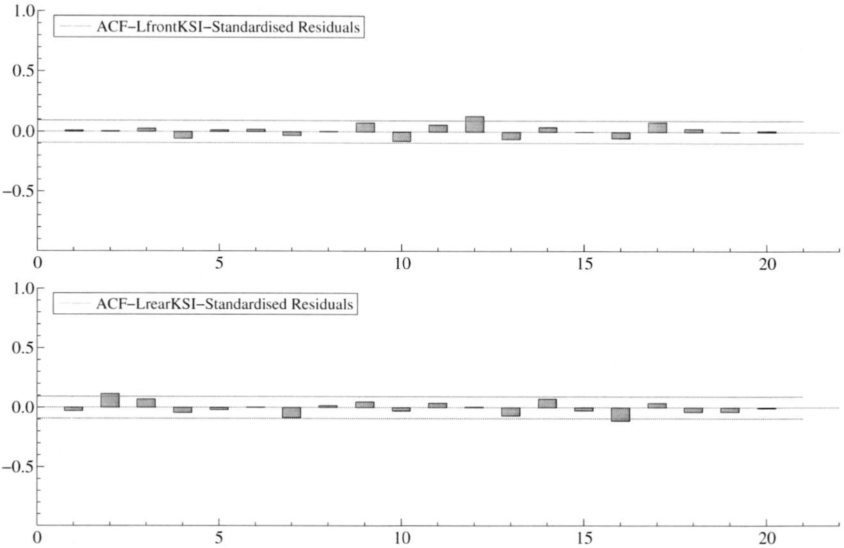

Fig. 6.5 Correlograms of rank one model for monthly numbers of front seat passengers (top) and rear seat passengers (bottom) killed or seriously injured in the UK in the period 1969–2007.

and

```
Analysis of variance matrices
Level disturbance variance matrix is 2 x 2 with imposed rank 1 and actual rank 1
Factors are determined by series LfrontKSI
Variance/correlation matrix
             LfrontKSI    LrearKSI
LfrontKSI    0.0002639       1.000
LrearKSI     0.0001810   0.0001241
Eigenvectors and eigenvalues
               LfrontKSI    LrearKSI
LfrontKSI         0.8247      0.5656
LrearKSI          0.5656     -0.8247
eigenvalues    0.0003881 -5.559e-021
percentage         100.0 -1.432e-015

Slope disturbance variance matrix is 2 x 2 with imposed rank 1 and actual rank 1
Factors are determined by series LfrontKSI
Variance/correlation matrix
             LfrontKSI    LrearKSI
LfrontKSI   7.084e-008       1.000
LrearKSI    1.195e-007  2.014e-007
Eigenvectors and eigenvalues
               LfrontKSI    LrearKSI
LfrontKSI        -0.5101      0.8601
LrearKSI         -0.8601     -0.5101
eigenvalues   2.723e-007  1.016e-023
percentage         100.0  3.731e-015
```

showing that *all* of the variation in the level and slope disturbance matrices is now explained by the first dimension, as expected. It follows that the state equations of the two level and slope components can be written as

$$\begin{aligned}\mu_{t+1}^{(1)} &= \mu_t^{(1)} + \nu_t^{(1)} + \xi_t^{(1)},\\ \mu_t^{(2)} &= \mu_t^{(1)} + \nu_t^{(2)},\\ \nu_{t+1}^{(1)} &= \nu_t^{(1)} + \zeta_t^{(1)},\\ \nu_t^{(2)} &= 1.68624\nu_t^{(1)} + 0.001602.\end{aligned}$$

Notwithstanding the fact that the residual diagnostic tests of the analyses presented in this section do not satisfy all of the model assumptions of independency and normality perfectly, we conclude that the impressive reduction in the UK number of front seat passengers KSI of 28.4% to 30.5% found in Durbin and Koopman (2001) and Commandeur and Koopman (2007) as a result of the introduction of the seat belt law in February 1983 is confirmed in the present analyses, even after adding 24 years of monthly observations to these time series data. However, the introduction of the UK seat belt laws for children and adults in the rear seat of cars in September 1989 and July 1991 apparently failed to have any significant impact on these types of road users.

6.5 Conclusions

We have presented an overview of uni- and multivariate state space time series analysis. An illustration of how the methodology based on state space can be implemented is given for the simultaneous analysis of two time series of traffic safety data. This account is far from complete and more details – such as how to deal with nonlinear models and non Gaussian error distributions – can be found in the references given.

References

Box, G. E. P., & Tiao, G. C. (1975). Intervention analysis with applications to economic and environmental problems. *Journal of the American Statistical Association, 70,* 70-79.

Commandeur, J. J. F., & Koopman, S. J. (2007). *An Introduction to State Space Time Series Analysis.* Oxford: Oxford University Press.

Durbin, J., & Koopman, S. J. (2001). *Time Series Analysis by State Space Methods.* Oxford: Oxford University Press.

Harvey, A. C. (1989). *Forecasting, structural time series models and the Kalman filter.* Cambridge: Cambridge University Press.

Harvey, A. C., & J. Durbin (1986). The effects of seat belt legislation on British road casualties: A case study in structural time series modelling. *Journal of the Royal Statistical Society A, 149 (3),* 187-227.

Kalman, R. E. (1960). A new approach to linear filtering and prediction problems. *Journal Basic Engineering, Transactions ASMA Series D, 82,* 35-45.

Koopman, S. J., Harvey, A. C., Doornik, J. A., & Shephard, N. (2007). *Stamp 8.0: Structural Time Series Analyser, Modeller and Predictor.* London: Timberlake Consultants.

Nocedal, J., & Wright, S. J. (1999). *Numerical Optimization.* New York: Springer Verlag.
Shumway, R. H., & Stoffer, D. S. (1982). An approach to time series smoothing and forecasting using the EM algorithm. *Journal of Time Series Analysis, 3,* 253-264.
Watson, M. W., & Engle, R. F. (1983). Alternative algorithms for the estimation of dynamic factor, MIMIC and varying coefficient regression. *Journal of Econometrics, 23,* 385-400.

Chapter 7
Continuous Time Modeling of Panel Data by means of SEM

Johan H.L. Oud and Marc J.M.H. Delsing

Abstract After a brief history of continuous time modeling and its implementation in panel analysis by means of structural equation modeling (SEM), the problems of discrete time modeling are discussed in detail. This is done by means of the popular cross-lagged panel design. Next, the exact discrete model (EDM) is introduced, which accounts for the exact nonlinear relationship between the underlying continuous time model and the resulting discrete time model for data analysis. In addition, a linear approximation of the EDM is discussed: the approximate discrete model (ADM). It is recommended to apply the ADM-SEM procedure by means of a SEM program such as LISREL in the model building phase and the EDM-SEM procedure by means of Mx in the final model estimation phase. Both procedures are illustrated in detail by two empirical examples: Externalizing and Internalizing Problem Behavior in children; Individualism, Nationalism and Ethnocentrism in the Flemish electorate.

7.1 Introduction

Continuous time modeling goes back to Newton (1643-1727) and Leibniz (1646-1716), who originated the tools of differential and integral calculus. Newton's laws of motion relate the position, speed, and acceleration of physical bodies by means of differential equations. Not less than two and a half centuries later, Simon (1952) introduced the use of differential equations into social science, followed by Coleman (1968) and Blalock (1969). Blalock illustrates his discussion by means of two

Johan H.L. Oud
Behavioural Science Institute, Radboud University Nijmegen, The Netherlands
e-mail: j.oud@pwo.ru.nl

Marc J.M.H. Delsing
Praktikon, Radboud University Nijmegen, The Netherlands
e-mail: M.Delsing@acsw.ru.nl

K. van Montfort et al., *Longitudinal Research with Latent Variables*,
DOI 10.1007/978-3-642-11760-2_7,

examples. In the first example, a system of two simple linear differential equations is used to describe and explain the arms race process between opposing nations. The second example was taken from Simon (1952) and formalizes Homans's theory about the human group (Homans, 1950), relating the variables interaction, friendship, and activity by means of differential equations. It should be noted that the applications in Newtonian mechanics as well as the examples provided by Simon and Blalock are deterministic and do not allow random error to enter the system.

In contrast to its popularity in physics and natural science, the use of continuous time methods in statistically orientated sciences such as economics and social science is still rare. Undoubtedly, one reason for the slow spread has been the difficulty of handling random phenomena in continuous time, in particular the definition of the random walk process on a continuous time scale as well as the associated stochastic integral. It took a century after the discovery of Brownian motion, the random walk behavior of particles in a liquid, before Norbert Wiener in 1928 succeeded to give this motion a rigorous mathematical definition. In honor of Wiener, the motion was later called Wiener process. Wiener was also the first to define integration of the Wiener process (Wiener stochastic integral), which in 1944 was generalized by the Japanese mathematician Itô (Itô stochastic integral). Nowadays, there is no reason to avoid the specification of random error in continuous time or the use of stochastic differential equations and their solution. The mathematical problems are solved and need not concern the research practitioner as will be shown in this chapter.

Just as in natural science, most phenomena studied in economics and social science evolve in continuous time. As emphasized by Bergstrom, the pioneer of continuous time modeling in econometrics, the economy does not cease to exist in between observations nor does it function only at quarterly or annual intervals corresponding to the observations (Phillips, 1993, p. 23). Bergstrom (1988) credited the British statistician Bartlett for being the first to deal with the problem of estimating the parameters of continuous time stochastic models from discrete time series. As Bartlett (1946) put it:

> The discrete time nature of our observations in many economic and other time series does not reflect any lack of continuity in the underlying series. Thus theoretically it should often prove more fundamental to eliminate this imposed artificiality. An unemployment index does not cease to exist between readings, nor does Yule's pendulum cease to swing.

Hereby, Bartlett for the first time criticized the unfortunate identification in conventional time series analysis of the dynamically relevant interval with the observation interval. Continuous time methods put the causal mechanisms on a continuous time scale, allowing the process to proceed in infinitesimally small steps, and so distinguish the underlying dynamics clearly from the discrete time measurement time points. This is especially important in social science, where measurement almost invariably occurs in discrete time, measurement time points are chosen rather arbitrarily, and observation intervals are often large. Particularly in the case of large intervals, approximating the continuous time process by a discrete time model formulated in terms of the observation interval leads to unacceptable results.

Traditionally, the application of continuous time methods is restricted to $N = 1$ research and estimation in the stochastic case is done by $N = 1$ time series estimation methods, especially filter techniques. To solve the problem of the data points in a time series being correlated, which violates the independence assumption of sampling theory, filter techniques purge the data from the predictable correlated parts to end up with uncorrelated "innovations". From 1990 onwards, Singer (1990, 1993, 1995, 1998) worked on the adaptation of these techniques for continuous time analysis of panel data. Singer's (1991) program LSDE (Linear Stochastic Differential Equations) performs maximum likelihood estimation of the continuous time model on the basis of the so-called exact discrete model (EDM). The EDM, claimed to be developed in 1961-1962 by Bergstrom (Bergstrom, 1988), will also be central in the present chapter. Many alternative but approximate estimation procedures, such as the approximate discrete model ADM (Bergstrom, 1966) or the multivariate latent differential equation MLDE (Boker, Neale, & Rausch, 2004) procedure, provide more or less accurate approximations of the underlying continuous time parameters on the basis of discrete time data. The EDM has the major advantage of linking the discrete time model parameters in an exact way to the underlying continuous time model parameters by means of nonlinear constraints. The EDM and estimation procedures using the EDM make sure that the parameters estimated are exactly equal to the parameters of the underlying differential equation model.

An alternative way to estimate the continuous time parameters for panel data through the EDM is Structural Equation Modeling (SEM). This was started by Oud (1978), employing the first published version of the SEM program LISREL (Jöreskog & Sörbom, 1976) described in Jöreskog's (1977) seminal publication about SEM. Later, Arminger (1986) and Oud, van Leeuwe, and Jansen (1993) used other SEM program versions for the same purpose. A similar approach was followed by Tuma and Hannan (1984), although they used related simultaneous equations procedures rather than SEM. Common to all these authors is that they were inspired by Coleman (1968) to employ the so-called "indirect" method in estimating the EDM. This consists of first estimating discrete time parameters by means of a SEM or similar program and then separately, in a second step, deriving the continuous time parameter values using the EDM. In general, the indirect method cannot be recommended. A simple example, where the indirect method breaks down, is in the case of unequal observation intervals (Tuma & Hannan, 1984). Here the imposition of simple equality constraints by the SEM program does not work and the direct application of the nonlinear constraints is called for.

Oud and Jansen (2000) showed how more recent nonlinear SEM software packages such as Mx (Neale, Boker, Xie, & Maes, 2006) can also be employed for maximum likelihood estimation of the continuous-time state space model parameters, but using the direct method: applying the nonlinear constraints of the EDM directly during estimation. A thorough comparison between the LSDE/EDM procedure using filter techniques and the direct SEM/EDM procedure was made by Oud and Singer (2008) in a series of Monte Carlo simulation studies. It turns out that in case the same model is analyzed in both procedures and the data are appropriate for both procedures, the estimation results from filter techniques and SEM are equal. In this chapter we will exclusively deal with SEM.

Although continuous time modeling is pertinent to an extremely broad subject field in social science, most problems with discrete time analysis and their solution by means of continuous time modeling are covered by the topic of reciprocal causal relationships. Reciprocal relationships are traditionally analyzed in discrete time by means of the cross-lagged panel design. In the next section, we will first go into this popular but, from the continuous time perspective, insidious analysis design. The motivation and basic principles of continuous time modeling will be clarified on the basis of the cross-lagged panel design. The full-fledged model and its estimation will be dealt with in the ensuing sections.

7.2 Analysis of Reciprocal Relationships in the Cross-Lagged Panel Design

The cross-lagged panel design studies and compares the effects that variables have on each other across time. Different from cross-sectional research, the causal direction in panel research is not based on instantaneous relationships between simultaneously measured variables x and y. Instead, different variables are used for opposite directions: x at time point 1 affecting y at time point 2, y at time point 1 affecting x at time point 2 (see Figure 7.1). The cross-lagged panel design is therefore supposed to be more suitable than cross-sectional research in answering, for example, whether parenting characteristics affect adolescents' adjustment or, conversely, whether adolescents' adjustment affects parenting characteristics, or whether both effects operate reciprocally (Neiderhiser, Reiss, Hetherington, & Plomin, 1999).

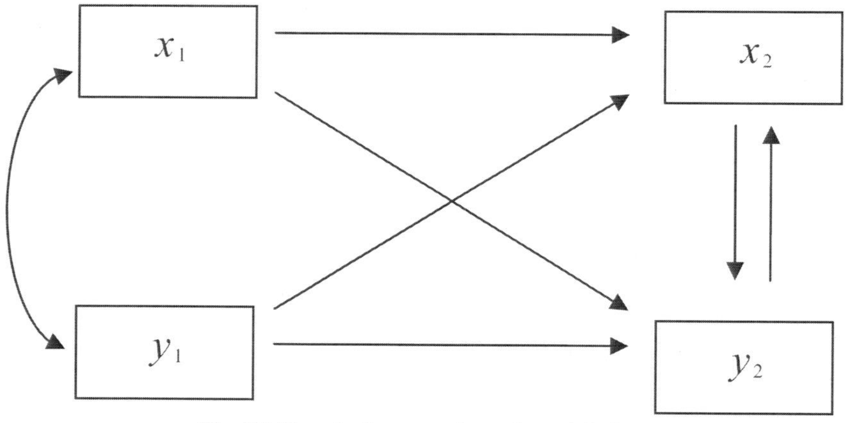

Fig. 7.1 Discrete-time cross-lagged panel design.

Having attracted attention in sociology earlier, the cross-lagged panel design is now becoming increasingly popular in psychology. Rueter and Conger (1998), for example, make clear that correlations between parental and children's behavior, which in the past were interpreted as unidirectional influences from parents to children, have in recent years assumed a reciprocal causal interpretation. This has led to a host of cross-lagged panel research to examine and test the direction of the effects. Other examples include cross-lagged reciprocal relationships between adolescent problem drug use, delinquent behavior, and emotional distress (Bui, Ellickson, & Bell, 2000), and between children's peer relations and antisocial behavior (Vuchinich, Bank, & Patterson, 1992).

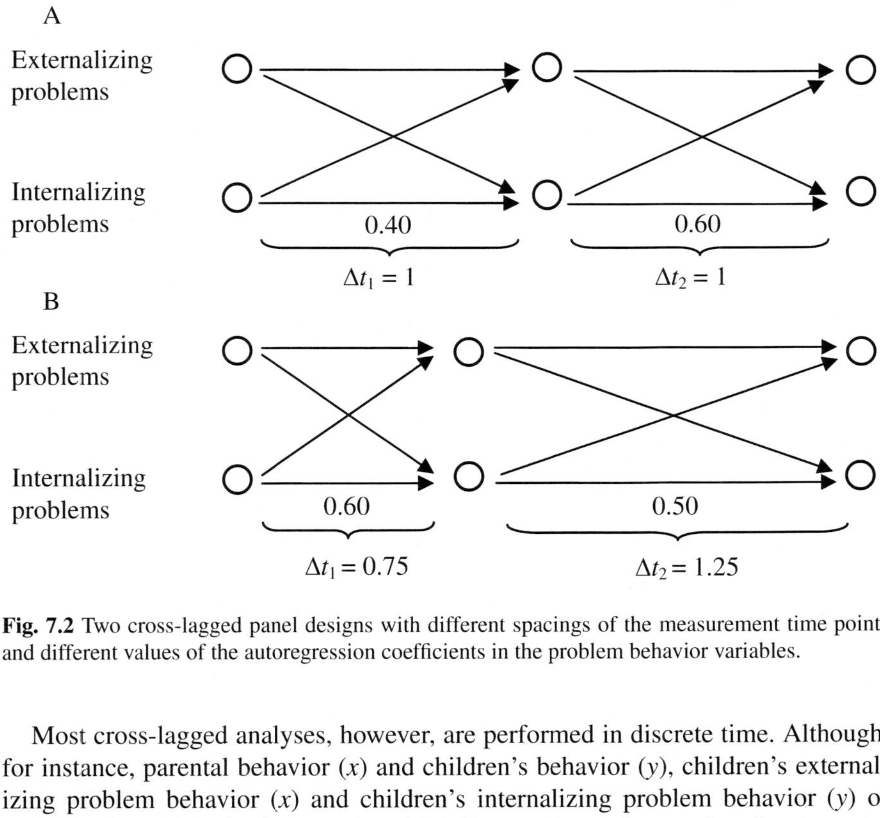

Fig. 7.2 Two cross-lagged panel designs with different spacings of the measurement time points and different values of the autoregression coefficients in the problem behavior variables.

Most cross-lagged analyses, however, are performed in discrete time. Although, for instance, parental behavior (x) and children's behavior (y), children's externalizing problem behavior (x) and children's internalizing problem behavior (y) or individualism (x) and ethnocentrism (y) influence themselves and each other continuously over time, measurements are typically taken not more than one or two times a year, resulting in a large observation interval. As a consequence, discrete time modeling becomes an oversimplification and often a distortion of reality. The path diagrams of the cross-lagged panel design in Figure 7.2 make this very clear. The oversimplification consists in the assumption that the arrows jump from one point in time to the next one and that nothing happens between measurements. In fact, the estimated cross-lagged coefficients (crossing arrows) and autoregression coefficients (horizontal arrows) over the observation interval Δt_i are complicated mixtures of the continuous time cross- and auto-effects in a constant interchange over, and heavily dependent on the length of, the chosen observation interval Δt_i. A variable with a high auto-effect, meaning that there is a strong tendency to sustain its value over time, tends also to retain the influence of other variables over a longer time interval than a variable with a low auto-effect. So, even a relatively small continuous time cross-effect can result in a relatively high cross-lagged effect in discrete time, if the variable influenced has a high auto-effect. But the converse can also be true: a relatively strong continuous time cross-effect having only small impact over a discrete time interval because of a rather low auto-effect in the dependent variable. Additionally, the result will be more strongly dependent on the auto-effect over the larger time interval ($\Delta t_2 = 1.25$ in diagram B) than over the shorter interval ($\Delta t_1 = 0.75$ in diagram B). So, the causal picture changes in discrete time, depending on the length of the chosen observation interval. Continuous time modeling is necessary to disentangle the continuous time cross-effects and auto-effects from the discrete time mixtures.

7.2.1 Relationship between Continuous and Discrete Time

The relationship between continuous and discrete time is governed by the matrix exponential

$$\mathbf{A}_{\Delta t_i} = \mathrm{e}^{\mathbf{A}\Delta t_i}. \tag{7.1}$$

Many paradoxical aspects of the relationship are explainable by the highly nonlinear character of the matrix exponential. Its power series definition will be given in (7.9) and a rather general computational form in (7.17). $\mathbf{A}_{\Delta t_i}$ is the discrete time autoregression matrix over observation interval $\Delta t_i = t_i - t_{i-1}$ $(i = 1,2,...)$ and $\mathbf{A}$ is the so-called drift matrix, which is the analogue of the autoregression matrix in continuous time. It is multiplied by the interval in the exponent of (7.1). Autoregression matrix $\mathbf{A}_{\Delta t_i}$ displays on the diagonal the autoregressions for each of the variables and off-diagonally the cross-lagged effects between the variables. Analogously, drift matrix $\mathbf{A}$ has the continuous time auto-effects on the diagonal and the continuous time cross-effects off-diagonally. It should be emphasized that (7.1), which specifies the exact relationship between $\mathbf{A}_{\Delta t_i}$ and $\mathbf{A}$, clearly shows that $\mathbf{A}_{\Delta t_i}$ changes as a function of the length of the observation interval, while $\mathbf{A}$ continues to be equal.

Table 7.1 gives a typical example of an $\mathbf{A}_{\Delta t_i}$ with corresponding exact $\mathbf{A}$, for $\Delta t_i = 1$ computed according to (7.1). The most conspicuous differences between the matrices are found in the diagonals of $\mathbf{A}_{\Delta t_i}$ (autoregressions 0.50, 0.40, and 0.30) and $\mathbf{A}$ (auto-effects –0.84, –1.05, and –1.60). Whereas the autoregressions in the diagonal of $\mathbf{A}_{\Delta t_i}$ are all positive, the corresponding auto-effects in $\mathbf{A}$ are all negative. This is a rather technical difference, which should be kept in mind when interpreting

Table 7.1 Discrete time autoregression matrix (left; $\Delta t_i = 1$) and corresponding continuous time drift matrix (right)

$$\begin{array}{c} \\ x_1 \\ x_2 \\ x_3 \end{array} \begin{array}{c} \begin{array}{ccc} x_1 & x_2 & x_3 \end{array} \\ \begin{bmatrix} 0.50 & 0.30 & 0.22 \\ 0.05 & 0.40 & 0.20 \\ 0.25 & 0.20 & 0.30 \end{bmatrix} \\ \mathbf{A}_{\Delta t_i} \end{array} \quad \begin{array}{c} \begin{array}{ccc} x_1 & x_2 & x_3 \end{array} \\ \begin{bmatrix} -0.84 & 0.64 & 0.44 \\ -0.09 & -1.05 & 0.69 \\ 0.76 & 0.40 & -1.60 \end{bmatrix} \\ \mathbf{A} \end{array}$$

differences. It is simply explained, however, when we take a closer look at the relationship between discrete and continuous time.

We start from the autoregression equation (7.2), specifying how by means of $\mathbf{A}_{\Delta t_i}$ each of the variables in vector $\mathbf{x}(t_i)$ is predictable by the variables in vector $\mathbf{x}(t_{i-1})$ at the previous time point:

$$\mathbf{x}(t_i) = \mathbf{A}_{\Delta t_i}\mathbf{x}(t_{i-1}). \tag{7.2}$$

For clarity, we do not yet specify an error component in (7.2), but this does not impact the relationship between autoregression matrix and drift matrix. From (7.2) we derive, dividing $\Delta\mathbf{x}(t_i) = \mathbf{x}(t_i) - \mathbf{x}(t_{i-1})$ by Δt_i:

$$\begin{aligned} \frac{\Delta\mathbf{x}(t_i)}{\Delta t_i} &= \mathbf{A}_*\mathbf{x}(t_{i-1}) \\ &\text{with } \mathbf{A}_* = (\mathbf{A}_{\Delta t_i} - \mathbf{I})/\Delta t_i. \end{aligned} \tag{7.3}$$

Difference equation (7.3) in terms of $\mathbf{A}_*$ approximates differential equation (7.4) in terms of continuous time matrix $\mathbf{A}$:

$$\frac{\mathrm{d}\mathbf{x}(t)}{\mathrm{d}t} = \mathbf{A}\,\mathbf{x}(t). \tag{7.4}$$

We assume the rather general conditions to be satisfied, which guarantee a unique solution of (7.4) for initial value $\mathbf{x}(t_{i-1}) = \mathbf{x}(t_0) = \mathbf{x}_0$ (Zadeh & Desoer, 1963, p. 294). Note that, although the differential equation model is specified for all t in some continuous time interval and also its solution is valid for all t in the interval, the solution is observed only at the discrete time points t_i. While the solution is given in autoregression form for arbitrary discrete time points t_i in (7.2), it is made explicit in terms of continuous time drift matrix $\mathbf{A}$ by matrix exponential (7.1).

Basically, differential equation model (7.4) can thus be viewed as a transformation of the popular autoregression model (7.2). First in (7.3) difference quotient $\Delta\mathbf{x}(t_i)/\Delta t_i$ is placed on the left-hand side, approximation $\mathbf{A}_*$ of $\mathbf{A}$ on the right-hand side, and subsequently shifting the time interval Δt_i towards zero makes $\mathbf{A}_*$ approach $\mathbf{A}$ more and more closely. As seen in (7.3), in the transformation from $\mathbf{A}_{\Delta t_i}$ into $\mathbf{A}_*$, each autoregression value in the diagonal of $\mathbf{A}_{\Delta t_i}$ is diminished by 1 and so becomes negative for autoregressions between 0 and 1. This explains why in general positive but less than 1 autoregressions in discrete time correspond to negative auto-effects in continuous time. By equation (7.1), it is further evident that in the case of zero off-diagonals in **A**, autoregressions between 0 and 1 must correspond to negative auto-effects. Zero off-diagonals cause the matrix exponential in (7.1) to reduce to scalar exponentials with negative values $-\infty < \mathrm{a} < 0$ in the diagonal of **A** leading to positive values $0 < \mathrm{e}^{\mathrm{a}\Delta t_i} < 1$ in the diagonal of $\mathbf{A}_{\Delta t_i}$ and vice versa. Note that in Table 7.1, the strength order of the positive autoregressions in the autoregression matrix (0.30 < 0.40 < 0. 50) is maintained in the negative drift matrix diagonals (–1.60 < –1.05 < –0.84). Depending on the off-diagonals, however, this is not necessarily the case.

Causally more interesting than the diagonals of the matrices in Table 7.1 are the paradoxical differences between discrete and continuous time that occur in the off-diagonal elements (effects between different variables). It turns out that the conclusions drawn in a discrete time analysis with respect to the cross-lagged coefficients in $\mathbf{A}_{\Delta t_i}$ may differ fundamentally from those to be drawn in a continuous time analysis on the basis of the corresponding cross-effects in **A**.

- *Equal discrete time coefficients become different in continuous time.*

For example, the two reciprocal cross-lagged coefficients with value 0.20 in the autoregression matrix – which in discrete time might lead to the conclusion that the strength of the causal effects between the variables x_2 and x_3 is equal in opposite directions – differ considerably in continuous time: 0.69 and 0.40.

- *The strength order of coefficients reverses going from discrete to continuous time.*

For example, in the autoregression matrix, the discrete time effect of x_3 on x_1 is greater than that of x_3 on x_2: 0.22 versus 0.20. However, in the corresponding drift matrix, it is the other way around: 0.44 for the first effect and 0.69 for the second effect.

- *Discrete time nonzero coefficients vanish or even change sign in continuous time.*

The effect of x_1 on x_2 with positive value 0.05 in discrete time gets the negative value of –0.09 in continuous time. So, even interpreting the sign of the effect between variables is not safe for the transition from discrete to continuous time.

7.2.2 Discrete Time Problems with Unequal and Equal Observation Intervals

Continuous time analysis is needed to draw correct conclusions about causal effects. Discrete time analysis gets into extreme trouble, however, in the case of unequal observation intervals. When different discrete time distances are used in the same

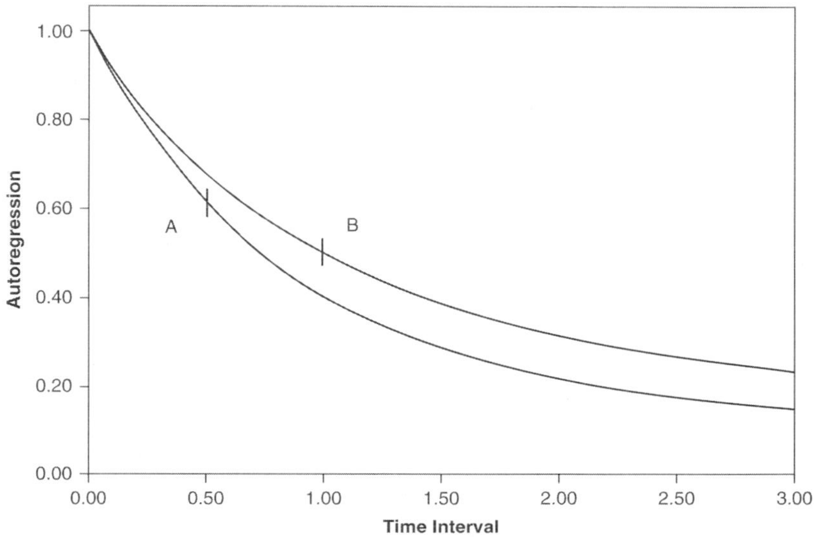

Fig. 7.3 Two different autoregression functions in two different studies A and B.

study or different researchers study the same causal effect in different distances, it becomes impossible to compare the strength of the causal effects found. This has long been recognized in discrete time analysis, notably by Gollob and Reichardt (1987). It seriously hampers cumulative progress of science, but cannot be solved in a discrete time context. This is exemplified in Figure 7.2 too. Supposing the other effects in the model to be equal over the two successive equal intervals in diagram A, one would not need continuous time modeling to conclude, on the basis of the autoregressions (0.40 and 0.60), that the auto-effect over the first interval is smaller than the one over the second interval. In diagram B with unequal intervals, however, no decision can be made as to which one represents a bigger true auto-effect: 0.60 over interval $\Delta t_1 = 0.75$ or 0.50 over longer interval $\Delta t_2 = 1.25$. This is because autoregressions depend on the time interval, and, in general, the smaller the interval, the larger the autoregression, reaching 1 for $t = 0$. To find out whether or not the auto-effects over the intervals are indeed equal, again continuous time analysis is needed to relate and compare the discrete time effects on the same underlying continuous time scale.

The possibly misleading results of a discrete time analysis in case of unequal observation intervals are clearly shown by the autoregression functions A and B in Figure 7.3. By definition, autoregression functions have value 1 at an interval of length 0 (no change) and generally this value decreases, when the observation interval becomes longer. Suppose A is valid in one study and B in a second study, while in study A an observation interval Δt_1 of 0.50 year is used, and in study B an observation interval Δt_2 of 1.00 year. Because autoregression function B exceeds

A everywhere, no other conclusion should be drawn than that the autoregression in study A is lower than in study B. Nevertheless, investigator A, finding the autoregression value of 0.61 at interval $\Delta t_1 = 0.50$, could erroneously conclude that the autoregression in his study is larger than in study B, where the lower value of 0.50 was found at interval $\Delta t_2 = 1.00$. Clearly, the correct answer can only be found in continuous time analysis by comparing the auto-effects in the two studies and generating the complete autoregression functions as in Figure 7.3.

Some discrete time analysts believe that the use of unequal observation intervals is the only culprit and that all problems would be solved by using and making comparisons for equal intervals only. Equal observation intervals are hardly less problematic than unequal observation intervals, however, as will become clear from the two reciprocal cross-lagged effect functions for variables x_1 and x_3 in Figure 7.4, both based on drift matrix **A** in Table 7.1. The cross-lagged effect functions specify the cross-lagged effects, not only for one specific interval ($\Delta t_i = 1$ in Table 7.1) and even not only for all discrete time observation intervals Δt_i in the study. Like autoregression functions, cross-lagged effect functions go through all infinitesimally increasing intervals Δt in continuous time, starting from $\Delta t = 0$. Unlike autoregression functions, which start at value 1, cross-lagged effect functions have starting value 0 (different variables cannot yet have any influence on each other over a zero time interval), build up the effect more or less rapidly until a maximum is reached somewhere (in Figure 7.4 maxima 0.250 and 0.240 are reached at the quite different intervals of $\Delta t = 1.02$ and $\Delta t = 1.64$, respectively), and eventually return to 0 in a stable model. Stability is defined by the eigenvalues of drift matrix **A**. If all eigenvalues have negative real parts, the model is stable. Eigenvalues of **A** can become complex in some situations, but in this chapter only real eigenvalues will be considered .

Autoregression functions as well as cross-lagged effect functions were computed by the matrix exponential in (7.5) which differs from (7.1) merely in allowing Δt to take all values in continuous time:

$$\mathbf{A}_{\Delta t} = \mathrm{e}^{\mathbf{A}\Delta t}. \tag{7.5}$$

Crucial is that, in discrete time research, autoregression matrices $\mathbf{A}_{\Delta t_i}$ are defined and estimated for the observation intervals Δt_i in the study only and are therefore unknown for intervals that are smaller than or unequal to multiples of Δt_i $(i = 1, 2, \ldots T)$, whereas $\mathbf{A}_{\Delta t} = \mathrm{e}^{\mathbf{A}\Delta t}$ in (7.5) is much more generally interpretable and computable for arbitrary continuous time intervals Δt. Basically, what we do in a continuous time analysis of discrete time data is first using (7.1) to find the continuous time drift matrix **A** that fits the empirical observation intervals, and next using (7.5) to generate the complete autoregression and cross-lagged effect functions on the basis of **A**.

A possible and by no means rare property of cross-lagged effect functions is shown in Figure 7.4. They are crossing at $\Delta t = 1.44$, both having the same value 0.239 at that interval. So, although according to **A** in Table 7.1 the effect of x_1 on x_3 is stronger than in the opposite direction from x_3 to x_1 (0.76 compared to 0.44) and

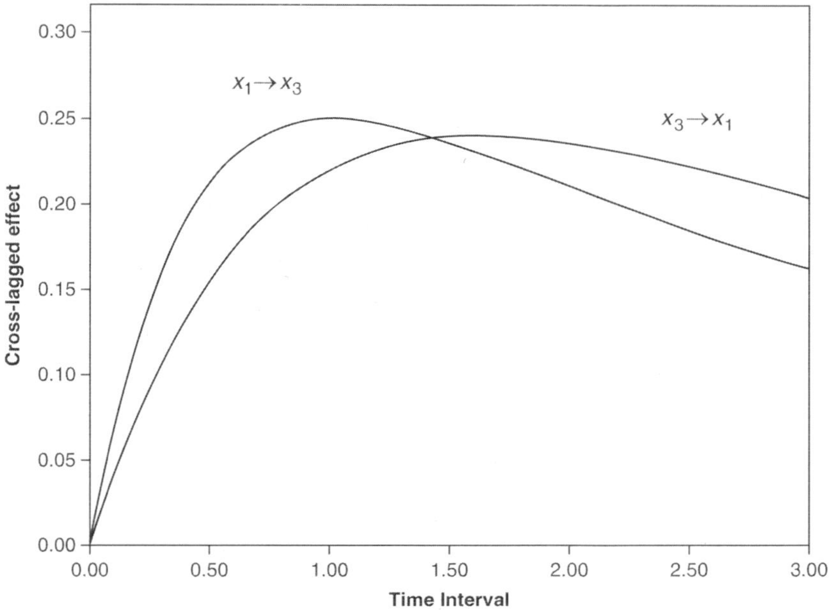

Fig. 7.4 Cross-lagged effect functions for the reciprocal effects between x_1 and x_3, based on the drift matrix in Table 7.1.

the cross-lagged effects in $\mathbf{A}_{\Delta t_i}$ of Table 7.1 indicate the same strength order at time interval $\Delta t = 1.00$ (0.25 compared to 0.22, both displayed also at $\Delta t = 1.00$ in Figure 7.4), the interplay between the variables over continuous time is such that the cross-lagged effects in Figure 7.4 become equally strong at $\Delta t = 1.44$ and even reverse the strength order for intervals $\Delta t > 1.44$. It is this possibility of crossing (non-monotone) cross-lagged effect functions (as well as non-monotone autoregression functions) that makes discrete time analysis useless for analyzing reciprocal relationships in the cross-lagged panel design. The implication of Figure 7.4 is that the relative strength of the reciprocal causal effects found between x_3 and x_1 depends on the observation interval chosen in the study. Investigators choosing their discrete time interval Δt_i between 0 and 1.44 years will come to the conclusion that x_1 has a stronger effect on x_3 (maximum difference of 0.058 reached at $\Delta t_i = 0.46$), whereas investigators choosing $\Delta t_i > 1.44$ years will arrive at the opposite conclusion (maximum difference of 0.042 reached at $\Delta t_i = 3.26$). No comparison problems would arise, at least not in the sense of contradictory results with regard to the strength order, if the cross-lagged effect functions in Figure 7.4, like the autoregression functions in Figure 7.3, were monotone (not crossing). Then it would not matter at what interval the comparison is made, because one would find the same order everywhere. However, the monotone or non-monotone character is seldom known beforehand and generally it is just the purpose of the research to find out.

What is worse is that it would not be of any help to choose and maintain the same observation interval. The cross-lagged effect functions in Figure 7.4 bring discrete time investigators using the same observation interval $\Delta t_i = 1.44$ also to a false conclusion, namely that the effect of x_1 on x_3 is equally strong as the effect of x_3 on x_1. This conclusion cannot be generalized to observation intervals $\Delta t_i \neq 1.44$, however, where different and, for $\Delta t_i > 1.44$, again false conclusions would be drawn, nor is it confirmed by the cross-effect coefficients in drift matrix **A**. Clearly, continuous time analysis, estimating the coefficients of the continuous time drift matrix **A** and displaying the consequences over the complete time axis by means of the full autoregression and cross-lagged effect functions, is the only solution to the problems of unequal as well as equal observation intervals.

7.2.3 Lagged and Instantaneous Effects Dilemma

We conclude the discussion of the cross-lagged panel design with another awkward problem often encountered by discrete time analysts, for which, again, there is no solution in discrete time. As seen in Figure 7.1, the analysis of the cross-lagged panel design allows the inclusion of two kinds of reciprocal effects between x and y: lagged reciprocal effects (i.e., x at time point 1 affecting y at time point 2, and y at time point 1 affecting x at time point 2) and instantaneous reciprocal effects (i.e., x at time point 2 affecting y at time point 2 and vice versa). One could choose the instantaneous coefficients, the lagged coefficients, or both to be present in the model, but the results are often different or even contradictory. This typically confronts the discrete time analyst with a dilemma. In the study by Vuchinich, Bank, and Patterson (1992), for example, the dilemma was whether to choose for instantaneous or lagged effects between parental disciplinary behavior and child antisocial behavior. The authors found significant instantaneous effects but no significant cross-lagged effects. The choice would become even more difficult, if these effects were to be estimated simultaneously, because then the results become highly dependent on the time interval Δt_i. In general, the longer the time interval between measurements, the higher the instantaneous coefficients become in comparison to the lagged coefficients. Most discrete time analysts feel that instantaneous and lagged effects should both be taken into consideration somehow. However, they do not and cannot know in discrete time how to connect and constrain these two types of effects to find the true underlying continuous time effects.

When analysts estimate the instantaneous and lagged effects simultaneously, autoregression equation (7.2) is in fact replaced by

$$\mathbf{x}(t_i) = \mathbf{A}_{ins}\mathbf{x}(t_i) + \mathbf{A}_{lag}\mathbf{x}(t_{i-1}). \tag{7.6}$$

Instantaneous matrix $\mathbf{A}_{ins}$ includes the instantaneous effects between the "current endogenous" variables in $\mathbf{x}(t_i)$. Lagged matrix $\mathbf{A}_{lag}$ includes the lagged effects from the "lagged endogenous" $\mathbf{x}(t_{i-1})$ on the "current endogenous" $\mathbf{x}(t_i)$. Equation (7.6) is known in econometrics as the "structural form" with (7.2) as the associated "reduced form". The clear relationships that exist between the two forms and the continuous time matrix exponential in (7.1), that is, between the coefficients in the reduced form or autoregression matrix $\mathbf{A}_{\Delta t_i}$, in the structural form matrices $\mathbf{A}_{ins}$ and $\mathbf{A}_{lag}$, and in the continuous time drift matrix $\mathbf{A}$, are shown in (7.7):

$$\mathbf{A}_{\Delta t_i} = (\mathbf{I} - \mathbf{A}_{ins})^{-1}\mathbf{A}_{lag} = \mathrm{e}^{\mathbf{A}\Delta t_i}. \tag{7.7}$$

By means of $\mathrm{e}^{\mathbf{A}\Delta t_i}$ in (7.7), which is the core of the exact discrete model EDM, nonlinear constraints are directly imposed on the coefficients in the autoregression matrix $\mathbf{A}_{\Delta t_i}$ for generating the exact drift matrix $\mathbf{A}$, skipping $\mathbf{A}_{ins}$ and $\mathbf{A}_{lag}$. In this way, the above mentioned dilemma of the choice between $\mathbf{A}_{ins}$ and $\mathbf{A}_{lag}$ is simply circumvented. The structural form is therefore not really indispensible in continuous time analysis. One might wonder, however, whether constraints could be imposed on the structural form matrices $\mathbf{A}_{ins}$ and $\mathbf{A}_{lag}$ for combining the instantaneous and lagged effects in an appropriate way to generate the underlying continuous time effects in $\mathbf{A}$ and thereby explicitly solving the discrete time dilemma. This has indeed been done in the so-called approximate discrete model ADM introduced by Bergstrom (1966; 1984, pp. 1172-1173), the same econometrician who originated the EDM. He showed, that by means of the simple linear constraints:

$$\begin{aligned}\mathbf{A}_{ins} &= \tfrac{1}{2}\mathbf{A}_{\dagger}\Delta t_i, \\ \mathbf{A}_{lag} &= \mathbf{I} + \tfrac{1}{2}\mathbf{A}_{\dagger}\Delta t_i,\end{aligned} \tag{7.8}$$

the ADM generates a quite reasonable approximation $\mathbf{A}_{\dagger}$ of exact $\mathbf{A}$. It immediately solves the dilemma of the discrete time analyst, because by means of (7.8) the two different matrices $\mathbf{A}_{ins}$ and $\mathbf{A}_{lag}$ are replaced by one and the same matrix $\mathbf{A}_{\dagger}$, which is the one to be interpreted and tested.

7.2.4 ADM and EDM

It is true that by using the ADM instead of the EDM one sacrifices exactness. However, although nonlinear SEM programs such as Mx, which include the exponential and matrix algebraic functions, can implement the EDM, the linearity of the constraints in (7.8) of the ADM also holds some attraction. Less nonlinearly oriented but more user-friendly SEM programs, which lack the exponential and matrix algebraic functions and therefore the possibility to apply the EDM, mostly allow implementation of the ADM. Oud (2007b) explains in detail how to apply the ADM-SEM procedure by means of LISREL (Jöreskog & Sörbom, 1996). In addition, LISREL and similar programs are particularly valuable in the modeling process, because they provide plenty of information about model fit and about modification results of individual parameters by means of the so-called modification indices. For this reason, it could be worthwhile in practice to first apply the ADM-SEM procedure in the model building phase by means of a program such as LISREL and then the EDM-SEM procedure in the final model estimation phase by means of Mx.

It should be noted that $\mathbf{A}_{\dagger}$, as an approximation of $\mathbf{A}$, compares favorably with other well-known approximations such as the relatively crude approximation $\mathbf{A}_{*} = (\mathbf{A}_{\Delta t_i} - \mathbf{I})/\Delta t_i$ in (7.3). This is seen by putting the exact nonlinear matrix exponential form $\mathbf{A}_{\Delta t_i} = \mathrm{e}^{\mathbf{A}\Delta t_i}$ and both approximate linear constraint forms in power series expansion

$$\begin{aligned} \mathbf{A}_{\Delta t_i} &= \mathrm{e}^{\mathbf{A}\Delta t_i} = \textstyle\sum_{k=0}^{\infty}(\mathbf{A}\Delta t_i)^k/k! \\ &= \mathbf{I} + \mathbf{A}\Delta t_i + \tfrac{1}{2}\mathbf{A}^2\Delta t_i^2 + \tfrac{1}{6}\mathbf{A}^3\Delta t_i^3 + \tfrac{1}{24}\mathbf{A}^4\Delta t_i^4 + \ldots \\ &\qquad\qquad\qquad\qquad\qquad\qquad\qquad\qquad \text{(exact)}, \\ \mathbf{A}_{\Delta t_i} &= (\mathbf{I} - \mathbf{A}_{ins})^{-1}\mathbf{A}_{lag} = (\mathbf{I} - \tfrac{1}{2}\mathbf{A}_{\dagger}\Delta t_i)^{-1}(\mathbf{I} + \tfrac{1}{2}\mathbf{A}_{\dagger}\Delta t_i) \\ &= \mathbf{I} + \mathbf{A}_{\dagger}\Delta t_i + \tfrac{1}{2}\mathbf{A}_{\dagger}^2\Delta t_i^2 + \tfrac{1}{4}\mathbf{A}_{\dagger}^2\Delta t_i^3 + \tfrac{1}{8}\mathbf{A}_{\dagger}^4\Delta t_i^4 + \ldots \\ &\qquad\qquad\qquad\qquad\qquad\qquad (\mathbf{A}_{\dagger}\ \text{approximation}), \\ \mathbf{A}_{\Delta t_i} &= \mathbf{I} + \mathbf{A}_{*}\Delta t_i \quad (\mathbf{A}_{*}\ \text{approximation}). \end{aligned} \tag{7.9}$$

Whereas the $\mathbf{A}_{*}$ approximation truncates the exact infinite series, the weights of the $\mathbf{A}_{\dagger}$ approximation $(\frac{1}{2}, \frac{1}{4}, \frac{1}{8}, \ldots)$ in the ADM are only seen to decrease less quickly than in the exact series $(\frac{1}{2}, \frac{1}{6}, \frac{1}{24}, \ldots)$ used in the EDM. In a simulation study with different estimation procedures, Oud (2007a) concluded that the ADM-SEM procedure did indeed yield more biased results than the EDM-SEM procedure, but that the overall quality in terms of the root mean squared error (RMSE) was hardly lower than in the EDM-SEM procedure. The ADM-SEM procedure compared also favorably with the approximate MLDE procedure of Boker, Neale, and Rausch (2004). In the examples to be presented below, we will first apply the ADM-SEM procedure, followed by the EDM-SEM procedure.

7.3 Linear Stochastic Differential Equation Model

The full linear stochastic differential equation model used in this chapter consists of two equations: a dynamic explanatory equation and a static measurement equation. The dynamic equation, shown in (7.10), extends the basic differential equation model in (7.4) by three important elements.

$$\frac{\mathrm{d}\mathbf{x}(t)}{\mathrm{d}t} = \mathbf{A}\mathbf{x}(t) + \mathbf{b} + \boldsymbol{\kappa} + \mathbf{G}\frac{\mathrm{d}\mathbf{W}(t)}{\mathrm{d}t}. \tag{7.10}$$

In addition to the drift matrix term $\mathbf{A}\mathbf{x}(t)$, introduced and discussed in detail in the previous section, the following new elements are found in (7.10): continuous time intercepts vector **b**, continuous time "trait" variables vector $\boldsymbol{\kappa}$, and finally the continuous time error process vector $\mathbf{G}\frac{\mathrm{d}\mathbf{W}(t)}{\mathrm{d}t}$, which makes the differential equation stochastic.

7.3.1 Continuous Time Intercepts

Nonzero intercepts in vector **b** aptly accommodate for the frequently observed nonzero mean trajectories $E[\mathbf{x}(t)]$. Importantly, **b** defines also the final means towards which a stable system eventually converges. In models with $\mathbf{b} = \mathbf{0}$ these final means are necessarily zero, with zero being the stable equilibrium position. An equilibrium position is the value of a constant curve satisfying the model. In models with $\mathbf{b} \neq \mathbf{0}$, it can first be proven, on the basis of (7.10), that the mean trajectories, starting from initial mean $E[\mathbf{x}(t_0)]$, follow nonzero curves (7.11) and next that convergence is towards nonzero $-\mathbf{A}^{-1}\mathbf{b}$.

$$E[\mathbf{x}(t)] = \mathrm{e}^{\mathbf{A}(t-t_0)}E[\mathbf{x}(t_0)] + \mathbf{A}^{-1}[\mathrm{e}^{\mathbf{A}(t-t_0)} - \mathbf{I}]\mathbf{b} \\ \text{with } E[\mathbf{x}(t \to \infty)] = -\mathbf{A}^{-1}\mathbf{b}. \tag{7.11}$$

The reason for the latter is the behavior of the matrix exponential $\mathrm{e}^{\mathbf{A}(t-t_0)}$ in a stable model. Because **A** in the exponent has negative eigenvalues and is multiplied by the interval length $t-t_0$, the matrix exponential eventually becomes zero (the concept of stability is equivalent to this property) and hence $E[\mathbf{x}(t)] \to -\mathbf{A}^{-1}\mathbf{b}$ for $t \to \infty$. The values in $-\mathbf{A}^{-1}\mathbf{b}$, in addition, are (stable) equilibrium positions. As a result of the commuting property $\mathrm{e}^{\mathbf{A}(t-t_0)}\mathbf{A}^{-1} = \mathbf{A}^{-1}\mathrm{e}^{\mathbf{A}(t-t_0)}$, choosing initial means $E[\mathbf{x}(t_0)] = -\mathbf{A}^{-1}\mathbf{b}$ in (7.11) leaves $E[\mathbf{x}(t)]$ unchanged.

So, the intercepts **b** enhance the flexibility of the model by allowing nonzero mean trajectories and nonzero final means. Flexibility is further enhanced by the possibility of subpopulation specific mean trajectories and final means within the same overall model. For this purpose (n x 1)-vector **b** is replaced by $\mathbf{B}_u\mathbf{u}$, which is also (n x 1) but the product of (n x r)-matrix $\mathbf{B}_u$ of regression coefficients and (r *x* 1)-vector **u** of exogenous variables. Suppose, for example, that boys and girls are assumed to follow a different development and to reach a different final position. As the first element of **u** we choose the unit variable, $u_1 = 1$ for all subjects in the population, and as the second element a dummy-variable, coded $u_2 = 0$ for boys and $u_2 = 1$ for girls. Let us call the first column of $\mathbf{B}_u$ $\mathbf{b}_1$ and the second column $\mathbf{b}_2$. By replacing **b** in (7.11) by $\mathbf{b}_1$ for boys and by $\mathbf{b}_1 + \mathbf{b}_2$ for girls, we then get two sets of n mean trajectories, $E[\mathbf{x}(t)]_{u_2=0}$ for boys and $E[\mathbf{x}(t)]_{u_2=1}$ for girls, and two sets of n final means, $E[\mathbf{x}(t \to \infty)]_{u_2=0} = -\mathbf{A}^{-1}\mathbf{b}_1$ for boys and $E[\mathbf{x}(t \to \infty)]_{u_2=1} = -\mathbf{A}^{-1}(\mathbf{b}_1 + \mathbf{b}_2)$ for girls. By using the same procedure to differentiate $E[\mathbf{x}(t_0)]_{u_2=0}$ for boys from $E[\mathbf{x}(t_0)]_{u_2=1}$ for girls, we additionally let boys and girls start from different positions. The procedure is easily extended for more than two subpopulations, more than two variables in **u** and, in addition to dummy-variables, also for metric variables in $\mathbf{u}$ such as income and, as shown by Oud and Singer (2008), even for changing exogenous variables (time-varying covariates) $\mathbf{u}(t)$. These kinds of models are often called conditional (e.g., see the chapter of Bollen and Zimmer in this volume). For metric variables u_i with many values (subpopulations) represented in the sample, which is typically the case with metric variables as, for example, income, the procedure outlined is often the only possible one. However, in the case of a limited number of subpopulations (e.g., boys and girls), an attractive alternative approach is performing a so-called multisample SEM analysis (Jöreskog & Sörbom, 1996), in which $\mathbf{b}$, $E[\mathbf{x}(t_0)]$, and possibly other parameters are allowed to vary in the subpopulations.

It should be noted that the intercepts in $\mathbf{b}$ are feeding the system continuously over time by a constant amount and therefore indeed result in different contributions from unequal intervals. In a discrete time model the intercepts contribute only at the observation time points chosen.

7.3.2 Continuous Time Trait Variables

Although, as discussed above, subpopulation intercepts allow a different mean trajectory and different final mean in each subpopulation, it is nevertheless paradoxical that a subject's current and future expected behavior should be exclusively determined by the population or subpopulation the subject happens to be modeled a member of. The flexibility of the model is further enhanced by the specification of random subject effects $\boldsymbol{\kappa}$ in (7.10): random intercept variables, called "trait" variables in the present chapter, which define for every subject an own subject-specific mean trajectory. The trait variables $\boldsymbol{\kappa}$, in distinction from the changing "state" variables $\mathbf{x}(t)$, have constant values across time as do the fixed intercepts $\mathbf{b}$. However, whereas the $\mathbf{b}$ are also constant across subjects, the normally distributed trait variables $\boldsymbol{\kappa}$ with mean $E(\boldsymbol{\kappa}) = \mathbf{0}$ and covariance matrix $\Phi_\kappa \neq \mathbf{0}$ have a different value for each subject and so model the subject specific deviations from the common mean defined by $\mathbf{b}$.

The constancy across time implies that $\boldsymbol{\kappa}$ already influences $\mathbf{x}(t)$ before the initial time point t_0, so that $\boldsymbol{\kappa}$ should be considered part of the initial state $\mathbf{x}(t_0)$ and, in general, $\boldsymbol{\kappa}$ and $\mathbf{x}(t_0)$ are correlated ($\Phi_{\mathbf{x}_{t_0},\kappa} \neq \mathbf{0}$). Both the variances of the trait variables in Φ_κ and their covariances with $\mathbf{x}(t_0)$ in $\Phi_{\mathbf{x}_{t_0},\kappa}$ are testable quantities. Both are expected to be nonzero, if subjects do indeed follow their subject-specific mean trajectory instead of coinciding with a single general mean trajectory. Supposing this is indeed the case, the distance between a subject-specific mean trajectory $E[\mathbf{x}(t)|\boldsymbol{\kappa}]$ and the (sub)population mean trajectory $E[\mathbf{x}(t)]$ is computed as

$$\begin{aligned} E[\mathbf{x}(t)|\boldsymbol{\kappa}] - E[\mathbf{x}(t)] &= \mathrm{e}^{\mathbf{A}(t-t_0)}\Phi_{\mathbf{x}_{t_0},\kappa}\Phi_\kappa^{-1}\boldsymbol{\kappa} + \mathbf{A}^{-1}[\mathrm{e}^{\mathbf{A}(t-t_0)} - \mathbf{I}]\boldsymbol{\kappa} \\ \text{with } E[\mathbf{x}(t\to\infty)|\boldsymbol{\kappa}] - E[\mathbf{x}(t\to\infty)] &= -\mathbf{A}^{-1}\boldsymbol{\kappa}. \end{aligned} \tag{7.12}$$

Here the first term on the right-hand side is a consequence of the regression of $\mathbf{x}(t_0)$ on $\boldsymbol{\kappa}$. As a result of the matrix exponential going again to zero for $t \rightarrow \infty$ in a stable model, the distance between the subject-specific and the (sub)population mean trajectories goes to a constant nonzero value: $-\mathbf{A}^{-1}\boldsymbol{\kappa}$.

It should be noted that the mean trajectories for (sub)populations or subjects are not only interesting as such, and in many cases even the main purpose of the study, but they also play a crucial role in the behavior of the estimated sample trajectories (conditional means $E[\mathbf{x}(t)|\mathbf{y}]$, where $\mathbf{y}$ is the total data vector of the subject). The reason is that these regress towards the mean trajectories (in a stable model) or egress from them (in an unstable model). Particularly, if a model contains trait variables, a subject's conditional mean regresses towards its own subject-specific mean trajectory (see (7.12)), whereas in a pure state model all subjects regress towards one and the same general mean trajectory (see (7.11)). Figure 7.5, taken from a youth delinquency study, shows the estimates of a mean trajectory, a subject-specific mean trajectory, and the subject's estimated sample trajectory. Outside of the measurement time points (recognizable by the kinks in the curve) the subject's sample trajectory is clearly seen to regress towards its subject-specific mean trajectory. The consequences are particularly dramatic for predictions. As the final values in the study are 2.12 for the mean trajectory and 3.88 for the subject-specific mean trajectory, the predicted final value for the subject differs no less than 1.76 from the one that would be found in a pure state model.

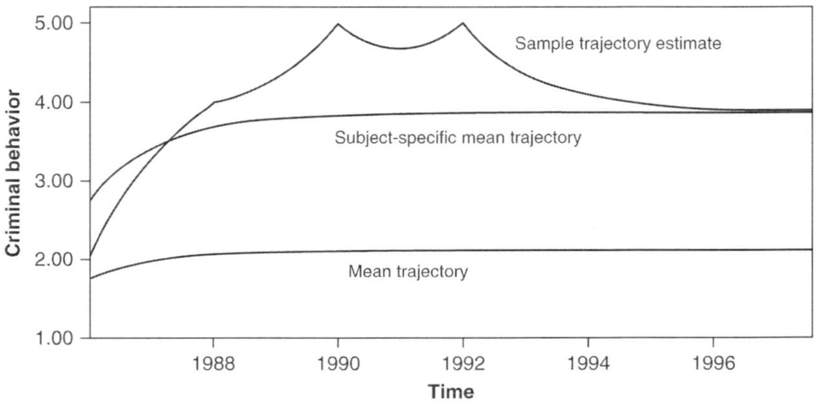

Fig. 7.5 Examples of mean trajectory, subject-specific mean trajectory and sample trajectory estimate.

7.3.3 Continuous Time Error Process

It hardly needs comment that the introduction of an error term into a longitudinal model enhances the applicability of the model. Historically, it took quite some time, however, to define continuous time error in a mathematically rigorous way. The continuous time error process $\mathbf{G}\frac{\mathrm{d}\mathbf{W}(t)}{\mathrm{d}t}$ in (7.10) builds upon the famous Wiener process $\mathbf{W}(t)$, the random walk through continuous time. At first sight $\mathbf{G}\frac{\mathrm{d}\mathbf{W}(t)}{\mathrm{d}t}$ looks quite complicated. Note, however, that discrete time error can be thought of analogously as the difference quotient $\frac{\Delta \mathbf{w}_t}{\Delta t}$ of a discrete time random walk $\mathbf{w}_t = \mathbf{w}_{t-\Delta t} + \mathbf{e}$. If the step sizes $\mathbf{e}$ for $\Delta t = 1$ are randomly drawn from a standard multinormal distribution $N(\mathbf{0}, \mathbf{I})$, successive non-overlapping increments $\Delta \mathbf{w}_t = \mathbf{w}_t - \mathbf{w}_{t-\Delta t}$ for $\Delta t \geq 1$ are independent with covariance matrix $\Delta t\mathbf{I}$. If one wants to model nonstandard error with larger or smaller variance than 1 for $\Delta t = 1$ and possibly correlated elements, but no change in the other properties, then the difference quotient for $\Delta t = 1$ could first be multiplied by lower triangular matrix $\mathbf{G}$, Cholesky factor of the desired covariance matrix $\mathbf{Q}$.

The properties defining the standard Wiener process $\mathbf{W}(t)$ are, in addition to its sample trajectories being continuous and starting at $\mathbf{W}(0) = \mathbf{0}$ (both with probability 1), precisely the conditions of independently and normally distributed increments $\Delta\mathbf{W}(t) = \mathbf{W}(t) - \mathbf{W}(t - \Delta t)$ with mean $\mathbf{0}$ and covariance matrix $\Delta t\mathbf{I}$ (Arnold, 1974, p. 46; Kuo, 2006, p. 7). The lower triangular matrix $\mathbf{G}$ in the continuous time error process $\mathbf{G}\frac{\mathrm{d}\mathbf{W}(t)}{\mathrm{d}t}$ is just there to allow increment variances to become lower or higher than 1 for $\Delta t = 1$ and to get nonzero correlations between elements. Product $\mathbf{Q} = \mathbf{G}\mathbf{G}'$ is the continuous time error covariance matrix, called "diffusion" matrix, and $\mathbf{G}$ the Cholesky factor of $\mathbf{Q}$. So, $\mathbf{Q}$ and $\mathbf{G}$ provide the same information and are easily expressed into each other.

The fame of the Wiener process is undoubtedly due to two peculiar facts that have given rise to a host of mathematical research. Its derivative $\frac{\mathrm{d}\mathbf{W}(t)}{\mathrm{d}t}$ or "white noise" cannot be defined as a derivative in the normal sense nor can the stochastic integral $\int_{t_0}^{t} \mathbf{F}(s)\mathrm{d}\mathbf{W}(s)$ in terms of a possibly time-varying function $\mathbf{F}(t)$ be defined as an ordinary integral. Solution (7.13) of stochastic differential equation (7.10) (see e.g., Arnold, 1974, pp. 128-134) is nevertheless seen to contain this type of integral for the error component. Defined in a proper way, however, its correct covariance matrix can be derived as given in (7.13):

$$\begin{aligned}
\mathbf{x}(t) &= \mathrm{e}^{\mathbf{A}(t-t_0)}\mathbf{x}(t_0) + \mathbf{A}^{-1}[\mathrm{e}^{\mathbf{A}(t-t_0)} - \mathbf{I}](\mathbf{b} + \boldsymbol{\kappa}) + \int_{t_0}^{t} \mathrm{e}^{\mathbf{A}(t-s)}\mathbf{G}\mathrm{d}\mathbf{W}(s) \\
&\text{with } \operatorname{cov} \int_{t_0}^{t} \mathrm{e}^{\mathbf{A}(t-s)}\mathbf{G}\mathrm{d}\mathbf{W}(s) = \int_{t_0}^{t} \mathrm{e}^{\mathbf{A}(t-s)}\mathbf{Q}\mathrm{e}^{\mathbf{A}'(t-s)}\mathrm{d}s \\
&\qquad = \operatorname{irow}\{\mathbf{A}_{\#}^{-1}[\mathrm{e}^{\mathbf{A}_{\#}(t-t_0)} - \mathbf{I}]\operatorname{row}\mathbf{Q}\} \\
&\qquad \text{for } \mathbf{Q} = \mathbf{G}\mathbf{G}' \text{ and } \mathbf{A}_{\#} = \mathbf{A} \otimes \mathbf{I} + \mathbf{I} \otimes \mathbf{A}.
\end{aligned} \tag{7.13}$$

Here $\otimes$ is the Kronecker product (postmultiplying each element of the former matrix by the latter matrix), "row" the rowvec operation (putting the elements of a matrix row-wise in a column vector), and "irow" the inverse operation (putting the elements back into the matrix again).

Note in solution (7.13) the predominant role of the matrix exponential having drift matrix $\mathbf{A}$ in the exponent, which appears in all three terms in the solution. Note also the similar structure of the integral expression $\mathbf{A}^{-1}[\mathrm{e}^{\mathbf{A}(t-t_0)} - \mathbf{I}]$ in the second term and $\mathbf{A}_{\#}^{-1}[\mathrm{e}^{\mathbf{A}_{\#}(t-t_0)} - \mathbf{I}]$ in the covariance matrix of the error term, replacing (nxn) matrix $\mathbf{A}$ in the second term by $(n^2 x\, n^2)$ matrix $\mathbf{A}_{\#}$ in the covariance matrix. $\mathbf{A}_{\#}$ is based again on drift matrix $\mathbf{A}$ and has all eigenvalues negative, if $\mathbf{A}$ has all eigenvalues negative. Therefore, analogously to final mean $-\mathbf{A}^{-1}\mathbf{b}$, the final error covariance matrix in a stable model is given by $\mathrm{irow}(-\mathbf{A}_{\#}^{-1}\mathrm{row}\,\mathbf{Q})$. If, for $\boldsymbol{\kappa} = \mathbf{0}$, the process starts with mean $-\mathbf{A}^{-1}\mathbf{b}$ and covariance matrix $\mathrm{irow}(-\mathbf{A}_{\#}^{-1}\mathrm{row}\,\mathbf{Q})$, the process is stationary, keeping the same mean and covariance matrix. Details about solution (7.13) can be found in Singer (1990) and Oud and Jansen (2000). It has the form that allows all parameters of the model to be estimated by means of the EDM as will be shown in the next section. Observe also, that the mean trajectory (7.11) is an immediate derivation from solution (7.13).

The rationale behind the approximate ADM procedure is different. The ADM is not based on the differential equation solution (7.13), where $\mathbf{x}(t)$ appears only on one side of the equation, but puts differential equation (7.10) first in integral form:

$$\begin{aligned}
&\textstyle\int_{t-\Delta t}^{t} \mathrm{d}\mathbf{x}(s) = \mathbf{A}\int_{t-\Delta t}^{t} \mathbf{x}(s)\mathrm{d}(s) + (\mathbf{b}+\boldsymbol{\kappa})\Delta t + \mathbf{G}[\mathbf{W}(t) - \mathbf{W}(t-\Delta t)].\\
&\text{or}\\
&\mathbf{x}(t) = \mathbf{x}(t-\Delta t) + \mathbf{A}\textstyle\int_{t-\Delta t}^{t} \mathbf{x}(s)\mathrm{d}(s) + (\mathbf{b}+\boldsymbol{\kappa})\Delta t + \mathbf{G}\Delta\mathbf{W}(t).
\end{aligned} \tag{7.14}$$

It next replaces the integral $\int_{t-\Delta t}^{t} \mathbf{x}(s)\mathrm{d}(s)$ on the right-hand side, having $\mathbf{x}(t)$ still inside of the integral, by the so-called trapezoid approximation $\frac{1}{2}[\mathbf{x}(t)+\mathbf{x}(t-\Delta t)]\Delta t$, which multiplies the length Δt of the integration interval by the average value at the end points. This gives rise to the approximate solution

$$\begin{aligned}
\mathbf{x}(t) \approx [\tfrac{1}{2}\mathbf{A}\Delta t]\mathbf{x}(t) + [\mathbf{I} + \tfrac{1}{2}\mathbf{A}\Delta t]\mathbf{x}(t-\Delta t) + (\mathbf{b}+\boldsymbol{\kappa})\Delta t + \mathbf{G}\Delta\mathbf{W}(t)\\
\text{with } \mathrm{cov}[\mathbf{G}\Delta\mathbf{W}(t)] = \mathbf{G}\mathbf{G}'\Delta t.
\end{aligned} \tag{7.15}$$

It explains the constraints imposed in (7.8) on the instantaneous and lagged coefficients for obtaining approximation $\mathbf{A}_{\dagger}$ of $\mathbf{A}$.

7.3.4 Measurement Equation

Latent variables abound in social science. It is probably no exaggeration to claim that the greater part of psychology and sociology draws on latent variables. Some of the latent variables, such as the trait variables in $\boldsymbol{\kappa}$ of (7.10) or the state variables $\mathbf{x}(t)$ in between measurement time points t_i, have no direct connection at all to the observed variables. For the latent state variables at the measurement points

t_i, however, we need to extend the model with a measurement equation, specifying how each of them is connected to the directly observed variables in $\mathbf{y}_{t_i}$:

$$\mathbf{y}_{t_i} = \mathbf{C}\mathbf{x}(t_i) + \mathbf{d} + \mathbf{v}_{t_i} \quad \text{with} \quad \text{cov}(\mathbf{v}_{t_i}) = \mathbf{R}. \tag{7.16}$$

Parameter matrix $\mathbf{C}$ specifies the loadings of the observed variables on the latent variables, parameter vector $\mathbf{d}$ the measurement intercepts or origins, and $\mathbf{R}$ the measurement error variances of the observed variables. If the state variables at the measurement time points are all observed, $\mathbf{y}_{t_i} = \mathbf{x}(t_i)$ and we specify $\mathbf{C} = \mathbf{I}$, $\mathbf{d} = \mathbf{0}$, $\mathbf{R} = \mathbf{0}$.

For identification reasons it is customary to fix, for each latent variable, one of the loadings in $\mathbf{C}$ at 1 and one of the measurement origins in $\mathbf{d}$ at 0. These values fix the measurement scale of the latent variable, 1 in $\mathbf{C}$ giving the latent variable, apart from measurement error variance, the same variance as the observed variable involved and 0 in $\mathbf{d}$ giving the same mean. Observe that the absence of a time index for the measurement parameter matrices and vector $\mathbf{C}$, $\mathbf{d}$, and $\mathbf{R}$ makes them time-invariant. Although time-invariance of the measurement model (measurement invariance) is no strict requirement, it is nevertheless extremely important for making sure that the latent variables keep the same meaning over time. One should have convincing reasons to deviate from measurement invariance for specific variables. Therefore, in the further development of the model, we will assume time-invariance.

7.4 Model Estimation by Means of SEM

In this section, based on the exact differential equation solution (7.13) and the approximate integral form (7.15), respectively, the full EDM and the full ADM will be formulated. Next, for estimation by means of a SEM program, all EDM or ADM parameter matrices will be put into inclusive SEM parameter matrices.

7.4.1 Full EDM

As will be clear from the subscripts t_i and Δt_i in the full EDM (7.17), the EDM is a discrete time model. The matrices with subscript Δt_i in (7.17) are defined for the discrete-time measurement time points only. Simultaneously, however, the EDM covers the continuous time model because of the nonlinear constraints imposed on the discrete time matrices in terms of the continuous time matrices from differential equation (7.10). It means that by applying the constraints on the discrete time matrices $\mathbf{A}_{\Delta t_i}, \mathbf{b}_{\Delta t_i}, \mathbf{H}_{\Delta t_i}, \mathbf{Q}_{\Delta t_i}$ during estimation, we simultaneously estimate the underlying continuous time parameter matrices $\mathbf{A}, \mathbf{b}, \Phi_{\kappa}, \Phi_{\mathbf{x}_{t_0},\kappa}$, $\mathbf{Q} = \mathbf{G}\mathbf{G}'$ (for convenience, vectors $\mathbf{b}_{\Delta t_i}$ and $\mathbf{b}$ are called matrices). The connection between the differential equation and the EDM is made by the exact solution (7.13) (choosing for

$t - t_0$ the observation intervals $\Delta t_i = t_i - t_{i-1}$, starting with $\Delta t_1 = t_1 - t_0$). All constraints on the discrete time matrices in EDM (7.17) are directly taken from exact solution (7.13).

For the computation of the matrix exponential $e^{\mathbf{A}\Delta t_i}$ in (7.17) the diagonalization method is used, which reduces the computation to scalar exponentials. After first diagonalizing $\mathbf{A} = \mathbf{MVM}^{-1}$ ($\mathbf{M}$ eigenvector matrix and $\mathbf{V}$ diagonal eigenvalue matrix of $\mathbf{A}$), next the scalar exponentials in diagonal matrix $e^{\mathbf{V}\Delta t_i}$ are computed, which finally is premultiplied by $\mathbf{M}$ and postmultiplied by $\mathbf{M}^{-1}$. SEM programs such as Mx do not allow to compute the matrix exponential directly, but allow matrix diagonalization and provide the scalar exponential function.

There are two options with regard to the trait covariance matrices Φ_κ and $\Phi_{\mathbf{x}_{t_0},\kappa}$: either you impose constraints on the discrete time analogues $\Phi_{\kappa\Delta t_i}$ and $\Phi_{\mathbf{x}_{t_0},\kappa\Delta t_i}$ separately in the forms shown in (7.17) or you constrain the coefficient matrix $\mathbf{H}_{\Delta t_i}$ of $\boldsymbol{\kappa}$ once, as a result of which both Φ_κ and $\Phi_{\mathbf{x}_{t_0},\kappa}$ come out in the right form automatically. The latter option is easiest and used here.

$$\begin{aligned}
\mathbf{x}_{t_i} &= \mathbf{A}_{\Delta t_i}\mathbf{x}_{t_i-\Delta t_i} + \mathbf{b}_{\Delta t_i} + \mathbf{H}_{\Delta t_i}\boldsymbol{\kappa} + \mathbf{w}_{t_i-\Delta t_i} \\
&\qquad\qquad \text{with } \operatorname{cov}(\mathbf{w}_{t_i-\Delta t_i}) = \mathbf{Q}_{\Delta t_i}, \\
\mathbf{A}_{\Delta t_i} &= e^{\mathbf{A}\Delta t_i} = \mathbf{M}e^{\mathbf{V}\Delta t_i}\mathbf{M}^{-1}, \\
\mathbf{b}_{\Delta t_i} &= \mathbf{A}^{-1}(e^{\mathbf{A}\Delta t_i} - \mathbf{I})\mathbf{b}, \\
\mathbf{H}_{\Delta t_i} &= \mathbf{A}^{-1}(e^{\mathbf{A}\Delta t_i} - \mathbf{I}), \\
\Phi_{\kappa\Delta t_i} &= \mathbf{H}_{\Delta t_i}\Phi_\kappa\mathbf{H}'_{\Delta t_i}, \\
\Phi_{\mathbf{x}_{t_0},\kappa\Delta t_i} &= \Phi_{\mathbf{x}_{t_0},\kappa}\mathbf{H}'_{\Delta t_i}, \\
\mathbf{Q}_{\Delta t_i} &= \operatorname{irow}\{\mathbf{A}_\#^{-1}[e^{\mathbf{A}_\#\Delta t_i} - \mathbf{I}]\operatorname{row}\mathbf{Q}\} \\
&\qquad \text{with } \mathbf{Q} = \mathbf{GG}' \text{ and } \mathbf{A}_\# = \mathbf{A}\otimes\mathbf{I} + \mathbf{I}\otimes\mathbf{A}.
\end{aligned} \tag{7.17}$$

Evidently, the EDM repeats equation (7.17) for successive observation intervals $\Delta t_i = \Delta t_1, \ldots, \Delta t_{T-1}$ (T the total number of observation time points). If the observation intervals are unequal, the discrete time matrices with subscript Δt_i are different across time but relate nonlinearly in terms of the common time-invariant continuous time matrices $\mathbf{A}$, $\mathbf{b}$, Φ_κ, $\Phi_{\mathbf{x}_{t_0},\kappa}$, $\mathbf{G}$. If the observation intervals are equal, simple equality constraints between the discrete time matrices of successive observation intervals would suffice too. In addition to the direct estimation method, therefore, the indirect method would become applicable: computing the estimates of the continuous time matrices on the basis of the five previously estimated discrete time matrices by applying the constraints in (7.17) in inverse direction. In particular, starting from

$$\mathbf{A} = \frac{1}{\Delta t_i}\ln\mathbf{A}_{\Delta t_i} = \frac{1}{\Delta t_i}\mathbf{M}\ln(\mathbf{V}_{\Delta t_i})\mathbf{M}^{-1} \tag{7.18}$$

by diagonalizing $\mathbf{A}_{\Delta t_i} = \mathbf{MV}_{\Delta t_i}\mathbf{M}^{-1}$ ($\mathbf{M}$ eigenvector matrix and $\mathbf{V}_{\Delta t_i} = e^{\mathbf{V}\Delta t_i}$ diagonal eigenvalue matrix of $\mathbf{A}_{\Delta t_i}$), $\mathbf{A}$ is found and then the other continuous time matrices easily follow. It cannot be emphasized enough that the indirect method is only applicable in the case of equal observation intervals. As in the case of unequal intervals no equality constraints can be imposed, each interval yields a different set of discrete time matrices, and, as a result of sampling fluctuations, each interval would also yield a different set of continuous time parameter matrices. So, in the case of unequal observation intervals, the direct method is the only suitable one.

In addition to the continuous time parameter matrices and the $T-1$ times repeated discrete time matrices in (7.17), the EDM as well as the ADM need one more parameter vector and one more parameter matrix for the initial time point t_0: initial means vector $\boldsymbol{\mu}_{\mathbf{x}_{t_0}}$ and initial covariance matrix $\Phi_{\mathbf{x}_{t_0}}$.

7.4.2 Full ADM

With regard to (7.19), the analogue of (7.17) for the ADM, which is directly taken from approximate integral form (7.15), the following observations apply. First, whereas the EDM (7.17) is formulated as a reduced form equation, the ADM (7.19) is in structural form. It means that the single autoregression matrix $\mathbf{A}_{\Delta t_i}$ in the EDM is replaced by two matrices in the ADM: instantaneous $\mathbf{A}^*_{\Delta t_i}$ and lagged $\mathbf{A}^{**}_{\Delta t_i}$. Both have been discussed earlier in Subsection 7.2.3, called there $\mathbf{A}_{ins}$ and $\mathbf{A}_{lag}$. The move from reduced form to structural form in combination with the replacement of exact drift matrix $\mathbf{A}$ by approximation $\mathbf{A}_\dagger$ leads to a dramatic simplification of the constraints on the discrete time matrices. The complicated nonlinear constraints in EDM (7.17) are replaced in ADM (7.19) by extremely simple linear expressions in terms of just the observation interval Δt_i or $\frac{1}{2}\Delta t_i$. Whereas the EDM constraints can only be applied by SEM programs such as Mx, which provide the exponential and matrix algebraic functions needed, the ADM constraints are applicable by almost any SEM program, in particular also by LISREL.

$$
\begin{aligned}
\mathbf{x}_{t_i} &= \mathbf{A}^*_{\Delta t_i}\mathbf{x}_{t_i} + \mathbf{A}^{**}_{\Delta t_i}\mathbf{x}_{t_i-\Delta_{t_i}} + \mathbf{b}^*_{\Delta t_i} + \mathbf{H}^*_{\Delta t_i}\boldsymbol{\kappa} + \mathbf{w}^*_{t_i-\Delta t_i} \\
&\qquad\qquad\qquad\qquad \text{with } \operatorname{cov}(\mathbf{w}^*_{t_i-\Delta t_i}) = \mathbf{Q}^*_{\Delta t_i}, \\
\mathbf{A}^*_{\Delta t_i} &= \tfrac{1}{2}\Delta t_i\mathbf{A}_\dagger, \\
\mathbf{A}^{**}_{\Delta t_i} &= \mathbf{I} + \tfrac{1}{2}\Delta t_i\mathbf{A}_\dagger, \\
\mathbf{b}^*_{\Delta t_i} &= \Delta t_i\mathbf{b}_\dagger, \\
\mathbf{H}^*_{\Delta t_i} &= \Delta t_i\mathbf{I}, \\
\Phi^*_{\kappa\Delta t_i} &= \Delta t_i^2\Phi_{\dagger\kappa}, \\
\Phi^*_{\mathbf{x}_{t_0},\kappa\Delta t_i} &= \Delta t_i\Phi_{\dagger\mathbf{x}_{t_0},\kappa}, \\
\mathbf{Q}^*_{\Delta t_i} &= \Delta t_i\mathbf{Q}_\dagger = \Delta t_i\mathbf{G}_\dagger\mathbf{G}'_\dagger.
\end{aligned} \tag{7.19}
$$

Again, if the observation intervals are equal, it becomes possible to extract the approximate continuous parameter matrices $\mathbf{A}_\dagger$, $\mathbf{b}_\dagger$, $\Phi_{\dagger\kappa}$, $\Phi_{\dagger\mathbf{x}_{t_0},\kappa}$, $\mathbf{G}_\dagger$ from the previously estimated set of structural form matrices by applying the simple constraints in

(7.19) in inverse direction. In fact, after applying during estimation, in addition to the equality constraints, the simple and only ADM constraints:

$$\mathbf{A}^{**}_{\Delta t_i} = \mathbf{I} + \mathbf{A}^{*}_{\Delta t_i}, \tag{7.20}$$

it comes down to dividing the left hand sides in (7.19) by $\frac{1}{2}\Delta t_i$, Δt_i, Δt_i^2, respectively, and finally computing the Cholesky factor $\mathbf{G}_{\dagger}$ of $\mathbf{Q}_{\dagger}$.

There is more, however, in the case of equal observation intervals. From structural form (7.19), by means of transformation matrix $\mathbf{D} = (\mathbf{I} - \mathbf{A}^{*}_{\Delta t_i})^{-1}$, we obtain the reduced form (7.21) which is in the form of (7.17). Because, as a result of the applied equality constraints, the reduced form matrices are equal across the successive time observation intervals, one-to-one relationships can be built between the ADM solutions, the reduced form solutions, and the EDM solutions, every reduced form solution giving rise to just one EDM solution and just one ADM solution and vice versa. As the confrontation with the data takes place via the common reduced form solution, the corresponding ADM and EDM solutions are equivalent, giving exactly the same model fit.

$$\begin{aligned}\mathbf{x}_{t_i} = \mathbf{D}\mathbf{A}^{**}_{\Delta t_i}\mathbf{x}_{t_i-\Delta_{t_i}} + \mathbf{D}\mathbf{b}^{*}_{\Delta t_i} + \mathbf{D}\mathbf{H}^{*}_{\Delta t_i}\boldsymbol{\kappa} + \mathbf{D}\mathbf{w}^{*}_{t_i-\Delta t_i} \\ \text{with } \operatorname{cov}(\mathbf{D}\mathbf{w}^{*}_{t_i-\Delta t_i}) = \mathbf{D}\mathbf{Q}^{*}_{\Delta t_i}\mathbf{D}'.\end{aligned} \tag{7.21}$$

In practice it means that one could start with the relatively simple ADM solution (7.19) by means of LISREL or some other user-friendly SEM program. Next, one could derive its reduced form using (7.21) and finally compute the corresponding EDM solution by means of the inverse constraints in (7.17) without any new SEM analysis. It should be noted that this is not possible in the case of unequal observation intervals, because then the indirect method is no option nor are the ADM and EDM solutions equivalent. Even then, however, it is often profitable to start with the relatively simple ADM solution to explore and evaluate the model and then use it as a reasonable initial solution for the final EDM analysis by means of the Mx program.

7.4.3 Putting ADM and EDM into SEM

A SEM model often can be specified in quite different ways and by different numbers of parameter matrices. Here we will put the ADM and EDM each into two equations with four parameter matrices: measurement parameter matrices $\boldsymbol{\Lambda}$ and $\boldsymbol{\Theta}$, and structural parameter matrices $\mathbf{B}$ and $\boldsymbol{\Psi}$:

$$\mathbf{y} = \boldsymbol{\Lambda}\boldsymbol{\eta} + \boldsymbol{\varepsilon} \quad \text{with } \operatorname{cov}(\boldsymbol{\varepsilon}) = \boldsymbol{\Theta}, \tag{7.22}$$

$$\boldsymbol{\eta} = \mathbf{B}\boldsymbol{\eta} + \boldsymbol{\zeta} \quad \text{with } \operatorname{cov}(\boldsymbol{\zeta}) = \boldsymbol{\Psi}. \tag{7.23}$$

For clarity, we limit the presentation to the case of four time points ($T = 4$: t_0, t_1, t_2, t_3) but this is easily reduced to three or extended to more than four time points. The model implied moment matrix $\mathbf{\Sigma} = \mathrm{f}(\mathbf{\Lambda}, \mathbf{\Theta}, \mathbf{B}, \mathbf{\Psi})$ is a function of the parameter matrices, the likelihood in turn is a function of $\mathbf{\Sigma}$ and sample moment matrix $\mathbf{S}$, and the maximum likelihood solution minimizes the discrepancy between $\mathbf{\Sigma}$ and $\mathbf{S}$ in the ML sense. Hence, for obtaining the maximum likelihood estimate of the ADM or EDM by means of a SEM program, it suffices to show how ADM and EDM are put into SEM parameter matrices $\mathbf{\Lambda}$, $\mathbf{\Theta}$, $\mathbf{B}$, and $\mathbf{\Psi}$. As the ADM is slightly simpler than the EDM, we start with the ADM.

For four time points the vectors $\mathbf{y}, \boldsymbol{\varepsilon}, \boldsymbol{\eta}, \boldsymbol{\zeta}$ in (7.22) and (7.23) look like:

$$\mathbf{y} = \begin{bmatrix} \mathbf{y}_{t_0} \\ \mathbf{y}_{t_1} \\ \mathbf{y}_{t_2} \\ \mathbf{y}_{t_3} \\ 1 \end{bmatrix}, \; \boldsymbol{\varepsilon} = \begin{bmatrix} \boldsymbol{\varepsilon}_{t_0} \\ \boldsymbol{\varepsilon}_{t_1} \\ \boldsymbol{\varepsilon}_{t_2} \\ \boldsymbol{\varepsilon}_{t_3} \\ 0 \end{bmatrix}, \; \boldsymbol{\eta} = \begin{bmatrix} \mathbf{x}_{t_0} \\ \mathbf{x}_{t_1} \\ \mathbf{x}_{t_2} \\ \mathbf{x}_{t_3} \\ 1 \\ \boldsymbol{\kappa} \end{bmatrix}, \; \boldsymbol{\zeta} = \begin{bmatrix} \mathbf{x}_{t_0} - \boldsymbol{\mu}_{\mathbf{x}_{t_0}} \\ \mathbf{w}_{t_1 - \Delta_{t_1}} \\ \mathbf{w}_{t_2 - \Delta_{t_2}} \\ \mathbf{w}_{t_3 - \Delta_{t_3}} \\ 1 \\ \boldsymbol{\kappa} \end{bmatrix}. \tag{7.24}$$

If the total number of variables in $\mathbf{y}$, the vector of observed variables including as the last variable the unit variable (1 for every subject in the sample), is $Tm + 1$, the total number of variables in $\boldsymbol{\eta}$, the vector of latent variables, is $(T+1)n + 1$ with n the number of state variables as well as trait variables. Hence, case $m = n$ (e.g., when all state variables are observed) is one example in which the total number of latent variables may exceed the total number of observed variables.

$$\mathbf{B} = \begin{bmatrix} \mathbf{0} & \mathbf{0} & \mathbf{0} & \mathbf{0} & \boldsymbol{\mu}_{\mathbf{x}_{t_0}} & \mathbf{0} \\ \mathbf{A}^{**}_{\Delta t_1} & \mathbf{A}^{*}_{\Delta t_1} & \mathbf{0} & \mathbf{0} & \mathbf{b}^{*}_{\Delta t_1} & \mathbf{H}^{*}_{\Delta t_1} \\ \mathbf{0} & \mathbf{A}^{**}_{\Delta t_2} & \mathbf{A}^{*}_{\Delta t_2} & \mathbf{0} & \mathbf{b}^{*}_{\Delta t_2} & \mathbf{H}^{*}_{\Delta t_2} \\ \mathbf{0} & \mathbf{0} & \mathbf{A}^{**}_{\Delta t_3} & \mathbf{A}^{*}_{\Delta t_3} & \mathbf{b}^{*}_{\Delta t_3} & \mathbf{H}^{*}_{\Delta t_3} \\ \mathbf{0} & \mathbf{0} & \mathbf{0} & \mathbf{0} & 0 & \mathbf{0} \\ \mathbf{0} & \mathbf{0} & \mathbf{0} & \mathbf{0} & \mathbf{0} & \mathbf{0} \end{bmatrix},$$

$$\mathbf{\Psi} = \begin{bmatrix} \mathbf{\Phi}_{\mathbf{x}_{t_0}} & & & & & \\ \mathbf{0} & \mathbf{Q}^{*}_{\Delta t_1} & & & & \\ \mathbf{0} & \mathbf{0} & \mathbf{Q}^{*}_{\Delta t_2} & & & \\ \mathbf{0} & \mathbf{0} & \mathbf{0} & \mathbf{Q}^{*}_{\Delta t_3} & & \\ \mathbf{0} & \mathbf{0} & \mathbf{0} & \mathbf{0} & 1 & \\ \mathbf{\Phi}_{\dagger \mathbf{x}_{t_0}, \kappa} & \mathbf{0} & \mathbf{0} & \mathbf{0} & \mathbf{0} & \mathbf{\Phi}_{\dagger \kappa} \end{bmatrix}, \tag{7.25}$$

$$\mathbf{\Lambda} = \begin{bmatrix} \mathbf{C} & \mathbf{0} & \mathbf{0} & \mathbf{0} & \mathbf{d} & \mathbf{0} \\ \mathbf{0} & \mathbf{C} & \mathbf{0} & \mathbf{0} & \mathbf{d} & \mathbf{0} \\ \mathbf{0} & \mathbf{0} & \mathbf{C} & \mathbf{0} & \mathbf{d} & \mathbf{0} \\ \mathbf{0} & \mathbf{0} & \mathbf{0} & \mathbf{C} & \mathbf{d} & \mathbf{0} \\ \mathbf{0} & \mathbf{0} & \mathbf{0} & \mathbf{0} & 1 & \mathbf{0} \end{bmatrix}, \; \mathbf{\Theta} = \begin{bmatrix} \mathbf{R} & & & & \\ \mathbf{0} & \mathbf{R} & & & \\ \mathbf{0} & \mathbf{0} & \mathbf{R} & & \\ \mathbf{0} & \mathbf{0} & \mathbf{0} & \mathbf{R} & \\ \mathbf{0} & \mathbf{0} & \mathbf{0} & \mathbf{0} & 0 \end{bmatrix}.$$

The ADM (7.19) is put into the SEM model matrices $\mathbf{B}$, $\boldsymbol{\Psi}$, $\boldsymbol{\Lambda}$, $\boldsymbol{\Theta}$ in the way shown in (7.25). Observe that, due to the specification of the matrices $\mathbf{H}^*_{\Delta t_i}$ in SEM matrix $\mathbf{B}$, the trait covariance matrices $\Phi_{\dagger\kappa}$ and $\Phi_{\dagger\mathbf{x}_{t_0},\kappa}$ appear directly in SEM matrix $\boldsymbol{\Psi}$.

For the EDM, an alternative specification of the trait variables in the latent vector $\boldsymbol{\eta}$ and its error vector $\boldsymbol{\zeta}$ is employed.

$$\boldsymbol{\eta} = \begin{bmatrix} \begin{bmatrix} \mathbf{x}_{t_0} \\ \boldsymbol{\kappa} \end{bmatrix} \\ \begin{bmatrix} \mathbf{x}_{t_1} \\ \boldsymbol{\kappa} \end{bmatrix} \\ \begin{bmatrix} \mathbf{x}_{t_2} \\ \boldsymbol{\kappa} \end{bmatrix} \\ \begin{bmatrix} \mathbf{x}_{t_3} \\ \boldsymbol{\kappa} \end{bmatrix} \\ 1 \end{bmatrix}, \quad \boldsymbol{\zeta} = \begin{bmatrix} \begin{bmatrix} \mathbf{x}_{t_0} - \boldsymbol{\mu}_{\mathbf{x}_{t_0}} \\ \boldsymbol{\kappa} \end{bmatrix} \\ \begin{bmatrix} \mathbf{w}_{t_1 - \Delta_{t_1}} \\ \mathbf{0} \end{bmatrix} \\ \begin{bmatrix} \mathbf{w}_{t_2 - \Delta_{t_2}} \\ \mathbf{0} \end{bmatrix} \\ \begin{bmatrix} \mathbf{w}_{t_3 - \Delta_{t_3}} \\ \mathbf{0} \end{bmatrix} \\ 1 \end{bmatrix}. \tag{7.26}$$

$$\mathbf{B} = \begin{bmatrix} \mathbf{0} & \mathbf{0} & \mathbf{0} & \mathbf{0} & \mathbf{0} & \mathbf{0} & \mathbf{0} & \mathbf{0} & \boldsymbol{\mu}_{\mathbf{x}_{t_0}} \\ \mathbf{0} & \mathbf{0} & \mathbf{0} & \mathbf{0} & \mathbf{0} & \mathbf{0} & \mathbf{0} & \mathbf{0} & \mathbf{0} \\ \mathbf{A}_{\Delta t_1} & \mathbf{H}_{\Delta t_1} & \mathbf{0} & \mathbf{0} & \mathbf{0} & \mathbf{0} & \mathbf{0} & \mathbf{0} & \mathbf{b}_{\Delta t_1} \\ \mathbf{0} & \mathbf{I} & \mathbf{0} & \mathbf{0} & \mathbf{0} & \mathbf{0} & \mathbf{0} & \mathbf{0} & \mathbf{0} \\ \mathbf{0} & \mathbf{0} & \mathbf{A}_{\Delta t_2} & \mathbf{H}_{\Delta t_2} & \mathbf{0} & \mathbf{0} & \mathbf{0} & \mathbf{0} & \mathbf{b}_{\Delta t_2} \\ \mathbf{0} & \mathbf{0} & \mathbf{0} & \mathbf{I} & \mathbf{0} & \mathbf{0} & \mathbf{0} & \mathbf{0} & \mathbf{0} \\ \mathbf{0} & \mathbf{0} & \mathbf{0} & \mathbf{0} & \mathbf{A}_{\Delta t_3} & \mathbf{H}_{\Delta t_3} & \mathbf{0} & \mathbf{0} & \mathbf{b}_{\Delta t_3} \\ \mathbf{0} & \mathbf{0} & \mathbf{0} & \mathbf{0} & \mathbf{0} & \mathbf{I} & \mathbf{0} & \mathbf{0} & \mathbf{0} \\ \mathbf{0} & \mathbf{0} & \mathbf{0} & \mathbf{0} & \mathbf{0} & \mathbf{0} & \mathbf{0} & \mathbf{0} & 0 \end{bmatrix},$$

$$\boldsymbol{\Psi} = \begin{bmatrix} \Phi_{\mathbf{x}_{t_0}} & & & & & & & & \\ \Phi_{\mathbf{x}_{t_0},\kappa} & \Phi_{\kappa} & & & & & & & \\ \mathbf{0} & \mathbf{0} & \mathbf{Q}_{\Delta t_1} & & & & & & \\ \mathbf{0} & \mathbf{0} & \mathbf{0} & \mathbf{0} & & & & & \\ \mathbf{0} & \mathbf{0} & \mathbf{0} & \mathbf{0} & \mathbf{Q}_{\Delta t_2} & & & & \\ \mathbf{0} & \mathbf{0} & \mathbf{0} & \mathbf{0} & \mathbf{0} & \mathbf{0} & & & \\ \mathbf{0} & \mathbf{0} & \mathbf{0} & \mathbf{0} & \mathbf{0} & \mathbf{0} & \mathbf{Q}_{\Delta t_3} & & \\ \mathbf{0} & \mathbf{0} & \mathbf{0} & \mathbf{0} & \mathbf{0} & \mathbf{0} & \mathbf{0} & \mathbf{0} & \\ \mathbf{0} & \mathbf{0} & \mathbf{0} & \mathbf{0} & \mathbf{0} & \mathbf{0} & \mathbf{0} & \mathbf{0} & 1 \end{bmatrix}, \tag{7.27}$$

$$\boldsymbol{\Lambda} = \begin{bmatrix} \mathbf{C} & \mathbf{0} & \mathbf{0} & \mathbf{0} & \mathbf{0} & \mathbf{0} & \mathbf{0} & \mathbf{0} & \mathbf{d} \\ \mathbf{0} & \mathbf{0} & \mathbf{C} & \mathbf{0} & \mathbf{0} & \mathbf{0} & \mathbf{0} & \mathbf{0} & \mathbf{d} \\ \mathbf{0} & \mathbf{0} & \mathbf{0} & \mathbf{0} & \mathbf{C} & \mathbf{0} & \mathbf{0} & \mathbf{0} & \mathbf{d} \\ \mathbf{0} & \mathbf{0} & \mathbf{0} & \mathbf{0} & \mathbf{0} & \mathbf{0} & \mathbf{C} & \mathbf{0} & \mathbf{d} \\ \mathbf{0} & \mathbf{0} & \mathbf{0} & \mathbf{0} & \mathbf{0} & \mathbf{0} & \mathbf{0} & \mathbf{0} & 1 \end{bmatrix}, \quad \boldsymbol{\Theta} = \begin{bmatrix} \mathbf{R} & & & & \\ \mathbf{0} & \mathbf{R} & & & \\ \mathbf{0} & \mathbf{0} & \mathbf{R} & & \\ \mathbf{0} & \mathbf{0} & \mathbf{0} & \mathbf{R} & \\ \mathbf{0} & \mathbf{0} & \mathbf{0} & \mathbf{0} & 0 \end{bmatrix}.$$

It could also be used for the ADM as the previous specification could be used for the EDM. The alternative specification highlights that the trait variables $\boldsymbol{\kappa}$ are just a special kind of state variables, namely constant across time. Hence, as shown in (7.26), the trait variables are added to the state vector $\mathbf{x}_{t_i}$ at each time point, leading to the larger total number of $2Tn+1$ variables in latent vector $\boldsymbol{\eta}$. The specification puts the model directly in a suitable form for the application of state space techniques, in particular the Kalman smoother or conditional mean $E[\mathbf{x}(t)|\mathbf{y}]$, for optimally estimating a subject's sample trajectory (Oud & Jansen, 1996; Oud, Jansen, van Leeuwe, Aarnoutse, & Voeten, 1999). Based on (7.17), the SEM specification of the EDM in (7.27), follows, apart from the trait variables, the same pattern as in the case of the ADM in (7.25). Whereas the state variables develop across time according to $\mathbf{A}_{\Delta t_i}$ and are influenced by the trait variables according to $\mathbf{H}_{\Delta t_i}$, the trait variables themselves, remaining constant over time, develop according to the identity matrix $\mathbf{I}$. The trait covariance matrix $\boldsymbol{\Phi}_{\kappa}$ and state-trait covariance matrix $\boldsymbol{\Phi}_{\mathbf{x}_{t_0},\kappa}$ are found in the second row of $\boldsymbol{\Psi}$. The measurement model matrices do not differ from the ones in (7.25), except that $\boldsymbol{\Lambda}$ has extra zero columns at the places of the unobserved trait variables.

7.4.4 Relating Models on Different Time Scales

Researchers in the same or different subject fields often use different observation intervals. As argued in Subsection 7.2.2, comparing longitudinal models with different observation intervals, a clear condition for cumulative progress in science, requires continuous time analysis. This is not all, however. Meaningful comparison also requires the results to be put on the same time scale. Because both the ADM and EDM are time-invariant, time scale shifts, $t' = t + d$, do not change the results. The ADM and EDM in (7.17) and (7.19) show, however, that changing the time scale unit, $t' = ct$ (for example, going from years t to months t': $t' = 12t$), indeed affects the parameter matrices, but in a quite simple way, not requiring any re-estimation of the parameter matrices. If one wants to compare one's results with another researcher, who used time scale $t' = ct$ ($\Delta t_i' = c\Delta t_i$) instead of one's own scale t, simply multiply four of the five parameter matrices by $1/c$ ($\mathbf{A}$, $\mathbf{b}$, $\boldsymbol{\Phi}_{\mathbf{x}_{t_0},\kappa}$, $\mathbf{G}$ in EDM; $\mathbf{A}_{\dagger}$, $\mathbf{b}_{\dagger}$, $\boldsymbol{\Phi}_{\dagger\mathbf{x}_{t_0},\kappa}$, $\mathbf{G}_{\dagger}$ in ADM) and one by $1/c^2$ ($\boldsymbol{\Phi}_{\kappa}$ in EDM ; $\boldsymbol{\Phi}_{\dagger\kappa}$ in ADM). The reason is that the discrete time matrices on the left hand side of the constraints equations in (7.17) and (7.19) keep the same value, while the change from Δt_i to $c\Delta t_i$ or from $\mathbf{A}^{-1}$ to $c\mathbf{A}^{-1}$ on the right hand side needs compensation by multiplying by $1/c$.

7.5 Relationships between Externalizing and Internalizing Problem Behavior

In this and the next section applications will be presented. The LISREL program will be used for the ADM-SEM procedure (see the commented LISREL script in the Appendix of Oud, 2007b) and the Mx program for the EDM-SEM procedure. All input and output files of the analyses performed in both sections are available in Chapter 7 at the book website `http://www.econ.upf.edu/~satorra/longitudinallatent/readme.html`. In this section, four LISREL maximum likelihood analyses and two Mx maximum likelihood analyses examine the relationships between externalizing and internalizing problem behavior in adolescents. It will be studied in continuous time whether and how strongly externalizing problem behavior influences internalizing problem behavior (failure perspective: Burke, Loeber, Lahey, & Rathouz, 2005; Capaldi, 1992), internalizing problem behavior influences externalizing problem behavior (acting out perspective: Carlson & Cantwell, 1980; Gold, Mattlin, & Osgood, 1989), or both variables influence each other reciprocally (mutual influence perspective: Overbeek, Vollebergh, Meeus, Luypers, & Engels, 2001). The two state variables in the model (externalizing problem behavior and internalizing problem behavior) are observed. So, the measurement model part includes only loadings 1, intercepts 0, and measurement error variances 0. In the next section, a model with an elaborate measurement model will be presented for relationships between three latent state variables.

The data analyzed are taken from a comprehensive Dutch study of family relationships and adolescent problem behavior (Nijmegen Family and Personality Study; Haselager & van Aken, 1999). Participants were 280 adolescents (139 boys, 141 girls) who were 14.5 years old on average (ranging from 11.4 to 16.0) at the first measurement wave. To assess adolescents' externalizing and internalizing problem behavior, participants completed the Nijmegen Problem Behavior List (NPBL; De Bruyn, Scholte, & Vermulst, 2005) at each of the three annual measurement waves. Further details regarding sample characteristics, measures, and procedure can be found in Delsing, Oud, van Aken, De Bruyn, and Scholte (2005).

Although the aim of the ADM is the estimation of the (approximate) underlying continuous time parameters, it is nevertheless clarifying to view in Figure 7.6 the discrete time part of the ADM in SEM form. The model contains the state variables *Ext* and *Int* and corresponding constant trait variables *Trait-Ext* and *Trait-Int*, for three time points leading to a total of eight variables in the SEM model (apart from the ninth unit variable, which is not depicted in Figure 7.6). The figure clearly shows the discrete time part of the ADM with instantaneous coefficients ($\mathbf{A}^{*}_{\Delta t_i}$) as well as lagged coefficients ($\mathbf{A}^{**}_{\Delta t_i}$). The ADM in SEM form is one of the rare SEM models with self-loop coefficients specified and estimated (diagonals in the instantaneous matrices $\mathbf{A}^{*}_{\Delta t_i}$, indicated in the figure by self-referencing arrows). In total, the continuous time part of the model contains 21 parameters to be estimated:

4 drift coefficients in $\mathbf{A}_{\dagger}$,
2 intercepts feeding mean development in $\mathbf{b}_{\dagger}$,

3 trait variances and covariance in $\Phi_{\dagger\kappa}$,
4 covariances between traits and initial states in $\Phi_{\dagger\mathbf{x}_{t_0},\kappa}$,
3 state variable diffusion coefficients in $\mathbf{G}_{\dagger}$,
2 initial state means in $\boldsymbol{\mu}_{\mathbf{x}_{t_0}}$,
3 initial state variances and covariance in $\Phi_{\mathbf{x}_{t_0}}$.

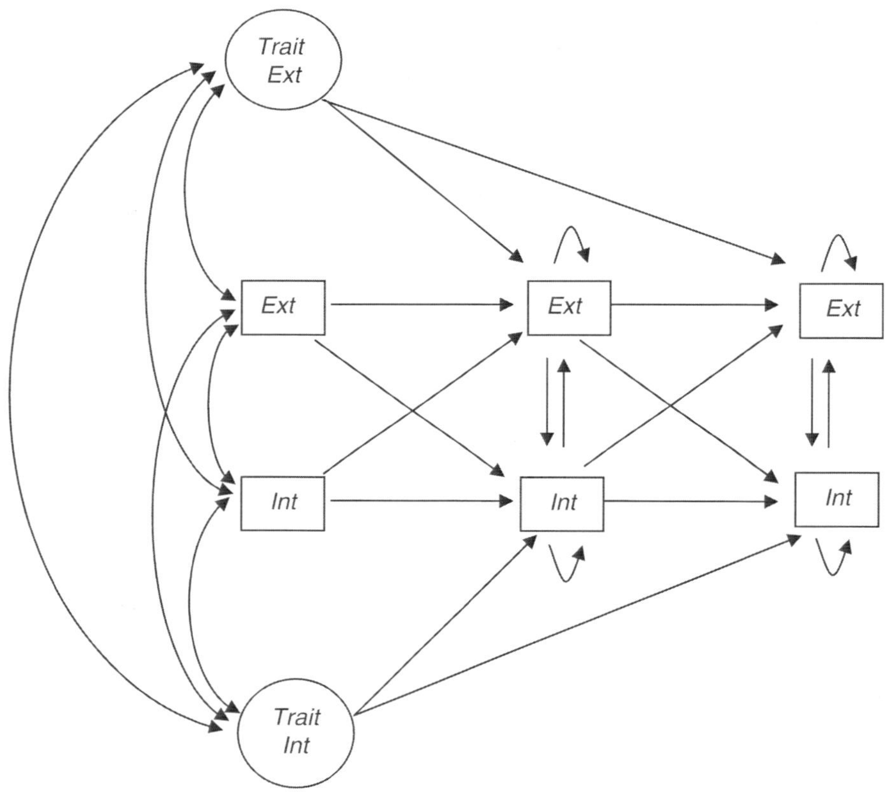

Fig. 7.6 The three-wave ADM-SEM cross-lagged panel model for adolescents' externalizing and internalizing problem behavior, including corresponding trait variables (subject-specific intercepts).

A full ADM as well as a full EDM can be proven to be identified for $T \geq 3$, assuming the measurement model part is identified. As the ADMs and EDMs in this section have only observed state variables, the model does not have free measurement parameters and so this part is identified automatically. Column 1 of Table 7.2 displays the estimate of the full ADM model (input file *ADM1.ls8* and output file *ADM1.out*). For convenience, the subscript † in the ADM parameter names is suppressed in Table 7.2. For equal observation intervals of length $\Delta t_i = 1$, many of the parameter estimates are immediately found in the LISREL parameter matrices **B** (BETA) and **Ψ** (PSI). Parameters not immediately found there but estimated as

Table 7.2 Estimates and model fit information for ADM1 (Full ADM), ADM2 (No Trait 1), ADM3 (No Trait 1, No *Int* → *Ext*) and EDM3 (No Trait 1, No *Int* → *Ext*); standardized drift coefficients

Parameter	ADM1	ADM2	ADM3	EDM3
$a_{11}(Ext)$	-0.790**	-0.302**	-0.317**	-0.320**
$a_{12}(Int \rightarrow Ext)$	0.347	-0.039		
$a_{21}(Ext \rightarrow Int)$	0.788**	0.616**	0.605**	0.704**
$a_{22}(Int)$	-1.056**	-1.134**	-1.110**	-1.251**
$\mu_{x_{1t_0}}$	17.943**	17.943**	17.943**	17.943**
$\mu_{x_{2t_0}}$	21.106**	21.106**	21.106**	21.106**
$\phi_{x_{1t_0}}$	27.323**	27.323**	27.261**	27.323**
$\phi_{x_{2t_0}}$	38.212**	38.212**	37.462**	38.212**
$\phi_{x_{21t_0}}$	10.990**	10.990**	10.774**	10.990**
b_1	7.817	5.979**	5.559**	5.606**
b_2	5.598	10.890*	10.730*	11.505*
g_{11}	4.922**	4.761**	4.762**	4.782**
g_{22}	6.228**	6.226**	6.219**	6.562**
g_{21}	1.347**	1.648**	1.557**	1.315**
ϕ_{κ_1}	10.965			
ϕ_{κ_2}	26.014	27.646	26.554	33.958
$\phi_{\kappa_{21}}$	-15.059			
$\phi_{x_{1t_0},\kappa_1}$	7.235			
$\phi_{x_{2t_0},\kappa_1}$	-5.364			
$\phi_{x_{1t_0},\kappa_2}$	-12.731*	-8.063*	-7.988*	-8.887*
$\phi_{x_{2t_0},\kappa_2}$	14.731	18.619*	17.999	20.694
χ^2	5.4	9.9	10.6	10.6
df	6	10	11	11
RMSEA	0.0	0.0	0.0	0.0

$^*p \leq .05$; $^{**}p \leq .01$.

so-called additional parameters are the four drift coefficients: auto-effects a_{11} and a_{22} (called PA(1) and PA(2), respectively, in the LISREL output) and cross-effects a_{12} and a_{21} (PA(3) and PA(4), respectively), and the three diffusion coefficients g_{11}, g_{22}, and g_{21} (PA(7), PA(9), and PA(8), respectively).

With regard to the main purpose of the study, assessing the existence and strength of the cross-effects a_{12} $(Int \rightarrow Ext)$ and a_{21} $(Ext \rightarrow Int)$ between internalizing and externalizing problem behavior, one should realize an important difference in interpretability between them and the auto-effects a_{11}, a_{22}. The auto-effects are scale free in the sense that they do not change under arbitrary linear transformations of the variables *Ext* and *Int* and so are directly interpretable. In particular, both *Ext* and *Int* show negative feedback ($-.790$ and -1.056), indicating stability or a quite strong tendency for an individual to converge to its subject-specific mean trajectory. The negative eigenvalues of the drift matrix confirm that the model as a whole is stable. The cross-effects are not scale free, however. Their value depends on the standard deviations of the independent and dependent variable involved. The cross-effects in Table 7.2 have therefore been standardized (PA(5) and (PA(6) in the LISREL output) through multiplication by the ratios of the initial standard deviations. However, as t-values are scale free, testing can best be done in terms of the unstandardized values of 0.293 for a_{12} (not significant) and 0.932 for a_{21} ($p < .01$) (the t-values computed by LISREL for the standardized values inappropriately also include the sampling variability of the standard deviations). The standardized values of 0.347 for a_{12} and 0.788 for a_{21} in combination with the testing results reported in Table 7.2 seem to reveal the existence of a strong unidirectional effect from externalizing to internalizing problem behavior with no or little effect in the opposite direction.

The specification of traits in a model has, in general, high impact on the estimates of the other parameters. Because in the full ADM no significant variances and covariances were found for the traits κ_1 and κ_2 except for the covariance between κ_2 and x_{1t_0}, we decided to retain only κ_2 in the next model ADM2 (files *ADM2.ls8* and *ADM2.out*), so that this model has subject-specific mean trajectories for internalizing problem behavior but only a single general mean trajectory for externalizing problem behavior. It is interesting that the exclusion of κ_1 from ADM2 led to the non-significant effect a_{12} $(Int \rightarrow Ext)$ in ADM1 turning slightly negative in ADM2 but with a non-significant t-value again that was even lower than in the ADM1. This was reason to next delete a_{12} from the model. The resulting ADM3 (files *ADM3.ls8* and *ADM3.out*) has all parameters significant except two which are related to κ_2. It retains in particular an impressively strong effect a_{21} $(Ext \rightarrow Int)$.

Having found a clear and particularly well fitting ADM (the extra constraints introduced into ADM2 and ADM3 do not deteriorate the fit shown by χ^2 and RMSEA), the obvious next step is to replace the approximate ADM by the exact EDM (see EDM3 in Table 7.2). As explained above for the case of equal observation intervals, one possibility would be to apply the indirect method by computing the reduced form matrices according to (7.21) and deriving the EDM from the ADM3 by (7.17) instead of using the direct method by running the Mx program. Note, that the (estimated) reduced form autoregression matrix $\mathbf{A}_{\Delta t_i}$ is already computed by LISREL in the first part of the matrix "Total effects of ETA on ETA" in *ADM3.out*. This indeed turns out to be exactly equal to the (estimated) autoregression matrix $\mathbf{A}_{\Delta t_i}$ computed by Mx in the first part of its BETA matrix (called "A" in GROUP 7 of Mx output file *EDM3.mxo*). Autoregressions for *Ext* and *Int* in both are equal to 0.72627 and 0.28614 and the cross-lagged effect in both is equal to 0.39352. One reason to apply the direct method by running the Mx program could be, however, that in addition to the EDM solution itself one gets also the correct standard errors.

The Mx analysis has therefore indeed been done (files *EDM3.mx* and *EDM3.mxo*) and the results are displayed in the last column of Table 7.2.

Although the reduced forms of ADM3 and EDM3 should be and are indeed equal in this case of equal observation intervals as is the fit of both models (the models are equivalent via their reduced form), the solutions themselves are close to each other but not equal; compare the last two columns in Table 7.2 (the EDM3 solution is found in GROUP 9 in the Mx output file *EDM3.mxo,* displaying all estimated parameter matrices; the standardized value of a_{21} is computed and found in GROUP 10). Our experience is, that the EDM often yields a somewhat more pronounced solution with the parameter estimates showing higher absolute values. This is clearly also the case here. For example, the standardized value of 0.704 for a_{21} $(Ext \rightarrow Int)$ in EDM3 points to an even stronger effect of externalizing problem behavior on internalizing problem behavior than found in ADM3. Our analyses leave little doubt that the failure perspective is the one confirmed by the data in this section and not the acting out or mutual influence perspectives discussed in the literature.

As has been stressed several times, the equivalence of the ADM and EDM is based on the equality of the observation intervals. If the observation intervals are unequal, ADM and EDM can give quite different reduced forms and a quite different fit. So, then no other choice is left than estimating the EDM independently from the ADM. To show how analyses with unequal observation intervals are performed

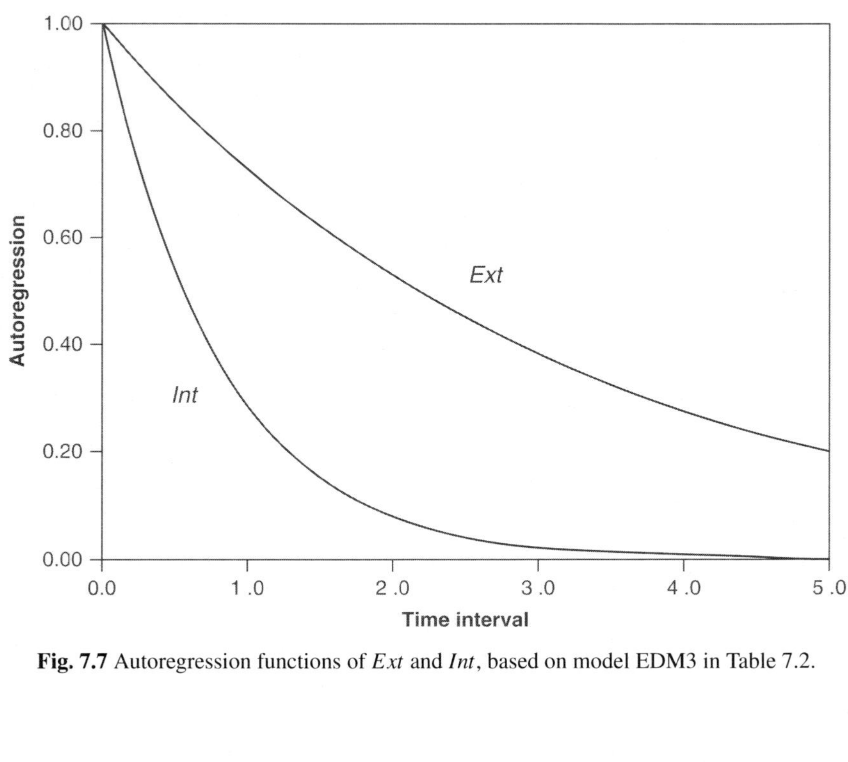

Fig. 7.7 Autoregression functions of *Ext* and *Int*, based on model EDM3 in Table 7.2.

and what could happen in the example with unequal intervals, instead of the true equal intervals $\Delta t_1 = \Delta t_2 = 1$, we simulated the second interval to be slightly larger than the first interval: $\Delta t_1 = 1$, $\Delta t_2 = 1.2$. So, the interval between the second and third wave is taken 1.2 years instead of the true 1 year. The analyses are performed by input files *ADM4.ls8* and *EDM4.mx* for LISREL and Mx, respectively, and the output is found in files *ADM4.out* and *EDM4.mxo*. Instead of the same χ^2-value of 10.6 for both ADM3 and EDM3, we now find $\chi^2 = 17.3$ for ADM4 and $\chi^2 = 11.2$ for EDM4. In both the fit deteriorates but ADM4 turns out to be much more sensitive to the wrong specification of the second observation interval than EDM4. Of course, autoregressions and cross-lagged effect are different for the unequal intervals in each analysis (whereas, as expected, the autoregressions were lower over the longer second interval, the cross-lagged effect turned out to be higher), but they also differ now between ADM4 and EDM4. Over the first interval the autoregressions in the ADM4 were 0.748 and 0.253 for *Ext* and *Int*, respectively, and in the EDM4 0.747 and 0.261; the cross-lagged effect in the ADM4 was 0.405 and in the EDM4 0.410. This clearly illustrates the necessity to estimate the EDM independently from the ADM in the case of unequal intervals.

We conclude the example with some of the consequences of the model estimated in continuous time: autoregression functions for *Ext* and *Int*, cross-lagged effect function for a_{21} $(Ext \rightarrow Int)$ and mean trajectories for *Ext* and *Int*, all based on the final EDM3 in Table 7.2. Autoregression and cross-lagged effect functions are

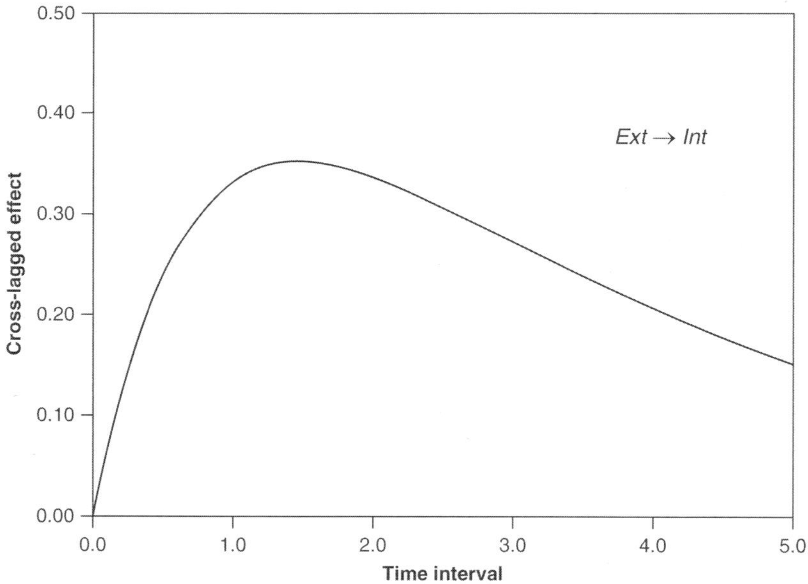

Fig. 7.8 Standardized cross-lagged effect function for *Ext* → *Int*, based on model EDM3 in Table 7.2.

computed using (7.5). An autoregression function describes the autonomous development of a variable, telling which proportion of its value present at the start predictably persists after increasing intervals. In discrete time analysis the autoregression is only computable on the basis of the discrete time intervals in the study. The continuous time analysis in Figure 7.7 reveals that the part of *Ext* predictable from its value at the start is everywhere higher than for *Int*, after an interval of $\Delta t =$ 2.2 years the predictable *Ext* part is still not less than half the original *Ext*, whereas only 7% of the original *Int* is left, and already after 6 months half of the original *Int* is lost.

The cross-lagged effect function in Figure 7.8 reveals that a unit (standard deviation) increase in *Ext* has its maximum impact of 35.2% of a standard deviation in *Int* after 1.5 years and that after 5 years still 15% is left. So, the cross-lagged effect function nicely clarifies what the meaning and impact across all intervals in continuous time is of the impressive cross-effect of 0.704 in model EDM3.

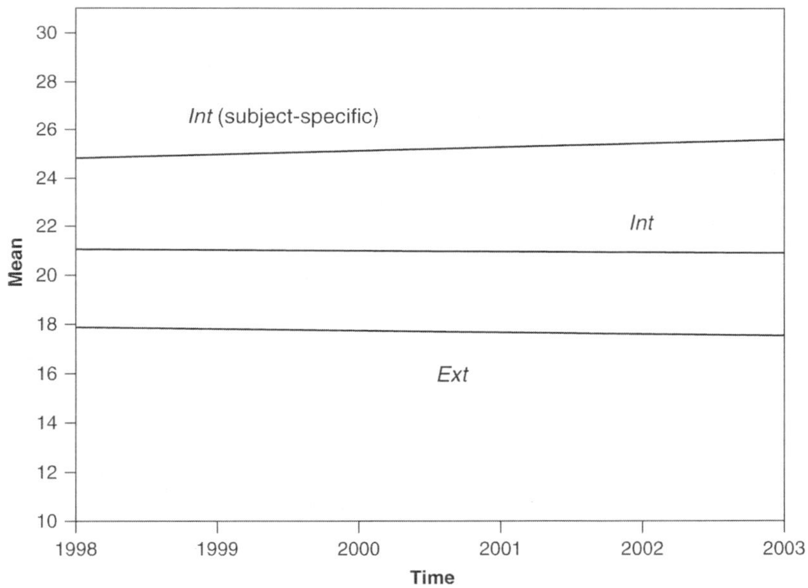

Fig. 7.9 Mean trajectories for *Ext* and *Int* and a subject-specific mean trajectory for *Int*, based on model EDM3 in Table 7.2.

General mean trajectories for *Ext* and *Int* in Figure 7.9 were computed using (7.11) and the subject-specific mean trajectory for *Int* using (7.12). As an illustration, the subject-specific mean trajectory was computed for a subject at one standard deviation above trait mean for *Int*, that is for $\kappa_2 = 6.18$. Because κ_1 was deleted from model EDM3 and therefore Φ_κ is not positive definite, (7.12) was applied by filling out in regression matrix $\Phi_{\mathbf{x}_{t_0},\kappa}\Phi_\kappa^{-1}$ the regressions of x_{1t_0} and x_{2t_0} on κ_2 only.

It turns out that the means hardly change. The general means, starting from 17.9 and 21.1 in 1998, going down only a tiny fraction over the data collection period between 1998 and 2000 towards 17.7 and 21.0, are predicted to be 17.6 and 20.9 in 2003 and converge for $t \to \infty$ to final values of 17.53 and 20.86. The subject-specific mean trajectory for *Int* of the subject with subject-specific intercept value of $b_2 + \kappa_2 = 17.7$, however, increases: from 24.9 in 1998 to 25.2 in 2000 and then in the prediction period, after reaching 25.6 in the middle of 2003, to final value 25.80. This final value is not much higher than the value in 2000 at the end of the data collection period.

7.6 Relationships between Individualism, Nationalism and Ethnocentrism in Flandres

The example in this section, taken from Toharudin, Oud, and Billiet (2008), is more comprehensive than the one in the previous section for two reasons. First, the state variables are latent and based on an elaborate measurement model for measuring the theoretical constructs Individualism (I), Nationalism (N) and Ethnocentrism (E). The constructs were repeatedly measured in three waves (1991, 1995, and 1999) in a panel of $N = 1274$ Flemish respondents and Dutch-speaking respondents in Brussels. Second, whereas the number of state variables was two in the previous section, in this example it is three, leading to a 3 x 3 drift matrix with six different causal connections between the latent variables. The purpose of the study was to find out, how the constructs develop and influence each other across time. On the basis of previous research a recursive causal structure was hypothesized: $I \to N, N \to E, I \to E$. Thus, in addition to the auto-effects, only three of the six possible cross-effects were hypothesized to be nonzero. In previous research, causal connections between the constructs were analyzed cross-sectionally only, or, if longitudinally, solely in correlational form without taking care of the causal direction of the effects, and never in continuous time. Again, the continuous time analysis of the data set started with an ADM analysis (LISREL input file *ADM-INE.ls8* and output file *ADM-INE.out*), followed by the corresponding EDM analysis (Mx input file *EDM-INE.mx* and output file *EDM-INE.mxo*).

First, attention will be paid to the measurement model. Individualism (I) or "unrestrained striving for personal interests", Nationalism (N) or "identification with the Flemish community in Belgium", and Ethnocentrism (E) or "negative attitude toward outgroups" were measured by 5, 4, and 8 items, respectively. Most of the items were 5-point-scale items, the answers consisting of different degrees of agreement/disagreement. Two item examples for each of the constructs are:

Individualism (I)
- -Everybody has to take care of himself first.
- -What counts is money and power.

Nationalism (*N*)
-Flanders must decide.
-Belgium has to disappear.

Ethnocentrism (*N*)
-Belgium should not have allowed in guest workers.
-Immigrants cannot be trusted.

Several items under Nationalism had slight differences in formulation between the first wave year 1991 and the two subsequent wave years 1995 and 1999. Measurement invariance (time invariance) analyses were performed, in which the loadings, measurement intercepts, and measurement error variances (**C**, **d**, and **R** in (7.16), respectively) of these *N*-items were compared between 1991 on the one hand and 1995 and 1999 on the other hand. It yielded that time invariance of *N*-item 4 for 1991 (called "4na91" in *ADM-INE.out)* in comparison with 1995 and 1999 had to be rejected. Consequently, the three measurement parameters of this item were allowed to deviate in 1991. Only one more deviation from time-invariance was allowed in the measurement model by freeing the measurement intercept of *I*-item 5 in 1999 (called "5in99" in *ADM-INE.out*). Freeing this single parameter, thereby increasing its value from 2.456 to 3.286, had the effect of decreasing the model χ^2 for ADM and EDM with the huge amount of 1209, implying a considerable improvement in model fit. It prevents the increase in this single item from unduly influencing the latent mean development of *I* between 1995 and 1999. All information about the loadings of the items can be found in the main body of matrix "LAMBDA-Y" in output file *ADM-INE.out*, about the measurement intercepts in the last column of this matrix, and about the measurement error variances in "THETA-EPS".

With regard to the dynamic model part and the initial state variances and means, first the ADM and EDM estimates will be presented and then details about the way the input files were formulated to obtain the estimates. Both solutions are given in Tables 7.3 and 7.4. In Table 7.3 the ADM solution is on the left hand side and the EDM solution on the right hand side. Because the estimates of the initial state variances and means are equal in both solutions, they are given only once in Table 7.4. Both tables also give *t*-values, providing precise information about the significance of the parameter estimates as well as about the standard errors (t = estimate/standard error). Although in both ADM and EDM, trait variables were specified (in the form of extra state variables as in (7.26) and (7.27)), all three trait variances were fixed at zero in the final analysis, because no positive estimates were found or expected to be found (see in *ADM-INE.out* the negative values under "Expected change for Psi" for the variances of "04w1TrI", "05w1TrN", and "06w1TrE"). We conclude that the initial variances are sufficient to differentiate trajectories for individual subjects from the mean trajectory and no extra trait variances are warranted. Diffusion coefficient matrix **G** and diffusion matrix $\mathbf{Q} = \mathbf{GG}'$ were specified diagonal because of the rather low modification indices and expected changes in *ADM-INE.out* for the off-diagonal elements.

Comparing the ADM and EDM solutions in Tables 7.3-7.4, it is striking how similar both solutions are with only very small differences in the third decimal of the parameter estimates. Also the differences in t-values are small, proving that the standard errors are very similar too. These almost equal results, obtained on the basis of different models by quite diverse programs, clearly confirm each other. It means also that the EDM can safely be evaluated by means of the ADM estimates, standard errors and other information given by the LISREL program. As expected, also the fit information is equal within precision limits. χ^2 = 7881 in LISREL and χ^2 = 7880 in Mx with df = 1304 do not seem to imply a particularly good fit. It should be noted, however, that the model with 51 observed variables is huge, the sample of N = 1274 big, and the (almost) strict time-invariance of the continuous-time model puts a lot of heavy constraints on the model, on the measurement part as well as on the dynamic part of the model. It is therefore no surprise that the popular

Table 7.3 ADM and EDM estimates; standardized coefficients in drift matrices $\mathbf{A}_\dagger$ and $\mathbf{A}$, t-values between parentheses

	$\mathbf{A}_\dagger$: *I*	$\mathbf{A}_\dagger$: *N*	$\mathbf{A}_\dagger$: *E*	$\mathbf{A}$: *I*	$\mathbf{A}$: *N*	$\mathbf{A}$: *E*
I	−0.069∗∗	−0.007	0.033∗∗	−0.070∗∗	−0.008	0.033∗∗
	(−9.41)	(−1.39)	(5.15)	(−9.23)	(−1.44)	(5.13)
N	0.013∗	−0.061∗∗	0.011∗	0.013∗	−0.061∗∗	0.012∗
	(2.31)	(−10.79)	(2.09)	(2.33)	(−10.17)	(2.11)
E	0.039∗∗	0.003	−0.062∗∗	0.040∗∗	0.003	−0.063∗∗
	(7.48)	(0.80)	(−11.98)	(7.53)	(0.82)	(−11.94)

	$\mathbf{b}_\dagger$	$\mathbf{b}$
I	0.061∗∗	0.061∗∗
	(3.28)	(3.32)
N	0.105∗	0.105∗
	(2.21)	(2.19)
E	0.094∗∗	0.095∗∗
	(7.58)	(7.62)

	$\mathbf{G}_\dagger$: *I*	$\mathbf{G}_\dagger$: *N*	$\mathbf{G}_\dagger$: *E*	$\mathbf{G}$: *I*	$\mathbf{G}$: *N*	$\mathbf{G}$: *E*
I	0.280∗∗			0.281∗∗		
	(11.37)			(10.83)		
N		0.683∗∗			0.685∗∗	
		(11.06)			(10.70)	
E			0.210∗∗			0.211∗∗
			(15.39)			(14.92)

*$p \leq .05$; **$p \leq .01$.

Table 7.4 Estimates of initial state (co)variances and means; t-values between parentheses, first line of ADM and second line of EDM

	I	N	E	
I	0.778**			2.460**
	(17.45)			(83.18)
	(16.57)			(82.96)
N	−0.183*	5.999**		4.231**
	(−2.41)	(17.38)		(51.64)
	(−2.41)	(16.51)		(50.49)
E	0.321**	0.106	0.508**	2.899**
	(13.44)	(1.87)	(17.10)	(116.82)
	(13.34)	(1.86)	(17.11)	(116.73)
		$\mathbf{\Phi}_{\mathbf{x}_{t_0}}$		$\boldsymbol{\mu}_{\mathbf{x}_{t_0}}$

$^*p \leq .05$; $^{**}p \leq .01$.

fit measure RMSEA (Browne and Cudeck, 1993) with value 0.068 indicates that the model fits reasonably.

Turning to the drift matrix **A**, which should give the answers to the main questions in the study, we first observe that the auto-effects are all three negative $(-0.069, -0.061, -0.062)$, indicating stability or a long-term tendency for the trajectories to converge to the mean trajectory. Stability is confirmed by the negative eigenvalues of the drift matrix. Interestingly, by accounting appropriately for the 4 year observation interval, the auto-effects are correctly comparable to the auto-effects of -0.320 and -1.251 in the previous example (Section 7.5) with a 1 year interval. Individualism, Nationalism and Ethnocentrism in the present example have a much weaker tendency to converge to their mean trajectory than externalizing problem behavior and internalizing problem behavior in the previous example (Section 7.5).

The cross-effects do not confirm the hypothesized recursive structure $I \rightarrow N$, $N \rightarrow E$, $I \rightarrow E$. In the place of non-significant and almost zero effect $N \rightarrow E$ come significant effects $E \rightarrow N$ and $E \rightarrow I$. The role of Nationalism is therefore quite different from what was expected. N turns out not to influence E, but, in contrast, to undergo a weak influence from E. So, N comes out as the dependent variable in the structure, weakly and nearly equally influenced by both other constructs (standardized coefficients of 0.013 and 0.011). In addition, a clear reciprocal relationship shows up between Individualism and Ethnocentrism: $I \rightarrow E$ but also $E \rightarrow I$ with standardized coefficients of 0.039 and 0.033, respectively. All standardized effects are small in strength, though, and, although significant, much smaller than the standardized effect of 0.704 found in the previous example (Section 7.5).

As mentioned above, in both the ADM and the EDM trait variables were specified as extra state variables. This can be seen by SEM matrix **B** (called "BETA" in

LISREL output *ADM-INE.out* and "A" in GROUP 42 of Mx output *EDM-INE.mxo*) containing 19 variables: in addition to the last variable, the unit variable, at each time point 3 state variables are followed by the extra 3 trait variables. Just as $\mathbf{B}$, $\mathbf{\Psi}$ is also a 19 x 19 matrix (called "PSI" in *ADM-INE.out* and "P" in GROUP 42 of *EDM-INE.mxo),* showing the initial covariance matrix $\Phi_{\mathbf{x}_{t_0}}$ in the first 3 x 3 diagonal block and the trait covariance matrix Φ_{κ} (in the final analysis fixed at zero) in the next 3 x 3 diagonal block.

For the ADM analysis, the equal observation intervals of $\Delta t_1 = \Delta t_2 = 4$ were reason to apply the simple ADM constraints in (7.20) between lagged and instantaneous matrices $\mathbf{A}^{**}_{\Delta t_1}$ and $\mathbf{A}^{*}_{\Delta t_1}$ in addition to the equality constraints between time points. In the LISREL output file *ADM-INE.out* one finds the lagged matrix $\mathbf{A}^{**}_{\Delta t_1}$ in BETA at variables 7-9 (dependent) and 1-3 (lagged independent) and the instantaneous matrix $\mathbf{A}^{*}_{\Delta t_1}$ at variables 7-9 (dependent and independent). The ADM constraints (7.20) are formulated in LISREL input file *ADM-INE.ls8* following "!ADM equality (auto)" and "!ADM equality constraints (cross)". As explained below formula (7.20), the estimates of drift coefficients in $\mathbf{A}_{\dagger}$, intercepts in $\mathbf{b}_{\dagger}$, and diffusion coefficients in $\mathbf{G}_{\dagger}$ can easily be obtained by hand from the estimated discrete time matrices using (7.19). For convenience, these simple computations have also been done by means of the LISREL program: division of $\mathbf{A}^{*}_{\Delta t_1}$ by $\frac{1}{2}\Delta t_1 = 2$ (multiplication by 0.5, yielding the drift coefficients in additional LISREL parameters PA(1)-PA(9)), division of $\mathbf{b}^{*}_{\Delta t_1}$ by $\Delta t_i = 4$ (multiplication by 0.25, yielding the drift coefficients in additional LISREL parameters PA(16)-PA(18)), and division of $\mathbf{Q}^{*}_{\Delta t_1}$ by $\Delta t_i = 4$, followed by the square-root of the result ($\sqrt{\frac{1}{4}q^{*}_{ii,\Delta t_1}} = 0.5\text{x}(q^{*}_{ii,\Delta t_1})^{0.5}$ yielding the diffusion coefficients in additional LISREL parameters PA(19)-PA(21)). All elements in trait matrices $\Phi_{\dagger\kappa}$ and $\Phi_{\dagger\mathbf{x}_{t_0},\kappa}$ were specified zero in the final analysis, but, if nonzero, they could have been computed by dividing $\Phi^{*}_{\kappa\Delta t_i}$ and $\Phi^{*}_{\mathbf{x}_{t_0},\kappa\Delta t_i}$ by $\Delta t_1^2 = 16$ and $\Delta t_1 = 4$, respectively.

The scale free character of t-values can be observed by comparison of the t-values for $\mathbf{A}^{*}_{\Delta t_1}$ and $\mathbf{A}_{\dagger}$ as well as for $\mathbf{b}^{*}_{\Delta t_1}$ and $\mathbf{b}_{\dagger}$. In both cases, the estimated values are different but the t-values are indeed equal. One would expect the t-values also to be equal for $\mathbf{Q}^{*}_{\Delta t_1}$ and $\mathbf{G}_{\dagger}$ as well as for the six unstandardized cross-effects in PA(2), PA(3), PA(4), PA(6), PA(7), PA(8) and their standardized values computed in PA(10)-PA(15). It turns out that the t-values of the three standard deviations in $\mathbf{G}_{\dagger}$ are not one time but exactly two times those of the variances in $\mathbf{Q}^{*}_{\Delta t_1}$. This is a consequence of the square root transformation covering negative as well as positive values. Because negative values have to be excluded for standard deviations, one should stick to half the values found for the diagonals in $\mathbf{G}_{\dagger}$ as reported in Table 7.3. The t-values computed by LISREL for the standardized coefficient values inappropriately also take into account the variability of the standard deviations used in standardization. The scale of the variables can be chosen arbitrarily, however. If the standard deviations would have been specified in PA(10)-PA(15) as fixed quantities, the t-values would have been equal to those computed for $\mathbf{A}^{*}_{\Delta t_1}$ and $\mathbf{A}_{\dagger}$ as reported in Table 7.3.

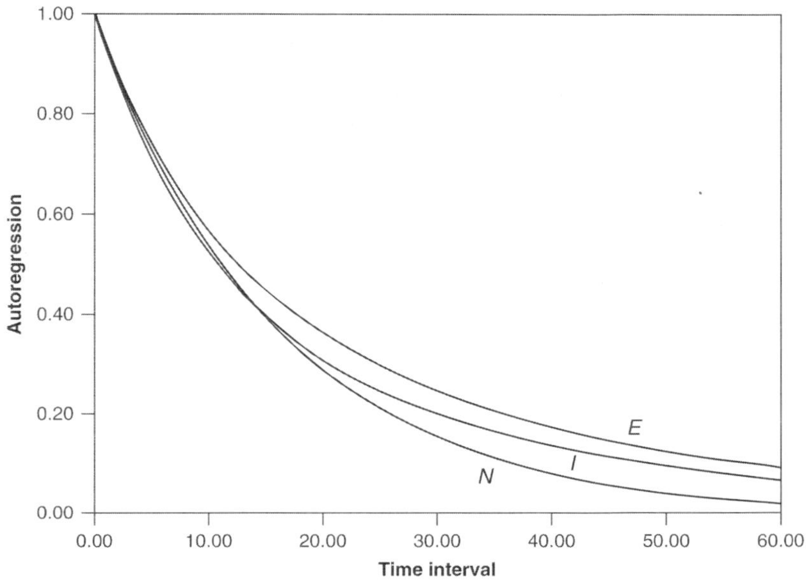

Fig. 7.10 Autoregression functions of I, N, and E, based on model EDM in Table 7.3.

For the EDM analysis a much more elaborate Mx input file was used than in the previous example (Section 7.5). The 45 GROUPS enable the estimation of the more general, time-varying model discussed in Oud and Jansen (2000). For estimating the present time-invariant model, only a subset of the GROUPS is needed, however. At the start of the input file *EDM-INE.mx*, it is clearly indicated in which GROUPS the time-invariant model matrices are specified. GROUP 44 in output file *EDM-INE.mxo* displays all the estimated matrices, reported in Tables 7.3-7.4. In GROUP 45 the standardized drift matrix is computed. Standard errors, on the basis of which the t-values reported in Tables 7.3-7.4 were computed, are displayed by Mx at the start of the output before GROUP 1. As in the case of the ADM, the t-values for the diffusion coefficients (standard deviations) in $\mathbf{G}$ have been halved. All discrete time matrices (see (7.17)) are found at the positions specified by (7.27) in $\mathbf{B}$ ("A" in GROUP 42) and in $\boldsymbol{\Psi}$ ("P" in GROUP 42).

To depict in continuous time the short-run and long-run implications of the model for Flandres, we conclude the example with the autoregression functions (Figure 7.10), cross-lagged effect functions (Figure 7.11), and mean trajectories (Figure 7.12), all based on the EDM estimates in Tables 7.3-7.4. Autoregression and cross-lagged effect functions have been computed using (7.5). The autoregressions and cross-lagged effects are computed not only for the actual observation intervals of 4 and 8 years, but interpolated and predicted for any interval over a rather extended prediction period. Although the differences between the autoregression functions

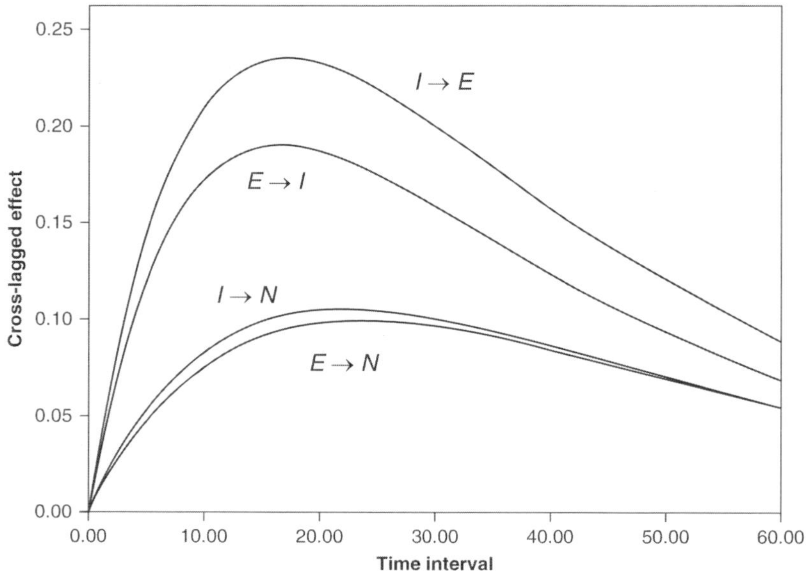

Fig. 7.11 Standardized cross-lagged effect functions for significant cross-effects of EDM in Table 7.3.

in Figure 7.10 are rather small (all three state variables have a strong tendency to persist over time, much stronger than, for instance, externalizing and internalizing problem behavior in Figure 7.7), the non-monotone character of the autoregression functions nicely illustrates the need for analyzing in continuous time. The strength order between N and I reverses after interval 14.2, N becoming the variable with the lowest autoregression. It means that a discrete-time analyst confronted, for example, with observation interval 16 or 20 would come to different conclusions than his colleague, working with interval 4 or 8. Continuous time analysis, however, prevents such erroneous conclusions by showing the complete picture.

In contrast to the the autoregression functions in Figure 7.10, the standardized cross-lagged effect functions in Figure 7.11 show monotonicity. For all intervals, at least over 0 to 60 years, the strength order between the four cross-lagged effect functions turns out to be the same as for the actual observation intervals of 4 and 8. Particularly, in the feedback loop between I and E, $I \longrightarrow E$ exceeds $E \longrightarrow I$ everywhere. With regard to the two relatively smaller effects, we observe that everywhere $I \longrightarrow N$ is slightly stronger than $E \longrightarrow N$. An interesting result of the continuous time analysis is further that all four cross-lagged effects reach their maximum quite some time after the empirical observation intervals of 4 and 8 years. The maximum of $I \longrightarrow E$ (0.235) is reached somewhat later, at interval 17.0, than the lower maximum of $E \longrightarrow I$ in the opposite direction (0.190), reached at 16.4. The maximum of $I \longrightarrow N$

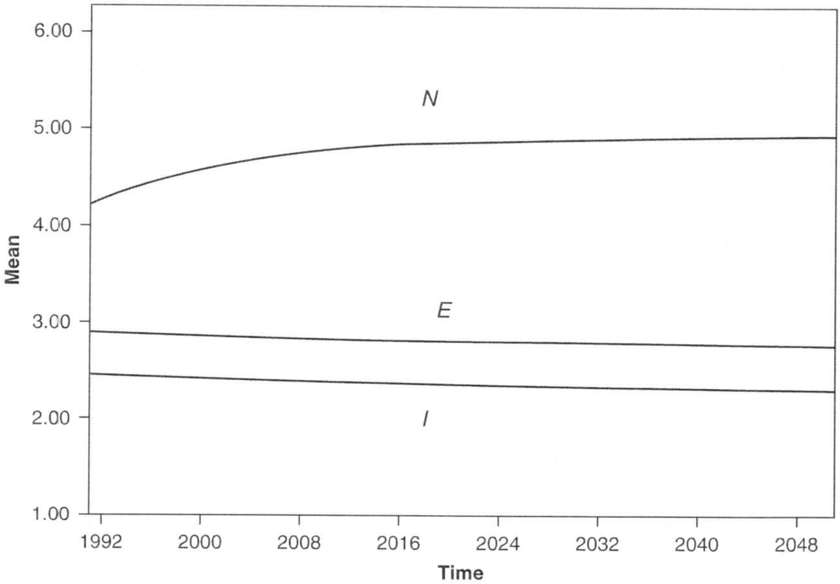

Fig. 7.12 Latent mean trajectories for I, N and E, based on Tables 7.3–7.4.

(0.105) at interval 22 is reached 1.2 years before the maximum of $E \rightarrow N$ (0.099) at 23.2.

Finally, Figure 7.12 shows that the mean values of Individualism and Ethnocentrism hardly changed in the data collection period 1991-1999 and are hardly expected to change in the prediction period. The mean of Individualism, starting at 2.46 in 1991 and decreasing to 2.43 in 1999, is expected to converge to final value 2.30. The mean of Ethnocentrism, starting at 2.90 in 1991 and decreasing to 2.87 in 1999, is expected to converge to final value 2.77. The mean of Nationalism, however, increased in the data collection period from 4.23 to 4.55, and a further limited increase for the near future is expected, but then the prediction levels off towards final value 4.91.

7.7 Conclusions

The development of externalizing and internalizing problem behavior in children or the attitude change in the Flemish electorate with regard to individualism, nationalism, and ethnocentrism are continuously evolving processes, rather than processes that show isolated, sudden changes at discrete points in time. The analyst, however, only observes at discrete points in time (for example, biennial, yearly or monthly observations). The typical approach in conventional (that is, discrete) time series modeling and panel data analysis is to ignore the continuous time nature of the processes underlying discrete time observations. Consequently, discrete time series and discrete panel data analyses are simplifications and often distortions of reality.

Comparability of results between different studies is the key to cumulative progress in science. Just because of the frequent model formulation and estimation in terms of the observation interval at hand, comparability is low in social and behavioral science. By means of the continuous-time approach using the exact discrete model (EDM) or approximate discrete model (ADM), explained in this chapter, the model parameters are made independent of the observation interval, and thus provide a common basis for accurate comparison of differently time-spaced models of the same or similar processes. As shown also in this chapter, if analysis results for the EDM or ADM from different authors use time scales in different units, they are easily translated into each other. Thus, results are made comparable without re-estimation being necessary.

Not all topics in continuous time analysis could be covered by the EDM-SEM and ADM-SEM procedures as expounded in the present chapter. We mention, in particular, time-varying models (Oud & Jansen, 2000) and models for oscillating movements (Oud, 2007a). In our conviction, however, the models presented in this chapter give a continuous time formulation to the typical kind of problems current longitudinal and panel research in social and behavioral science is involved with. A final but important topic not dealt with in the present chapter is the handling of incomplete data. In a longitudinal SEM context this can be solved in most cases by the expectation-maximization (EM) procedure using the Kalman smoother, explained in Oud and Jansen (1996) and applied in continuous-time modeling by Oud and Jansen (2000), or the individual likelihood procedure (Neale, 2000; Wothke, 2000) as implemented, for instance, in Mx.

References

Arnold, L. (1974). *Stochastic differential equations.* New York: Wiley.

Arminger, G. (1986). Linear stochastic differential equations for panel data with unobserved variables. In N. B. Tuma (Ed.), *Sociological methodology* (pp. 187-212). Washington: Jossey-Bass.

Bartlett, M. S. (1946). On the theoretical specification and sampling properties of autocorrelated time-series. *Journal of the Royal Statistical Society (Supplement), 7,* 27-41.

Blalock, H. M., Jr. (1969). *Theory construction: From verbal to mathematical formulations.* Englewood Cliffs, NJ: Prentice-Hall.

Bergstrom, A. R. (1966). Nonrecursive models as discrete approximations to systems of stochastic differential equations. *Econometrica, 34,* 173-182.

Bergstrom, A. R. (1984). Continuous time stochastic models and issues of aggregation over time. In Z. Griliches & M. D. Intriligator (Eds.), *Handbook of econometrics: Vol. 2* (pp. 1145-1212). Amsterdam: North-Holland.

Bergstrom, A. R. (1988). The history of continuous-time econometric models. *Econometric Theory, 4,* 365-383.

Boker, S., Neale, M., & Rausch, J. (2004). Latent differential equation modeling with multivariate multi-occasion indicators. In K. van Montfort, J. Oud, & A. Satorra (Eds.), *Recent developments on structural equations models: Theory and applications* (pp. 151-174). Dordrecht, The Netherlands: Kluwer Academic Publishers.

Browne, M. W. & R. Cudeck (1993). Alternative ways of assessing model fit. In K. A. Bollen & J. Scott Long (Eds.), *Testing structural equation models* (pp. 136-162). Newbury Park: Sage.

Bui, K. V. T., Ellickson, P. L., & Bell, R. M. (2000). Cross-lagged relationships among adolescent problem drug use, delinquent behavior, and emotional distress. *Journal of Drug Issues, 30,* 283-303.

Burke, J. D., Loeber, R., Lahey, B. B., & Rathouz, P. J. (2005). Developmental transitions among affective and behavioral disorders in adolescent boys. *Journal of Child Psychology and Psychiatry, 46,* 1200-1210.

Capaldi, D. M. (1992). Co-occurrence of conduct problems and depressive symptoms in early adolescent boys: II. A 2-year follow-up at grade 8. *Development and Psychopathology, 4,* 125-144.

Carlson, G. A., & Cantwell, D. P. (1980). Unmasking masked depression in children and adolescents. *American Journal of Psychiatry, 137,* 445-449.

Coleman, J. S. (1968). The mathematical study of change. In H.M. Blalock, Jr. & A. Blalock (Eds.), *Methodology in social research* (pp. 428-478). New York: McGraw-Hill.

De Bruyn, E. E. J., Scholte, R. H. J., & Vermulst, A. A. (2005). *Psychometric analyses of the Nijmegen problem behavior list (NPBL): A research instrument for assessing problem behavior in community samples using self- and other reports of adolescents and parents.* Nijmegen, The Netherlands: Institute of Family and Child Studies, Radboud University Nijmegen

Delsing, M. J. M. H., Oud, J. H. L., van Aken, M. A. G., De Bruyn, E. E. J., & Scholte, R. J. H. (2005). Family loyalty and adolescent problem behavior: The validity of the family group effect. *Journal of Research on Adolescence, 15,* 127-150.

Gold, M., Mattlin, M., & Osgood, D. W. (1989). Background characteristics and response to treatment of two types of institutionalized delinquent boys. *Criminal Justice and Behaviour, 16,* 5-33.

Gollob, H. F., & Reichardt, C. S. (1987). Taking account of time lags in causal models. *Child Development, 58,* 80-92.

Haselager, G. J. T., & van Aken, M. A. G. (1999). *Codebook of the research project Family and Personality: Vol. 1. First measurement wave.* Nijmegen, The Netherlands: University of Nijmegen, Faculty of Social Science.

Homans, G. C. (1950). *The human group.* New York: Harcourt, Brace & World.

Jöreskog, K. G. (1977). Structural equation models in the social sciences: Specification, estimation and testing. In Krishnaiah, P. R. (Ed.), *Applications of statistics* (pp. 265-287). Amsterdam: North-Holland.

Jöreskog, K. G., & Sörbom, D. (1976). *LISREL III: Estimation of linear structural equation systems by maximum likelihood methods: A FORTRAN IV program.* Chicago: National Educational Resources.

Jöreskog, K. G., & Sörbom, D. (1996). *LISREL 8: User's reference guide.* Chicago: Scientific Software International.

Kuo, H.-H. (2006). *Introduction to stochastic integration.* New York: Springer.

Neale, M.C. (2000). Individual fit, heterogeneity, and missing data in multigroup structural equation modeling. In T.D. Little, K.U. Schnabel, & J. Baumert (Eds.), *Modeling longitudinal and multilevel data* (pp. 219-240). Mahwah NJ: Lawrence Erlbaum.

Neale, M. C., Boker, S. M., Xie, G., & Maes, H. H. (2006). *Mx: Statistical Modeling* (7th ed.). Richmond, VA: Department of Psychiatry.

Neiderhiser, J. M., Reiss, D., Hetherington, E. M., & Plomin, R. (1999). Relationships between parenting and adolescent adjustment over time: Genetic and environmental contributions. *Developmental Psychology, 35,* 680-692.

Oud, J. H. L. (1978). *Systeem-methodologie in sociaal-wetenschappelijk onderzoek [Systems methodology in social science research].* Doctoral dissertation. Nijmegen: Alfa.

Oud, J. H. L. (2007a). Comparison of four procedures to estimate the damped linear differential oscillator for panel data. In K. van Montfort, J. Oud, & A. Satorra (Eds.), *Longitudinal models in the behavioral and related sciences* (pp. 19-39). Mahwah, NJ: Lawrence Erlbaum Associates.

Oud, J. H. L. (2007b). Continuous time modeling of reciprocal effects in the cross-lagged panel design. In S.M. Boker & M.J. Wenger (Eds.), *Data analytic techniques for dynamical systems in the social and behavioral sciences*. Mahwah, NJ: Lawrence Erlbaum Associates.

Oud, J. H. L., & Jansen, R. A. R. G. (1996). Nonstationary longitudinal LISREL model estimation from incomplete panel data using EM and the Kalman smoother. In U. Engel & J. Reinecke (Eds.), *Analysis of change: Advanced techniques in panel data analysis* (pp. 135-159). Berlin: Walter de Gruyter.

Oud, J. H. L., & Jansen, R. A. R. G. (2000). Continuous time state space modeling of panel data by means of SEM. *Psychometrika, 65,* 199-215.

Oud, J. H. L., Jansen, R. A. R. G., van Leeuwe, J.F.J., Aarnoutse, C. A. J., & Voeten, M. J. M. (1999). Monitoring pupil development by means of the Kalman filter and smoother based upon SEM state space modeling. *Learning and Individual Differences, 10,* 121-136.

Oud, J. H. L., van Leeuwe, J. F. J, & Jansen, R. A. R. G., (1993). Kalman filtering in discrete and continuous time based on longitudinal LISREL models. In J. H. L. Oud & A. W. van Blokland-Vogelesang (Eds.), *Advances in longitudinal and multivariate analysis in the behavioral sciences* (pp.3-26). Nijmegen: ITS.

Oud, J. H. L., & Singer, H. (2008). Continuous time modeling of panel data: SEM versus filter techniques. *Statistica Neerlandica, 62*, 4-28.

Overbeek, G. J., Vollebergh, W. A. M., Meeus, W. H. J., Luijpers, E. T. H., & Engels, R. C. M. E. (2001). Course, co-occurence and longitudinal associations between emotional disturbance and delinquency from adolescence to young adulthood: A six-year three-wave study. *Journal of Youth and Adolescence, 30*, 401-426.

Phillips, P. C. B. (1993). The ET Interview: A. R. Bergstrom. In P. C. B. Phillips (Ed.), *Models, methods, and applications of econometrics* (pp. 12-31). Cambridge, MA: Blackwell.

Rueter, M. A., & Conger, R. D. (1998). Reciprocal influences between parenting and adolescent problem-solving behavior. *Developmental Psychology, 34,* 1470-1482.

Simon, H. A. (1952). A formal theory of interaction in small groups. *American Sociological Review, 17,* 202-211.

Singer, H. (1990). *Parameterschätzung in zeitkontinuierlichen dynamischen Systemen [Parameter estimation in continuous time dynamic systems].* Konstanz: Hartung-Gorre.

Singer, H. (1991). *LSDE - A program package for the simulation, graphical display, optimal filtering and maximum likelihood estimation of linear stochastic differential equations: User's guide*. Meersburg: Author.

Singer, H. (1993). Continuous-time dynamical systems with sampled data, errors of measurement and unobserved components. *Journal of Time Series Analysis, 14*, 527-545.

Singer, H. (1995). Analytical score function for irregularly sampled continuous time stochastic processes with control variables and missing values. *Econometric Theory, 11*, 721–735.

Singer, H. (1998). Continuous panel models with time dependent parameters. *Journal of Mathematical Sociology*, 23, 77-98.

Toharudin, T., Oud, J. H. L., & Billiet, J. B. (2008). Assessing the relationships between Nationalism, Ethnocentrism, and Individualism in Flanders using Bergstrom's approximate discrete model. *Statistica Neerlandica, 62*, 83-103.

Tuma, N. B., & Hannan, M. (1984), *Social dynamics: Models and methods.* New York:Academic Press.

Vuchinich, S., Bank, L., & Patterson, G. R. (1992). Parenting, peers, and the stability of antisocial behavior in preadolescent boys. *Developmental Psychology, 38,* 510-521.

Wothke, W. (2000). Longitudinal and multigroup modeling with missing data. In T.D. Little, K.U. Schnabel, & J. Baumert (Eds.), *Modeling longitudinal and multilevel data* (pp. 219-240). Mahwah NJ: Lawrence Erlbaum.

Zadeh, L. A., & Desoer, C. A. (1963*). Linear system theory: The state space approach.* New York: McGraw-Hill.

Chapter 8
Five Steps in Latent Curve and Latent Change Score Modeling with Longitudinal Data

John J. McArdle and Kevin J. Grimm

Abstract This paper describes a set of applications of one class of longitudinal growth analysis - latent curve (LCM) and latent change score (LCS) analysis using structural equation modeling (SEM) techniques. These techniques are organized in five sections based on Baltes & Nesselroade (1979). (1) Describing the observed and unobserved longitudinal data. (2) Characterizing the developmental shape of both individuals and groups. (3) Examining the predictors of individual and group differences in developmental shapes. (4) Studying dynamic determinants among variables over time. (5) Studying group differences in dynamic determinants among variables over time. To illustrate all steps, we present SEM analyses of a relatively large set of data from the National Longitudinal Survey of Youth (NLSY). The inclusion of all five aspects of latent curve modeling is not often used in longitudinal analyses, so we discuss why more efforts to include all five are needed in developmental research.

8.1 Introduction

Many debates in developmental research conclude with a suggestion that the collection of longitudinal data is a necessary ingredient for the study of developmental phenomena. Methodological researchers have defined these issues in extensive detail, but most rely on "the explanation of inter-individual differences (or similarities) in intra-individual change patterns" (e.g., Wohlwill, 1973; Baltes & Nesselroade, 1979). During the last two decades, many methodologists have contributed to the

John J. McArdle
Department of Psychology, University of Southern California, Los Angeles
e-mail: jmcardle@usc.edu

Kevin J. Grimm
Department of Psychology, University of California, Davis, One Shields Avenue, Davis, CA 95616, USA
e-mail: kjgrimm@ucdavis.edu

K. van Montfort et al., *Longitudinal Research with Latent Variables*,
DOI 10.1007/978-3-642-11760-2_8, © Springer-Verlag Berlin Heidelberg 2010

knowledge base, and the classic models for "growth curve analysis" seem to have been revived as an important research technique (e.g., see Rogosa & Willett, 1985; McArdle & Epstein, 1987; Meredith & Tisak, 1990). The term growth curve analysis denotes the processes of describing, testing hypotheses, and making scientific inferences regarding growth and change patterns in a wide range of time-related phenomena. Of course, these *curves* are not limited to the phases where the organism *grows*, but it can also be used to describe and analyze phases where the organism *declines, accelerates, decelerates, oscillates,* or even *remains stable*.

This paper describes a set of applications of one class of longitudinal growth models - *latent change score (LCS)* analysis using *structural equation modeling (SEM)* techniques. These techniques can be presented in many ways, but we organize this information in five sections, as steps of developmental data analysis, based on a sequential rationale inspired by Baltes & Nesselroade (1979):

Step 1 – *Describing the Observed and Unobserved Longitudinal Data* – We consider some useful ways to summarize longitudinal data, including statistical information from both the complete and incomplete cases.

Step 2 – *Characterizing the Developmental Shape of Individuals and Groups* – We try to describe both the group and individual characteristics of development and demonstrate the general ease and flexibility of the SEM approach.

Step 3 – *Examining the Predictors of Individual and Group Differences in Developmental Shapes* – We recognize individual differences in growth may be the result of combinations of other measured variables. We describe how SEM can be used in both multilevel and multiple-group forms to provide empirical evidence for hypotheses concerning the correlates of individual longitudinal patterns.

Step 4 – *Studying Dynamic Determinants among Variables over Time* – We show how the time-dependent nature of the latent variables can be represented in SEM and used to study lead-lag relations using simple dynamic expressions.

Step 5 – *Studying Group Differences in Dynamic Determinants Among Variables over Time* – We show how the multi-group and latent mixture dynamic models can be fit to examine heterogeneous lead-lag relationships for different groups of individuals.

As the reader will notice, we first use latent curve modeling (LCM) to begin the analyses, but we then emphasize the direct use of latent change scores (LCS) for more clarity in the model alternatives. This clarification may assist the researcher in considering the alternative change models available. This point is important because the LCS allows us to rather easily join seemingly different concepts about change from classical models based on time-series and auto-regression or latent growth curve analyses.

As an illustration for the five steps, we present SEM analyses of data from the well-known and publicly available *National Longitudinal Survey of Youth* (NLSY)

– Children and Young Adults. In this study, the children of female respondents were repeatedly measured biennially from 1986 through 2000. The longitudinal data of the NLSY includes measures of achievement (e.g., Peabody Individual Achievement Test; PIAT; Dunn & Markwardt, 1970) and behavior problems (e.g., Behavior Problems Index; BPI; Zill, 1990). These analytic illustrations are used to convey the main presumptions and techniques as well as the benefits and limitations of these approaches in developmental research.

Our main goal is to present an overview of the general developmental methodology, and to demonstrate the practical and flexible utility of these methods for developmental research. We do not provide extensive mathematical and statistical details, but the computer input and output scripts for each step of the SEM analyses are available from our website `http://kiptron.usc.edu/`as well as from `http://www.econ.upf.edu/~satorra/longitudinallatent/readme.html` . Most importantly, the inclusion of all five aspects of latent curve modeling is often overlooked in longitudinal analyses, so we end by discussing why all five steps are needed in developmental investigations.

8.2 Step 1: Describing the Observed and Unobserved Longitudinal Data

The first step in any useful data analysis is an adequate description of the data. However, the collection and presentation of longitudinal data can be difficult, so the unique aspects of these data should be emphasized.

8.2.1 The National Longitudinal Survey of Youth – Children and Young Adults

The data examined here come from children who were measured at least once between age 8 and 14, so the overall $N = 6{,}790$. As previously mentioned, data collection occurred biennially with measurements occurring in every even year from 1986 through 2000. Figure 8.1 is a display of individual growth data for the (a) PIAT reading comprehension and (b) BPI antisocial behavior measure by age for a sub-sample of $n = 100$ randomly selected participants. The y-axis indexes the participants' scores and the x-axis is an index of the participants' age-at-testing. The connected lines in this figure are graphic descriptions of the change pattern for Reading Comprehension scores for each individual, so each line is termed a *growth curve* or *trajectory*. The plot allows us to see the overall trends for changes in achievement and antisocial behavior through childhood and adolescence as well as how the data are incomplete.

(A)

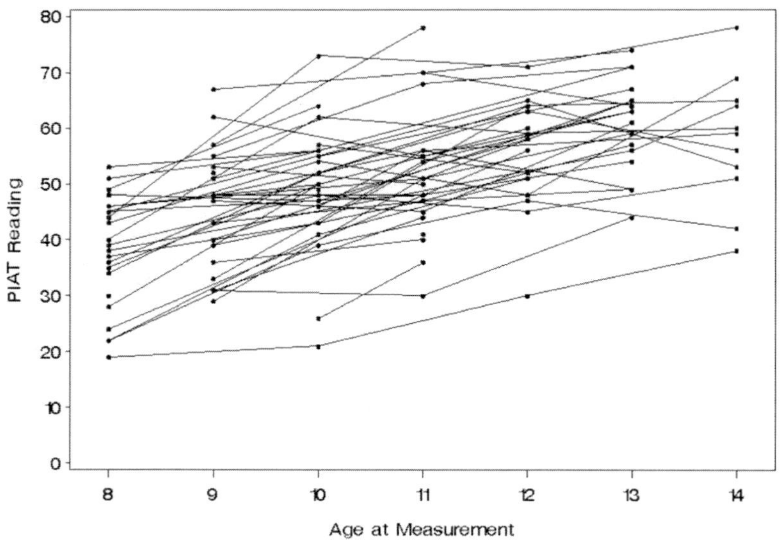

(B)

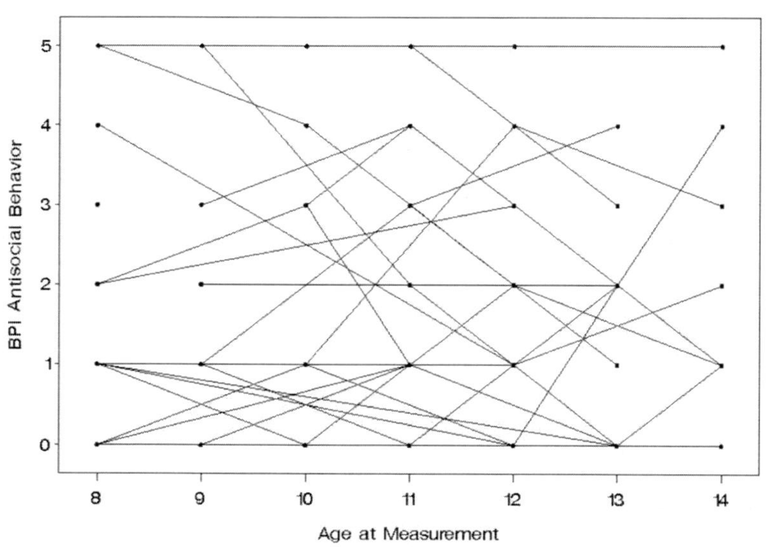

Fig. 8.1 Longitudinal plots of (A) Reading Comprehension from the Peabody Individual Achievement Test and (B) Antisocial Behavior from the Behavior Problems Index for a random sample of 100 participants.

8.2.2 Describing the Observed Data

The sample sizes, means, standard deviations, and correlations of these raw measures from age 8 to 14 are listed in Table 8.1. The means and standard deviations show a simple pattern from age 8 to 14 with increases in performance coupled with increases in variation. The correlations over time, the unique statistical information of longitudinal data, also present a relatively simple pattern of results with most correlations suggesting a relatively high level of the stability of individual differences (e.g., $r > .5$). We use SEM to test hypotheses about these longitudinal statistics.

8.2.3 Results from Dealing with Incomplete Information

The summary information presented in Table 8.1 is not limited to only those participants with complete data at all ages (no participants have complete data). To deal with this problem we present a description of the patterns of complete and incomplete data in Table 8.1c. The incomplete data patterns can be represented as the proportion of data or coverage for each covariance of these scores – at any time no more than 42% of the participants have data (age 8 and 9) and in some cases only 1% of the information is available (at ages 13 and 14).

In Table 8.1a and 8.1b we also use brackets to list an "incomplete data" estimate of the sample means, standard deviations, and correlations. These estimates are based on what is typically termed *full information maximum likelihood* (FIML; Little, 1995; McArdle, 1994; Cnaan, Laird, & Slasor, 1997). This approach allows us to examine the initial summary statistics "as if all persons were measured at all occasions" and, hopefully, deal with any selection bias in the longitudinal sampling strategy. These newly estimated statistics are not exactly the same as the pair-wise estimates, but they are not altered very much, indicating these data meet the minimal conditions of "missing at random" (MAR; Little, 1995). Most importantly, these estimated statistics do not suffer from some common statistical problems of pair-wise estimates based on different sample sizes, and use all available information from every person. As a result, we do not need to select a subset of persons because they have complete data, nor do we have to make our timing basis based on a data collection strategy (e.g., 1986, 1988, ..., 2000). Instead, we choose to examine age-at-testing (see equation (8.1) below) as opposed to wave-of-testing or year-of-testing, based on our developmental interest.

Let us be clear at the start that an age-based approach by itself does not guarantee that all model assumptions are met (e.g., see McArdle et al., 2002; cf., Sliwinski & Buschke, 1999; Miyazaki & Raudenbush, 2000). In fact, this age-based approach is not often used in standard developmental research, where it is much more likely to find time (e.g., the occasion of measurement) as the focal axis of development. However, in this specific case, the individuals were sampled from an ongoing developmental process that is likely to have strong age related components, and there was no common point of intervention except for the natural differences due to grade and maturity. The standard MAR assumptions that are needed do require an important belief on the part of the analyst – that the ways in which data are incomplete are somehow reflected in the data that are complete. While we think this is a reasonable

Table 8.1 Observed and unobserved summary statistics for the Peabody Individual Achievement Test Reading Comprehension (Read) and Behavior Problems Index Antisocial Behavior (Anti) scores from the National Longitudinal Survey of Youth data at eight time points (N = 6970; MLE-MAR estimates in brackets; Step 1, see Figures 8.1A and 8.1B

(a) Observed and unobserved means and standard deviations over age

Variable	N	Mean [MLE]	SD [MLE]	Skewness	Kurtosis	Min	Max
Read – Age 8	2847	31.1 [30.7]	9.8 [9.8]	.42	-.23	0	70
Read – Age 9	2833	36.7 [36.6]	10.3 [10.4]	.03	-.40	0	78
Read – Age 10	2660	41.5 [41.2]	10.6 [10.7]	-.09	.36	0	84
Read – Age 11	2566	44.7 [45.0]	11.4 [11.4]	-.13	.43	0	84
Read – Age 12	2226	48.1 [48.1]	11.4 [11.4]	-.15	.35	0	81
Read – Age 13	2047	50.3 [50.5]	12.1 [12.3]	-.19	.42	0	84
Read – Age 14	1734	52.1 [52.4]	12.0 [12.1]	-.24	.47	0	84
Anti – Age 8	3046	1.49 [1.53]	1.52 [1.53]	.97	.23	0	6
Anti – Age 9	2987	1.52 [1.52]	1.59 [1.59]	1.04	.36	0	6
Anti – Age 10	2722	1.52 [1.51]	1.61 [1.61]	.98	.17	0	6
Anti – Age 11	2644	1.54 [1.50]	1.63 [1.63]	.95	.02	0	6
Anti – Age 12	2287	1.60 [1.56]	1.62 [1.63]	.88	-.11	0	6
Anti – Age 13	2140	1.66 [1.60]	1.67 [1.68]	.81	-.35	0	6
Anti – Age 14	1798	1.70 [1.64]	1.72 [1.74]	.75	-.54	0	6

(b) Observed and unobserved correlations (each entry includes pairwise r and [MLE-MAR r])

	1.	2.	3.	4.	5.	6.	7.	8.	9.	10.	11.	12.	13.	14.
1. Read– Age 8	1.00													
2. Read– Age 9	.57	1.00												
	[.46]													
3. Read– Age 10	.63	.64	1.00											
	[.64]	[.58]												
4. Read– Age 11	.56	.69	.67	1.00										
	[.57]	[.69]	[.68]											
5. Read– Age 12	.57	.56	.68	.72	1.00									
	[.58]	[.60]	[.70]	[.65]										
6. Read– Age 13	.48	.62	.60	.66	.77	1.00								
	[.51]	[.62]	[.62]	[.67]	[.75]									
7. Read– Age 14	.53	.57	.62	.69	.67	.63	1.00							
	[.54]	[.59]	[.63]	[.69]	[.68]	[.71]								
8. Anti – Age 8	-.19	-.23	-.22	-.21	-.21	-.18	-.22	1.00						
	[-.20]	[-.20]	[-.22]	[-.24]	[-.22]	[-.17]	[-.24]							
9. Anti – Age 9	-.29	-.22	-.05	-.20	-.30	-.15	-.19	.58	1.00					
	[-.23]	[-.21]	[-.16]	[-.21]	[-.23]	[-.17]	[-.22]	[.53]						
10. Anti – Age 10	-.19	-.29	-.21	-.11	-.24	-.16	-.21	.59	.59	1.00				
	[-.17]	[-.25]	[-.20]	[-.19]	[-.22]	[-.21]	[-.21]	[.60]	[.64]					
11. Anti – Age 11	-.15	-.19	-.20	-.21	-.22	-.21	-.24	.52	.64	.57	1.00			
	[-.22]	[-.19]	[-.23]	[-.22]	[-.23]	[-.22]	[-.28]	[.50]	[.64]	[.57]				
12. Anti – Age 12	-.19	-.17	-.20	-.11	-.21	-.35	-.21	.50	.54	.59	.54	1.00		
	[-.18]	[-.13]	[-.20]	[-.15]	[-.20]	[-.21]	[-.22]	[.50]	[.54]	[.60]	[.58]			
13. Anti – Age 13	-.27	-.14	-.25	-.17	-.26	-.20	-.10	.45	.54	.49	.58	.48	1.00	
	[-.29]	[-.16]	[-.22]	[-.19]	[-.20]	[-.21]	[-.20]	[.45]	[.53]	[.52]	[.59]	[.51]		
14. Anti – Age 14	-.17	-.14	-.17	-.24	-.20	-.05	-.22	.45	.38	.53	.61	.60	.57	1.00
	[-.19]	[-.15]	[-.19]	[-.23]	[-.21]	[-.20]	[.24]	[.46]	[.51]	[.54]	[.60]	[.62]	[.61]	

Table 8.1 (Continued)

(c) Covariance coverage (proportion of participants with available data at each age and combination of ages)

	1.	2.	3.	4.	5.	6.	7.	8.	9.	10.	11.	12.	13.	14.
1. Read – Age 8	.42													
2. Read – Age 9	.02	.42												
3. Read – Age 10	.28	.02	.39											
4. Read – Age 11	.05	.28	.02	.38										
5. Read – Age 12	.22	.04	.25	.02	.33									
6. Read – Age 13	.04	.21	.04	.23	.02	.30								
7. Read – Age 14	.17	.03	.20	.03	.20	.01	.26							
8. Anti – Age 8	.38	.02	.28	.05	.22	.04	.17	.43						
9. Anti – Age 9	.02	.38	.02	.27	.04	.20	.03	.02	.42					
10. Anti – Age 10	.27	.02	.35	.02	.24	.04	.18	.28	.02	.38				
11. Anti – Age 11	.05	.27	.02	.34	.02	.22	.03	.05	.28	.02	.37			
12. Anti – Age 12	.21	.04	.24	.02	.29	.01	.19	.22	.04	.24	.02	.32		
13. Anti – Age 13	.04	.20	.04	.23	.02	.27	.01	.04	.21	.04	.23	.02	.30	
14. Anti – Age 14	.17	.03	.19	.03	.19	.01	.23	.17	.03	.19	.03	.20	.01	.25

set of assumptions, these will never be completely correct, and we try to point out critical junctures where a failure to meet MAR assumptions may be important.

8.3 Step 2: Characterizing Developmental Shapes for Groups and Individuals

The second step in a longitudinal data analysis is the attempt to highlight the key features of the data in terms of a *model*. In contemporary behavioral science research, one common approach to growth curve analysis is to write a *trajectory* equation for each group and individual. One such *trajectory* equation for *repeated measurements* of an observed variable, *Read*, at multiple times ($t = 1$ to T) for the same person ($n = 1$ to N), written in the mixed-model form of

$$Read[t]_n = g_{0n} + g_{1n} \cdot B[t] + e[t]_n \,. \tag{8.1}$$

This model includes three *unobserved or latent scores* representing the individual's (1) level (g_{0n}), (2) slope (g_{1n}) representing *linear change over time* and (3) independent errors of measurements ($e[t]_n$). To indicate the form of the systematic change, we use a set of group coefficients or *basis weights* (e.g., slope loadings) which define the timing or *shape of the trajectory over time* (e.g., $B[t] = t - 1$). It is typical to estimate the fixed group means for intercept and slopes (μ_0, μ_1) but also the implied random variance and covariance terms (σ_0^2, σ_1^2, σ_{01}) describing the distribution of individual deviations (d_{0n}, d_{1n}) around the group means. We also follow a traditional convention and assume there is a single random error variance within each time (σ_e^2), and the error terms are assumed to be normally distributed and uncorrelated with all other components. This final assumption about a single error variance mimics the assumptions of most other repeated measures models (e.g., Mixed-Effects ANOVA).

One important issue emerges when we recognize that there is nothing actually pre-defined about the basis of time ($B[t]$), and this allows us to investigate many alternative forms of the time axis (e.g., McArdle & Bell, 2000). For example, it may be more appropriate in this case to study multiple ages (e.g., age = 8 to 14) on the same person and write

$$Read[age]_n = g_{0n} + g_{1n} \cdot B[age] + e[age]_n \tag{8.2}$$

because using age as the basis of timing allows a more interpretable set of trajectories.

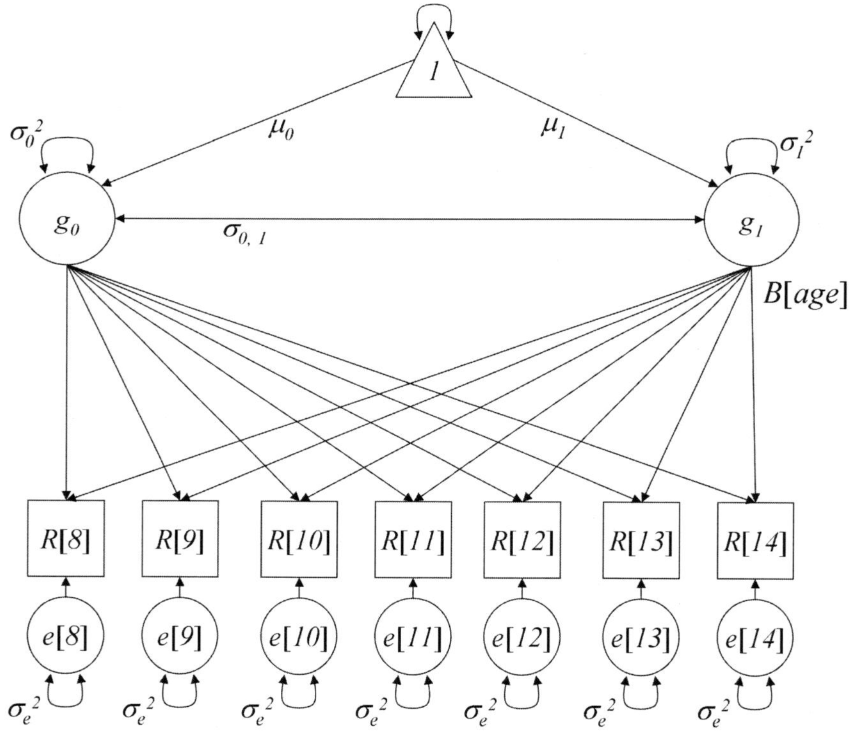

Fig. 8.2 Path diagram of a latent growth curve for Reading Comprehension.

A path diagram of this growth curve is presented in Figure 8.2 and is an exact translation of the necessary matrix algebra of these models (See Grimm & McArdle, 2005; McArdle, 2005; McArdle & McDonald, 1984). These diagrams can be conceptually useful devices for understanding the basic modeling concepts. In this path diagram the observed variables are drawn as squares, unobserved variables as circles, the required constant is included as a triangle, and parameters are labeled such that invariant parameters (e.g., residual variances) have the same label. Model parameters representing "fixed" or "group" coefficients are drawn as one headed arrows while "random" or "individual" features are drawn as two-headed arrows. In this case the level and slope are often assumed to be random variables with "fixed" means (μ_0, μ_1) but "random" variances (σ_0^2, σ_1^2) and covariance (σ_{01}). Of course, this is essentially a model based on means and covariances with MAR assumptions about the incomplete data.

8.3.1 Basic Linear Growth Models Results

Some initial growth curve modeling results for the NLSY Reading data are presented in Table 8.2. In these longitudinal models any change score (g_{1n}) is assumed to be constant *within* an individual but is not assumed to be the same *between* individuals. We do not estimate the unobserved scores but estimate several parameters that characterize the key features of the unobserved scores.

Table 8.2 Selected results from five latent growth models fitted to NLSY longitudinal data (N = 6970; Step 2).

Parameter	*2a*: No Growth	*2c*: Linear	*2d*: Latent
Fixed Effects			
Basis b[8]	=0	=0	=0
Basis b[9]	=0	=1	.28
Basis b[10]	=0	=2	.48
Basis b[11]	=0	=3	.66
Basis b[12]	=0	=4	.80
Basis b[13]	=0	=5	.92
Basis b[14]	=0	=6	=1
Level μ_0	42.1*	32.6*	30.7*
Slope μ_1	—	3.7*	21.7*
Random Effects			
Error σ_e^2	114.0*	42.7*	39.9*
Level σ_0^2	56.2*	61.1*	61.1*
Slope σ_1^2	—	1.0*	35.6*
Correlation ρ_{01}	—	.24*	.15*
Fit Indices			
Parameters	3	6	11
Degrees of Freedom	32	29	24
Log Likelihood	-66677	-61710	-61416
χ^2	10635	699	110
RMSEA	0.22	0.06	0.02

Note: The fit statistics for the time-based linear model (Model 2b) are not presented here because the data are considered to be different because of their organization.

The first model labeled 2a is the *no-growth* model fitted with only three parameters: a level mean ($\mu_0 = 42.1$), a level variance ($\sigma_0^2 = 56.1$), and an error variance ($\sigma_e^2 = 114.0$). The model yields a likelihood (L^2 = -66677) which shows the no-growth baseline is a poor fit compared to the totally unrestricted or saturated model ($\chi^2 = 10635$, $df = 32$), in which means, variances, and covariances are estimated for all observed variables. This model is typically used as a baseline against which to judge the fit of more informative models. The second growth model (2b) uses a fixed set of basis coefficients that change linearly with the *number of occasions* representing the time passed since the participant was enrolled in the study. Therefore, $B[t] = t - 1$ where t is the occasion number. Based on the data collection paradigm of the NLSY, a one-unit change in t represents a two-year change (e.g., measurements occurred every two-years from 1986 to 2000). This model has three additional parameters compared to the no-growth model: a slope mean (μ_1), variance (σ_1^2), and a level-slope covariance (σ_{01}). This model yielded a new likelihood (L^2 = -61856), which was a distinct improvement over the no-growth model ($\Delta - 2LL = 9644$ for 3 additional parameters). The resulting estimates describe a function that begins at 35.2 at the first occasion and increased by 7.3 units every two years. The variance estimates of the level and slope parameters were significant (σ_0^2 = 76.4.1, $\sigma_1^2 = 3.8$) indicating inter-individual differences in reading ability upon entering the study and in the linear change over time. Additionally, the level-slope correlation was .20 indicating a small positive relationship between children's reading performance upon entering the study and their linear rate of change. The error variance was estimated to be 41.6.

The second *linear growth* model (2c) was fit uses a fixed set of basis coefficients or slope loadings that change linearly with *age* and formed by taking $B[age] = (age-8)$, or the fixed values of $B[age] = [0, 1, 2, 3, 4, 5, 6]$). This linear scaling is only one of many that could be used, but was chosen to permit a practical interpretation of the slope parameters in terms of a *per-year change* and centers the level to represent 8 years of age. Therefore, the parameters related to the level reflect parameters associated with age 8. This linear growth model has three additional parameters compared to the *no-growth* model: a slope mean (μ_1) and variance (σ_1^2), and a level-slope covariance (σ_{01}). This model yields a new likelihood (L^2 = -61710) that represents a relatively large distance from the unrestricted model ($\chi^2 = 699$ on $df = 29$) but was an improvement over the no-growth model (*2c* vs *2a*: $\Delta\chi^2 = 9936$ on $\Delta df = 3$). The resulting means describe a function that started relatively low at age 8 ($\mu_0 = 32.6$) but increased by 3.7 units per year between ages 8 and 14 ($\mu_1 = 3.6$). The variance estimates of the level and slope parameters were significant ($\sigma_0^2 = 61.1$, $\sigma_1^2 = 1.0$) indicating inter-individual differences in the linear growth parameters. Additionally, the level-slope correlation was .24 indicating a small positive relationship between children's reading performance at age 8 and their linear rate of change. The error variance has been reduced ($\sigma_e^2 = 42.7$) compared to the *no-growth* model, which also indicates an improvement in fit.

The time-based and age-based models are not nested, which makes directly comparing their fit somewhat problematic. But this mimics a traditional problem that emerges when rescaling any X-variable in a traditional regression – if different transformations of X scores are used to predict Y scores the resulting parameters and fit can change. In the SEM framework the raw data always has the same likelihood and degrees of freedom, but different latent variable models based on $B[t]$ can have different likelihoods and degrees of freedom. One way to compare these models is based on likelihood comparison fit statistics, such as the AIC or BIC. In both cases here, the age-based model fit better. More importantly, the age-based model has important substantive interpretations. In this observational study we are observing the phenomena of changes in reading scores as they are unfolding. While the beginning of the study may be of paramount importance for the researchers, it is not likely that anything happened at this point to consider it important in the participants' lives (e.g., unlike a surgical procedure).

8.3.2 Nonlinearity Using Latent Basis Curves

An attractive nonlinear alternative of the linear growth model was proposed by Rao (1958) and Tucker (1958, 1966) in the form of summations of "latent curves" (see Meredith & Tisak, 1990). The use of this latent growth curve offers a simple way to investigate the shape of a growth curve - we allow the basis coefficients ($B[age]$) to take on a form based on the empirical data. In this approach we estimate the basis coefficients (e.g., $B[9-13]$) with the exception of two ($B[8]$ and $B[14]$) for identification purposes. In this latent basis model we end up with an optimal shape for the group curve and individual differences with one change component (see McArdle & Epstein, 1987; McArdle & Bell, 2000).

The fourth model fitted (2d) was this kind of latent basis growth model. For identification purposes, we fixed $B[8] = 0$ and $B[14] = 1$, but the remaining basis coefficients were estimated from the data. This results in a large improvement in the model likelihood ($L^2 = -61416$), which was much closer to the unrestricted model ($\chi^2 = 110$ on $df = 24$), and substantially better than the nested baseline ($\Delta\chi^2 = 10525$ on $\Delta df = 8$) and nested linear models ($\Delta\chi^2 = 589$ on $\Delta df = 5$). The error variance has also been reduced ($\sigma_e^2 = 39.9$). The estimated latent means were $\mu_0 = 30.7$ and $\mu_\iota = 21.7$, their variances were $\sigma_0^2 = 61.1$ and $\sigma_1^2 = 35.6$, and the intercept/slope correlation was $\rho_{01} = 0.15$. The estimated basis coefficients were .28, .48, .66, .80, and .92 for ages 9 – 13. The coefficients indicated a decelerating growth function. Additional nonlinear models, including multiphase (Cudeck & Klebe, 2002) and structured curves (Browne & du Toit, 1991) can be fit to these data, but are not described here. These additional nonlinear models may be able to adequately represent the data with fewer parameters. We refer the reader to Oud & Jansen (2000), Cudeck & Klebe (2002), Browne & du Toit (1991), and Ram & Grimm (2007) for further details regarding nonlinear models.

8.4 Step 3: Modeling Individual Differences in Developmental Scores and Patterns

The estimated means of the level and slope in the previous analyses allow us to plot the group trajectory over time. Similarly the estimated variance parameters allow us to consider the size of the between group differences at each age. However, no prior information obtained in model fitting tells us about the *sources of this variance*. To further explore the differences between persons we expand the basic latent growth model to include impacts on the latent parameters. There are several techniques to evaluate the sources of inter-individual differences and we consider three common methods including the growth model with an extension variable, the multiple group growth model, and the growth mixture model.

8.4.1 The Growth Model with an Extension Variable

Let us assume a variable termed X indicates some measurable difference between persons (e.g., sex, educational level). If we measure this variable at one occasion we might like to examine its influence in the context of a growth model for our outcome of interest (e.g., reading achievement). One popular model is based on the use of "adjusted" growth parameters as popularly represented in the *analysis of covariance*. In growth curve terms, this model is written with fixed (group) coefficients (γ) with some effect on the measured scores at each occasion ($Read[t]$), and the X is an independent observed (or assigned) predictor variable and written as

$$\begin{aligned} g_{0n} &= \nu_0 + \gamma_0 \cdot X_n + d_{0n}\,, \\ g_{1n} &= \nu_1 + \gamma_1 \cdot X_n + d_{1n}\,, \end{aligned} \tag{8.3}$$

where we have intercepts (ν) and regression slopes (γ) for the effect of X on the two latent components (g_0 and g_1) with residuals (d_0 and d_1). In this case the latent growth parameters ($\mu_{0:x}$, $\mu_{1:x}$, $\sigma_{0:x}$, $\sigma_{1:x}$, $\sigma_{0,1:x}$) are considered to be conditional on the expected values of the measured X variable. In the early factor analytic literature this relation between an observed X and a common factor score was termed an "extension analysis" (Horn, 1973). The apparent complexity of the covariance model leads to a simpler and increasingly popular way to add an external variable – we can write a *growth model with an extension variable* where the X variable has a direct effect on the parameters of the growth curve.

8.4.2 Results Growth Model with an Extension Variable

A variety of additional variables have been measured in the NLSY, including demographic (e.g., gender, mother's and father's education.), self reported health behaviors (e.g., smoking, drinking, physical exercise, etc.) and other problems (e.g., general health, illness, medical procedures, etc.). In the analyses presented here we consider two variables: gender (effect coded as -0.5 for males and $+0.5$ for females) and the mother's age at the child's birth (centered at 24 years of age).

We add gender and mother's age at child's birth as predictors of the level and slope. Table 8.3 is a list of results including the two variables as predictors of the level and slope. The model (3a) has a misfit ($\chi^2 = 125$ on $df = 34$) and this is an improvement when compared to the model in which the regression parameters were fixed at zero ($\chi^2 = 276$ on $df = 38$; $\Delta\chi^2 = 151$ on $\Delta df = 4$). The parameter estimates suggest the following interpretations. (0) The latent basis coefficients ($B[t]$) were unaffected by the inclusion of the predictors. (1) There were accurate (significant) differences between males and females on both the level and slope with females predicted to have a greater reading level at age 8 ($\gamma_0 = 1.9$), but a slightly slower rate of change from age 8 to 14 ($\gamma_1 = -1.3$). (2) The mother's age at the child's birth was also predictive of the level ($\gamma_0 = 0.19$) and slope ($\gamma_1 = 0.20$) of the growth model for reading comprehension. Older mothers at the child's birth were predicted to have children with a greater level of reading ability at age 8 as well as a faster rate of change from age 8 to 14.

Table 8.3 Results from latent growth models with extension variables fit to the NLSY longitudinal data (Step 3)

Parameters	*3a*: Level		*3a*: Slope
Fixed Effects			
Basis $B[t]$	= 0, .27*, .48*, .66*, .80*, .92*, = 1		
Intercept v_0	30.8*		22.3*
Regression from gender γ_g	1.9*		-1.3*
Regression from mother's age at birth γ_a	.19*		.20*
Random Effects			
Residual δ_d^2	59.6*		34.6*
Error σ_e^2		39.9*	
Correlation $\rho_{d0,ds}$		.15	
Fit indices			
Parameters		20	
Degrees of Freedom		34	
Log Likelihood		-97446	
χ^2		125	
RMSEA		.02	

8.4.3 Group Differences from a Multiple Group Perspective

The initial representation of group differences uses a set of estimated parameters to summarize between group differences. This idea is clearly represented by coding a set of variables (X) to characterizing the group differences and then examining the effect of this set (X) on the model parameters. However, this method is limited in a number of important ways. For example, some reasonable forms of group differences in the growth processes (e.g., different developmental shapes) are not possible within the standard framework. For example, different groups of people could have different "amplitude" or be in different "phases" in their growth pattern. These group differences in the features of growth are not separated within the basic level and change parameters although they may be realistic features of development.

An SEM treatment of this kind of a model uses concepts derived from multiple-group factor analysis (e.g., Jöreskog & Sörbom, 1979; McArdle & Cattell, 1994). In these kinds of models, each group (g = 1 to G) is assumed to follow a latent growth model where the basis coefficients ($B[t]^{(g)}$) are allowed to vary across groups. Since the groups need to be independent (each person can only be in one group) this kind of grouping is most easily done for discrete categorical variables (i.e., sex, but not educational level or maternal age at birth). A multiple group growth model (see McArdle, 1989) with age as the time-basis can be written as

$$Read[age]_n^{(g)} = g_{0n}^{(g)} + g_{1n}^{(g)} \cdot B[age]^{(g)} + e[age]_n^{(g)}. \tag{8.4}$$

This multiple group growth model permits the examination of the presumed invariance of the latent basis functions (i.e., $B[age]^{(1)} = B[age]^{(2)} = \ldots\ B[age]^{(g)} = \ldots\ B[age]^{(G)}$). The rejection of this model implies that each independent group has a different shape of growth. If invariance is found we can also examine the equality of the variances of the latent level and slope ($\sigma_0^{(g)} = \ldots\ \sigma_0^{(G)}$ and $\sigma_1^{(g)} = \ldots\ \sigma_1^{(G)}$) and their covariance ($\sigma_{01}^{(g)} = \ldots\ \sigma_{01}^{(G)}$). Further analyses could include the fixed effects (μ_0, μ_1), error deviations ($\sigma_e^{(g)}$), and functions of all the other parameters. These multiple group hypotheses represent additional types of group differences than was possible with the *growth modeling with an extension variable* approach.

8.4.4 Results for Group Differences in Growth of Reading for Males and Females

To illustrate this kind of analysis here, we fit multiple group growth models with gender as the grouping variable. Table 8.4 contains the parameter estimates and fit statistics for three models. In these cases the two groups were created, so the unrestricted likelihood for these data was based on two sets of mean and covariance matrices; one for males and one for females.

The first model (4a) allows both groups to have completely different latent growth curves. The model now includes 11 parameters for each group, and the 22 estimates are listed in the first two columns. This resulted in a reasonable fit to both data sets (χ^2 = 131 on df = 41). A few small differences in estimates can be seen between the two groups, but one key difference appears to be the smaller slope

Table 8.4 Numerical results from multiple group latent growth models fitted to male and female NLSY longitudinal data (Step 3)

Growth Model Parameters	*4a*: Latent Growth for Gender		*4b*: Loading Invariance over Both Groups		*4c*: Total Invariance over Both Groups
Fixed Effects	Males $n = 3448$	Females $n = 3342$	Males $n = 3448$	Females $n = 3342$	
Basis $b[8]$	=0	=0	=0		=0
Basis $b[9]$	.27*	.29*	.28*		.28*
Basis $b[10]$	.48*	.48*	.48*		.48*
Basis $b[11]$	.67*	.64*	.66*		.66*
Basis $b[12]$	.80*	.79*	.80*		.80*
Basis $b[13]$	.93*	.90*	.92*		.92*
Basis $b[14]$	=1	=1	=1		=1
Level μ_0	29.8*	31.5*	29.7*	31.6*	30.7*
Slope μ_1	22.1*	21.4*	22.4*	21.1*	21.7*
Random Effects					
Error σ_e^2	38.3*	41.4*	38.3*	41.5*	39.9*
Level σ_0^2	66.3*	54.2*	66.2*	54.1*	61.6*
Slope σ_1^2	41.7*	28.2*	42.6*	27.3*	35.6*
Correlation ρ_{01}	.21*	.11*	.20*	.11*	.15*
Fit Statistics					
Parameters	22		17		11
Degrees of Freedom	48		53		59
Log Likelihood	-61362		-61366		-61416
χ^2	131		138		237
RMSEA	.02		.02		.03

variance for the females. The second model (4b) adds the restriction that the latent basis coefficients, while free to vary, must be identical across males and females. This model was similar in fit to the free model ($\chi^2 = 138$ on $df = 53$; $\Delta\chi^2 = 7$ on $\Delta df = 5$), and this indicates the shapes of the curves may be considered the same across gender.

The third model (4c) adds the restriction that all parameters, while free to vary, must be identical across males and females. This model showed a loss in fit ($\chi^2 = 237$ on $df = 59$) compared to the previous model (4b vs. 4c: $\Delta\chi^2 = 99$ on $\Delta df = 6$), indicating some of the latent means and/or covariances are different. As previously seen in the model with gender as an extension variable, the growth factor means were somewhat different between males and females. Additionally, it appears that the slope variances were also somewhat different.

8.4.5 Mixture Models for Latent Groups

Another fundamental problem is the discrimination between models of (a) multiple curves for one group of people from (b) *multiple groups of people with different curves*. It is possible for us to have, say, three clusters of people, each with a distinct growth curve, but when we aggregate information over all people we end up with a complex growth pattern with multiple growth factors for a single population as opposed to a simple growth pattern for three groupings of people. This is the essence of a latent grouping of people, and parallels the "person centered approach" to multivariate data analysis (e.g., Cattell, 1980; Magnussen, 2003).

The recent series of models termed *growth mixture models* have been developed for this purpose (Muthén & Muthén, 2000; Muthén & Shedden, 1999; Nagin, 1999; Wedel & DeSarbo, 1995). In these analyses the distribution of the latent parameters are assumed to come from a "mixture" of two or more overlapping distributions. Current techniques in mixture models have largely been developed under the assumption of a small number of discrete or probabilistic "classes of persons" based on mixtures of multivariate normals. More formally, we can write a model as a probability weighted sum of curves where the probability of class membership (π_{cn}) is defined for the person in c = 1 to C classes. With a age-based growth curve as the within-class model we can write the growth mixture model as

$$\begin{aligned} Read[age]_n &= \sum_{c=1}^{C} \pi_{cn} \left(g_{0n}^{(c)} + g_{1n}^{(c)} \cdot B[age]^{(c)} + e[age]_n^{(c)} \right) \\ \text{where } &\sum_{c=1}^{C} \pi_{cn} = 1 \text{ and } 0 \leq \pi_{cn} \leq 1 \,. \end{aligned} \tag{8.5}$$

In this kind of growth mixture analysis we estimate the threshold parameter for the latent distribution (τ_p, for the *pth* parameter) while simultaneously estimate separate model parameters for the resulting latent groups.

The growth mixture models may be seen as a *model-restricted fuzzy-set cluster analysis* – a multiple group model without exact knowledge of group membership for each individual. The concept of an unknown or latent grouping can be successively based on the logic of multiple group factorial invariance. The resulting estimates yield a likelihood which can be compared to the results obtained from a model with one less class, so the mixture model distribution can be treated as a hypothesis to be investigated. As in standard discriminant analysis, we can also estimate the probability of assignment of individuals to each class in the mixture. In growth mixture modeling, it is important to fully examine how the latent classes differ from one another. Building on the work of multiple group growth models, described above, we examine differences in the basis coefficients (i.e., $B[age]^{(1)} = \ldots\ B[age]^{(c)} = \ldots\ B[age]^{(C)}$). The rejection of this model implies that each latent class has a different shape of growth. If invariance is found we can also examine the equality of the variances of the latent level and slope ($\sigma_0^{(c)} = \ldots\ \sigma_0^{(C)}$ and $\sigma_1^{(c)} = \ldots\ \sigma_1^{(C)}$) and their covariance ($\sigma_{01}^{(c)} = \ldots\ \sigma_{01}^{(C)}$). Further analyses could include the fixed effects (μ_0, μ_1), error deviations ($\sigma_e^{(g)}$), and functions of all the other parameters.

8.4.6 Results from Latent Mixture Models

These latent growth mixture models were fit using the NLSY reading data and some of the results are described here. In a first latent mixture model (4d) we estimated a two-class model with free parameters for both groups. This model required 23 parameters and led to another likelihood ($L^2 = -61072$). We recognize the statistical basis of this comparison is still somewhat controversial, but if we consider the threshold as an implied parameter in some previous models, we can get some sense of the gain in fit. The threshold parameter is a point estimate of the position on the outcome distribution where the individuals would be separated in classification into one group or another. In this case, the threshold ($\tau = -1.29$) is a z-score that suggests the total group can be considered a mixture of two classes of different sizes, n_1 = 1,463 and n_2 = 5,327, with different growth patterns between groups but the same growth pattern within groups. By contrast to the one-class model ($L^2 = -61416$) this 2-class model appears to be an improvement; however, numerical instability (and convergence problems) was found (i.e.; for one of the classes as the level variance was near zero). In a second model the level variance and level/slope covariance was fixed at zero in the first class. The result was a model with most participants categorized into the second class ($n = 6121$; $\tau = -2.07$).

In second set of latent mixture model (4e) we allowed the possibility of two latent classes ($C = 2$) with different parameters for the latent means and variance but assumed the same growth basis. This model resulted in a model with convergence problems for the same reasons as the previous model (4d). Finally, the latent means were allowed to vary between latent classes, but the remaining parameters were forced to be equal across latent classes. This final mixture model resulted in convergence problems as the estimated within-class level-slope correlation was greater than 1. Therefore, the results from these growth mixture models did not provide any evidence of latent classes with divergent growth patterns. It's important to remember there was variability in the growth factors (Model 2c), but the results from these mixture models confirms that this variation was distributed normally.

8.5 Step 4: Studying Dynamic Determinants across Multiple Variables

In recent research we have considered some ways to improve the clarity of the basic dynamic change interpretations with conventional SEM analytic techniques. These dynamic change hypotheses have led to the development of a set of alternative models, based on classical principles of dynamic change, but represented in the form of *latent change scores* (e.g., McArdle, 2001; McArdle & Nesselroade, 1994). This alternative representation makes it relatively easy to represent a dynamic hypothesis about the change within a variable, and about the time-ordered determination of one variable upon another.

8.5.1 Modeling Latent Change Scores

The introduction of multiple variables at each longitudinal occasion of measurement leads naturally to questions about time-dependent relationships among growth. A classical SEM for multiple variables over time is based on a *latent variable cross-lagged regression model* (see Cook & Campbell, 1977; Rogosa, 1978). This model can be written for latent scores with over-time auto-regressions (φ_y, φ_x) and cross-regressions (δ_{yx}, δ_{xy}) for time-lagged predictors, but the standard applications of this model do not include systematic growth components (i.e., individual slopes). For this reason, recent SEM analyses have examined *parallel growth curves*, including the correlation of various components (McArdle, 1988, 1989; Willett & Sayer, 1994). A popular alternative used in multilevel and mixed effects modeling is based on the analysis of covariance with $X[t]$ as *time-varying covariates*. In this model the regression coefficient (e.g., $X[t] \rightarrow Y[t]$) is usually assumed to be the same at all occasions. These last two models are easy to implement using existing computer software (e.g., Sliwinski & Buschke, 1999; Sullivan et al., 2000; Verbeke et al., 2000), but the typical applications are limited to a few basic forms of dynamic hypotheses.

To expand our SEM for other dynamic concepts we now reconsider the trajectory equations from a different starting point. First, we assume we have a pair of observed scores ($Y[t]$ and $Y[t-1]$) measured over a defined interval of time ($\Delta t = 1$), and write a model with latent scores ($y[t]$ and $y[t-1]$), and corresponding errors of measurement ($e[t]$ and $e[t-1]$). We can now define a new latent variable that represents the change in the latent scores for y. The *latent change score* is defined as in equation (8.6a). This latent change score is not the same as an observed change score ($\Delta Y[t]_n$) because the latent score is considered separate from the model based error component. Now we can write the trajectory over time in the observed variables as with (8.6b).

$$\Delta y[t]_n = y[t]_n - y[t-1]_n\,, \tag{8.6a}$$

$$Y[t]_n = g_{0n} + \left(\sum_{t=2}^{T} \Delta y[t]_n\right) + e[t]_n\,. \tag{8.6b}$$

Of course, the main alteration in this approach is that in this LCS representation we do not directly define the basis coefficients ($B[t]$; as in equation (8.1)), but instead we directly define *change as an accumulation of the first differences among latent variables*. This deceptively simple algebraic device allows us to define the trajectory equation as an accumulation of the latent changes ($\Delta y[t]$) up to time t *based on any model of change*.

One benefit of this LCS approach is that all of the previous latent growth models can be re-conceptualized in terms of first differences, and some new models emerge (as in McArdle & Nesselroade, 1994; McArdle, 2001, 2009; McArdle & Hamagami, 2001). We first re-iterate traditional models and then present some new models. We can start with the simple baseline model of *no change* by stating (8.7a), so that this difference model represents a trajectory with (8.7b).

$$\Delta y[t]_n = 0, \tag{8.7a}$$

$$Y[t]_n = g_{0n} + e[t]_n. \tag{8.7b}$$

Thus, the baseline model allows systematic individual differences at all occasions, and random error at all occasions, but no systematic changes over time.

In contrast, we can write (8.8a), so that this change model represents a trajectory with (8.8b).

$$\Delta y[t]_n = g_{1n}, \tag{8.8a}$$

$$Y[t]_n = g_{0n} + \left(\sum_{t=2}^{T} g_{1n}\right) + e[t]_n. \tag{8.8b}$$

So,

$$Y[1]_n = g_{0n} + e[1]_n,$$
$$Y[2]_n = g_{0n} + g_{1n} + e[2]_n,$$
$$Y[3]_n = g_{0n} + g_{1n} + g_{1n} + e[3]_n,$$

or, in general,

$$Y[t]_n = g_{0n} + g_{1n}(t-1) + e[t]_n,$$

and so the trajectory is linear over time.

As another alternative, we can consider a model where the changes are directly proportional to the previous latent score by writing (8.9a) and this change model represents a trajectory with (8.9b).

$$\Delta y[t]_n = \beta \cdot y[t-1]_n \tag{8.9a}$$

$$Y[t]_n = g_{0n} + \left(\sum_{t=2}^{T} \beta \cdot y[t-1]_n\right) + e[t]_n \tag{8.9b}$$

So

$$Y[1]_n = g_{0n} + e[1]_n,$$
$$Y[2]_n = g_{0n} + (\beta \cdot y[1]) + e[2]_n,$$
$$Y[3]_n = g_{0n} + (\beta \cdot y[1] + \beta \cdot y[2]) + e[3]_n,$$

and so on.

This accumulated trajectory is an exponentially accelerating function over time. As yet another alternative, we can write a composite change expression model where we consider both a systematic constant change (g_{1n}) and a proportional change (β) over time. The change equation for this dual change score model can be written as (8.10a) and this change model represents a trajectory with (8.10b).

$$\Delta y[t]_n = g_{1n} + \beta \cdot y[t-1]_n, \tag{8.10a}$$

$$Y[t]_n = g_{0n} + \left(\sum_{t=2}^{T} g_{1n} + \beta \cdot y[t-1]_n \right) + e[t]_n. \tag{8.10b}$$

So

$$\begin{aligned} Y[1]_n &= g_{0n} + e[1]_n, \\ Y[2]_n &= g_{0n} + (g_{1n} + \beta \cdot y[1]) + e[2]_n, \\ Y[3]_n &= g_{0n} + (g_{1n} + \beta \cdot y[1] + g_{1n} + \beta \cdot y[2]) + e[3]_n, \end{aligned}$$

or, in general,

$$Y[t]_n = g_{0n} + g_{1n}(t-1) + \left(\sum_{t=2}^{T} \beta \cdot y[t-1] \right) + e[t]_n.$$

This accumulating of the composite change model (8.10a) leads to a potentially complex nonlinear growth trajectory (8.10b). Depending on the sign and size of the coefficients, this nonlinear growth trajectory follows an increasing or decreasing, accelerating or decelerating exponential form (e.g., $Y[t]_n = c_{0n} + c_{1n} \cdot (1 - e^{\pi \cdot t}) + e[t]_n$).

Of course, this use of latent change scores is a generic approach that can be extended to many other forms of change models. For example, McArdle (2001) examined the proportional change model with an independent residual (i.e., an autoregressive model) as well as a model of changes in the common factor scores. Hamagami & McArdle (2007) investigated the forms of changes based on second order difference operators. A key feature of this latent change score approach to defining trajectories over time is that we are not limited to the models discussed here. Instead, the latent change score approach opens up possibilities for other parametric analyses of repeated observations.

An immediate benefit of this approach is seen when we deal with multiple variables over time. In a simple case, we can first organize the model into a set of *bivariate dynamic change score* equations that relate the latent changes in each variable to the previous states of those variables and a constant change component. If we use the simple starting points of the models considered above, one set of dynamic equations can be written as

$$\begin{aligned} \Delta y[t]_n &= g_{1n} + \beta_y \cdot y[t-1] + \gamma_{yx} \cdot x[t-1] \\ \Delta x[t]_n &= h_{1n} + \beta_x \cdot x[t-1] + \gamma_{xy} \cdot y[t-1] \end{aligned} \tag{8.11}$$

Where g_{1n} and h_{1n} are the constant change components for y and x, β_y and β_x are the proportional change parameters describing how each variable influences itself over time, and γ_{yx} and γ_{xy} are the coupling parameters describing how each variable influences each other over time. It may be useful to note that all the desirable latent slope parameters are not jointly identifiable, so we typically estimate only the latent means (μ_{g1} and μ_{h1}; see Figure 8.3). Also, to simplify the expressions, we start with an explicit repetition of all model parameters across each time (i.e., β_x, β_y, γ_{yx}, and γ_{xy} do not depend on t), and we recognize this is not a necessary feature of real data. This simplified form of a bivariate trajectory model is depicted as a path diagram in Figure 8.3.

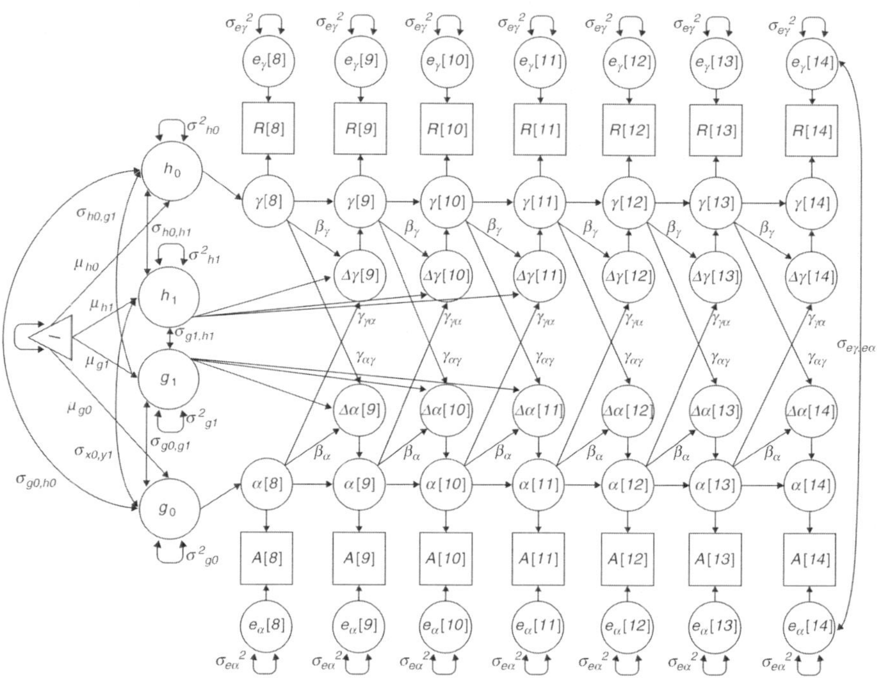

Fig. 8.3 Path diagram of a bivariate latent difference score model for Reading Comprehension and Antisocial Behavior.

In this simplified form of a bivariate dynamic system we assume a dual change score model within each variable but also permit *coupling* parameters (γ) across different variables. This model is used to estimate the time-dependent effect of latent $x[t]$ on $\Delta y[t+1]$ (γ_{yx}) as well as coupling parameter representing the time-dependent effect of latent $y[t]$ on $\Delta x[t+1]$ (γ_{xy}). This model subsumes all aspects of the previous cross-lagged, correlated growth, and time-varying covariate models as special cases. These latent change score models can lead to more complex nonlinear trajectory equations (e.g., non-homogeneous equations) and the use of latent change scores allow for the analysis of a variety of dynamic models using standard SEM (for more detailed explanations, see McArdle, 2001, 2009).

8.5.2 Results from Fitting Latent Change Score Models

The latent change score dynamic models were fitted to the reading comprehension variable and to the antisocial behavior scores. Several alternative Reading/Antisocial Behavior bivariate coupling models based on Figure 8.3 were fitted to the data. In the first model (5a), the coupling parameters (γ) were fixed to zero so the changes in reading and antisocial behaviors were not time-dependent. In the second model (5b), the coupling parameter from reading to changes in antisocial behavior was estimated whereas the coupling parameter from antisocial behavior to changes in reading was fixed to zero. In the third model (6c), the coupling parameter from reading to changes in antisocial behaviors was fixed to zero and the coupling parameter from antisocial behavior to changes in reading was estimated. These two models (5b and 5c) test whether reading was a leading indicator of changes in antisocial behavior (5b) and whether antisocial behavior was a leading indicator of changes in reading (5c). The final model (5d) was the bidirectional coupling model in which both coupling parameters were estimated.

The fitting of a sequence of alternative models was needed to interpret the replicability of the *coupling* across the reading and antisocial variables. Table 8.5 contains parameter estimates and fit statistics for the four bivariate dynamic models fit to reading and antisocial behaviors to determine whether one or more of the coupling parameters (γ) were different from zero. In the first model (5a), the coupling parameters were fixed at zero and led to a likelihood of $L^2 = -92162$. This model can be used as a baseline for comparison for the models in which coupling parameters were estimated. In the second model (5b) the parameter representing the effect of antisocial behavior on changes in reading was fixed to zero; however the effect of *reading* on changes in antisocial behaviors was estimated. This model resulted in a slight improvement in fit ($\Delta\chi^2 = 8$ on $df = 1$) compared to the *no coupling* model (5a). Similarly, the third model in which the coupling parameter from antisocial behavior to changes in reading was estimated and the parameter from reading to changes in antisocial behaviors was fixed to zero resulted in an improvement in fit ($\Delta\chi^2 = 8$ on $df = 1$) compared to the *no coupling* model. Finally, the bidirectional coupling model (5d) was fit and was an improvement over the *no coupling* model ($\Delta\chi^2 = 17$ on $df = 2$) and the two unidirectional coupling models (5b and 5c; $\Delta\chi^2 = 9$ on $df = 1$). Therefore, Model 5d, in which reading and antisocial behaviors were both dynamically related, was the most reasonable representation of the time-dependent relationships. The resulting interpretation is a dynamic process where scores on reading achievement have a tendency to impact changes in antisocial behavior in a positive manner and antisocial behavior has a tendency to effect subsequent change in reading achievement negatively. Therefore, children who have

Table 8.5 Results of bivariate latent change score dynamic models fitted to PIAT Reading Comprehension and BPI Antisocial Problem Behaviors (Step 4)

Model Parameters	*5a: No Coupling*		*5b: Read → ΔAnti*		*5c: Anti → ΔRead*		*5d: Bidirectional Coupling*	
	Read	Anti	Read	Anti	Read	Anti	Read	Anti
Fixed Effects								
Initial Mean μ_0	30.8*	1.5*	30.8*	1.5*	30.8*	1.5*	39.8*	1.5*
Slope Mean μ_1	11.7*	-.2*	11.8*	-.3*	13.6*	-.2*	13.7*	-.3*
Proportion β	-.19*	.16*	-.19*	.13*	-.19*	.15*	-.19*	.12*
Coupling γ	—	—	—	.004*	-1.32*	—	-1.32*	.004*
Random Effects								
Error Variance σ_e^2	39.9*	1.0*	39.9*	1.0*	39.7*	1.0*	39.7*	1.0*
Initial Variance σ_0^2	61.0*	1.5*	61.0*	1.5*	61.1*	1.5*	61.1*	1.5*
Slope Variance σ_1^2	5.5*	.06*	5.5*	.05*	5.6*	.05*	5.7*	.04*
Correlation ρ_{01}	.75*	-.90*	.75*	-.82*	.51*	-.88*	.52*	-.79*
Correlation ρ_{r0a0}	-.30*		-.30*		-.32*		-.31*	
Correlation ρ_{r1a1}	.22*		.05		-.31		-.40	
Correlation ρ_{r0a1}	.26*		.10		.26*		.09	
Correlation ρ_{r1a0}	-.30*		.29*		.38		.38	
Correlation ρ_{erea}	.01		-.01		-.00		-.00	
Fit Statistics								
Parameters	19		20		20		21	
Degrees of Freedom	100		99		99		98	
Log Likelihood	-92162		-92158		-92158		-92154	
χ^2	236		228		228		219	
RMSEA	.01		.01		.01		.01	

a greater reading comprehension scores tend to show slightly more positive changes in antisocial behaviors (negatively valenced) and children displaying more antisocial behaviors tend to show less positive changes in reading comprehension.

The estimated model parameters were dependent on the scalings used, but the trajectory expectations allow us to interpret the results in a relatively "scale-free" form – Figure 8.4 gives a summary of this state-space plot as a *vector field* (for details, see Boker & McArdle, 1995; McArdle et al., 2001). Any pair of coordinates is a starting point (e.g., intercept for reading and antisocial behavior) and the directional arrow is a display of the expected pair of 1-year changes from this point. This figure shows an interesting dynamic property – *the change expectations of a dynamic model depend on the starting point*. From this perspective, we can also interpret the negative level-level correlation ($\rho_{r0,a0} = -.31$), which describes the placement of the individuals in the vector field, and the slope-slope correlation ($\rho_{r1,a1} = -.40$), which describes the location of the subsequent change scores for individuals in the vector field. The resulting "flow" shows a dynamic process where reading comprehension and antisocial behavior scores have a tendency to impact changes in each other from age 8 to 14.

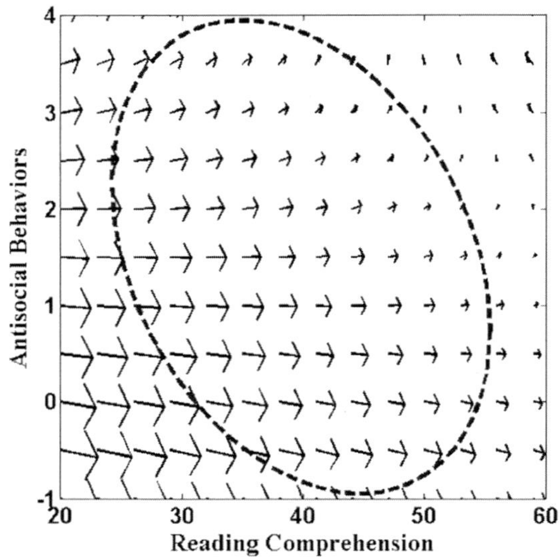

Fig. 8.4 Vector field for the pattern of results from the bivariate latent change score model for Reading Comprehension and Antisocial Behavior.

8.6 Step 5: Studying Group Differences in Dynamic Determinants across Multiple Variables

The final step was to examine whether there were group differences in the dynamic time-dependent associations. That is, we want to determine whether there were group differences in the lead-lag relationships (*Read* $\rightarrow$ Δ*Anti*; *Anti* $\rightarrow$ Δ*Read*). The models for this step are a combination of the previous two steps (Group Differences & Dynamics). That is, the bivariate dual change score model with bidirectional coupling is brought into a multiple group and latent mixture framework to study differences in lead-lag relationships for observed and unobserved groups.

8.6.1 Results for Multiple Group Dynamic Models with Gender

As in the univariate multiple group models for reading we begin with a model in which all of the parameters were separately estimated for males and females. This model yields a fit ($L^2 = -91876$) which is reasonable ($\chi^2 = 356$ on $df = 196$) given the observed data for males and females. The estimated parameters were similar for males and females; however there were some small interesting differences. For example, the males tended to be more antisocial than females at age 8 and the effect of reading on changes in antisocial behavior appeared to be stronger for females than males. The dynamic parameters (β, γ) were then constrained to be equivalent for males and females and the resultant fit ($L^2 = -91878$; $\chi^2 = 360$ on $df = 200$) was similar ($\Delta\chi^2 = 4$, $\Delta df = 4$) to when the dynamic parameters were independently estimated for males and females indicating the lead-lag relationships that exist between reading and antisocial behavior were the same for males and females. Next, the variance/covariance parameters were set equal for males and females, which resulted in a substantial loss in fit ($\chi^2 = 615$ on $df = 213$). Therefore, there were variance/covariance parameters that were significantly different for males and females. From the previous model, it appeared the level and slope variances for reading and antisocial behaviors were greater for males than females. Additionally, males had greater level of antisocial behavior at age 8 and females tended to have higher levels of reading achievement at age 8.

8.6.2 Results for Dynamic Mixture Models

The first dynamic mixture model was a two-class bivariate dual change score model with bidirectional coupling. In this model all of the parameters were separately estimated for the two-classes. This model required 43 parameters and yielded a likelihood ($L^2 = -89639$) and likelihood based fit statistics (BIC = 179659). Comparing the likelihood and BIC from this two-class mixture model to the likelihood (-92154) and BIC (184493) from one-class model indicated an improvement. The threshold parameter (τ) was estimated to be 0.63 indicating the sample could be considered a mixture of two classes of different sizes, $n_1 = 4665$ and $n_2 = 2637$, with different dynamic relationships. The first class showed a dynamic pattern that was similar to the overall model with reading comprehension having a small positive (0.004) effect on changes in antisocial behaviors while antisocial behaviors had a large negative effect (-1.64) on changes in reading comprehension. The second class, on the other hand, had no significant coupling parameters indicating that reading comprehension and antisocial behaviors did not have a time-dependent relationship for this class of participants. This separation of individuals into people who did show a specific coupling from persons who seem uncoupled is an important theoretical issue that requires careful consideration and replication. Although an initial set of values can be estimated using this latent change mixture model approach, it also seems obvious that replicated results across multiple studies would give us a much stronger basis to form homogeneous groupings of people.

8.7 Discussion

This chapter serves to provide some methodological and analytical methods the examination of longitudinal data using the general rubric of growth curve modeling techniques in a structural equation modeling framework. We recognize that SEM is just one framework for longitudinal data analysis and represents a limited class of longitudinal data analytic techniques (e.g., Nesselroade & Baltes, 1979; Collins & Sayer, 2001). However, the analyses presented here include some of the most up-to-date combinations of longitudinal models dealing with the developmental-dynamic processes with unobserved heterogeneity. The five steps we outlined here represent one way to organize some of the inherent complexity of longitudinal data analysis, but these techniques are central to answering questions that are often posed and initiate the collection of longitudinal data.

These five steps form a sequence with increasing levels of practical and theoretical knowledge, so it is useful to consider them in the order presented here. The inclusion of all five aspects of latent curve modeling is often overlooked in longitudinal analyses. That is, latent curves models (#2) are often fit without first describing the basic data (#1). Group differences (#3) are presented without a full evaluation of various growth curves that may be appropriate for the data (#2). In many recent cases, inferences about latent curve dynamics across variables (#4 and #5) are offered using simpler models that are incapable of providing this information (e.g., #3). For these reasons, a longitudinal researcher should consider the issues within each step before moving on to the next step. Of course, it is easy to envision situations where it would be best to apply the steps in a different sequence, or even to elaborate on one step based on the research questions. Obviously, models of the complexity of Steps 4 and 5 may only be useful in the more advanced stages of research. Further steps beyond these are possible, and should deal with dynamic models from a time-series perspective (e.g., Nesselroade et al., 2002), models based on differential equations (e.g., Oud & Jansen, 2001), selection effects due to survival (e.g., McArdle et al., 2005), and deal with experimental group dynamics (e.g., McArdle, 2007).

The structural-dynamic models discussed here represent only a sample of the mathematical and statistical models appropriate for longitudinal data and the choice of longitudinal models should be based on the specific research question under investigation (see Grimm, 2007). Indeed, some of the most difficult problems for future work on latent curves will be focused on the rather elusive meaning of the latent model parameters themselves (Zeger & Harlow, 1987; McArdle & Nesselroade, 2003). The choice of an appropriate substantive-vs-methodological interface (see Wohlwill, 1991) creates problems that remain among the most difficult challenges for future work. In this sense, the five step sequence advocated here is mainly intended as a practical way to organize the otherwise daunting task of developmental analyses of multivariate multiple occasion data.

Acknowledgements The work described here has been supported since 1980 by the National Institute on Aging (Grant#AG-07137). Kevin Grimm was supported by a National Science Foundation REECE Program Grant (DRL-0815787) and the National Center for Research on Early Childhood Education, Institute of Education Sciences, U.S. Department of Education (R305A06021). We are especially grateful to the assistance of our close friend and colleague, John R. Nesselroade. This research was also influenced by discussions with many others, including Paul Baltes, Steven

Boker, Emilio Ferrer, Paolo Ghisletta, Fumiaki Hamagami, John Horn, Bill Meredith, and Carol Prescott.

References

Baltes, P. B., & Nesselroade, J. R. (1979). History and rationale of longitudinal research. In J. R. Nesselroade & P. B. Baltes (Eds.), *Longitudinal research in the study of behavior and development* (pp. 1-39). New York: Academic Press.

Boker, S. M., & McArdle, J. J. (1995). Statistical vector field analysis applied to mixed cross-sectional and longitudinal data. *Experimental Aging Research, 21*, 77-93.

Browne, M., & du Toit, S. H. C. (1991). Models for learning data. In L. Collins & J. L. Horn (Eds.), *Best methods for the analysis of change* (pp. 47-68). Washington, DC: APA Press.

Bryk, A. S., & Raudenbush, S. W. (1992). *Hierarchical linear models: Applications and data analysis methods*. Newbury Park, CA: Sage.

Cattell, R. B. (1980). The separation and evaluation of personal and environmental contributions to behavior by the person-centered model (PCER). *Multivariate Behavioral Research, 15*, 371-402.

Cnaan, A., Laird, N. M., & Slasor, P. (1997). Using the general linear mixed model to analyze unbalanced repeated measures and longitudinal data. *Statistics in Medicine, 16*, 2349-2380.

Cook, T. D., & Campbell, D. T. (1979). *Quasi-experimentation design and analysis issues for field settings*. Skokie, IL: Rand-McNally.

Cudeck, R., & Klebe, K. J. (2002). Multiphase mixed-effects models for repeated measures data. *Psychological Methods, 7*, 41-63.

Dunn, L. M., & Markwardt, F. C. (1970). *Peabody Individual Achievement Test manual*. Circle Pines, MN: American Guidance Service.

Grimm, K. J. (2007). Multivariate longitudinal methods for studying developmental relationships between depression and academic achievement. *International Journal of Behavioral Development, 31*, 328-339.

Grimm, K. J., & McArdle, J. J. (2005). A note on the computer generation of structural expectations. In F. Dansereau & F. Yammarino (Eds.), *Multi-level issues in strategy and research methods* (Volume 4 of Research in multi-level issues) (pp. 335-372). Amsterdam: JAI Press/Elsevier.

Hamagami, F., & McArdle, J. J. (2007). Dynamic extensions of latent difference score models. In S. M. Boker & M. J. Wenger, *Data analytic techniques for dynamical systems. Notre Dame series on quantitative methodology* (pp. 47-85). Mahwah, NJ: Lawrence Erlbaum Associates.

Horn, J. L. (1973). On extension analysis and its relation to correlations between variables and factor scores. *Multivariate Behavioral Research, 8*, 477-489.

Jöreskog, K. G., & Sörbom, D. (1979). *Advances in factor analysis and structural equation models*. Cambridge, MA: Abt Books.

Laird, N. M., & Ware, J. H. (1982). Random effects models for longitudinal data. *Biometrics, 38*, 963-974.

Littell, R. C., Miliken, G. A., Stoup, W. W., & Wolfinger, R. D. (1996). *SAS system for mixed models*. Cary, NC: SAS institute.

Little, R. J. A. (1995). Modeling the dropout mechanism in repeated-measures studies. *Journal of the American Statistical Association, 90*, 1112-1121.

Magnussen, D, (2003). Person-Centered methodology. In S. C. Peck & R. W. Roeser (Eds.), *Person-centered approaches to studying development in context*. San Francisco: Jossey-Bass.

McArdle, J. J. (1988). Dynamic but structural equation modeling of repeated measures data. In J. R. Nesselroade & R. B. Cattell (Eds.), *Handbook of multivariate experimental psychology: Vol. 2* (pp. 561-614). New York: Plenum.

McArdle, J. J. (1989). Structural modeling experiments using multiple growth functions. In P. Ackerman, R. Kanfer & R. Cudeck (Eds.), *Learning and individual differences: Abilities, motivation, and methodology* (pp. 71-117). Hillsdale, NJ: Lawrence Erlbaum Associates.

McArdle, J. J. (1994). Structural factor analysis experiments with incomplete data. *Multivariate Behavioral Research, 29,* 409-454.

McArdle, J. J. (2001). A latent difference score approach to longitudinal dynamic structural analyses. In R. Cudeck, S. du Toit & D. Sörbom (Eds.), *Structural equation modeling: Present and future* (pp. 342-380). Lincolnwood, IL: Scientific Software International.

McArdle, J. J. (2005). The development of the RAM rules for latent variable structural equation modeling. In J. J. McArdle & A. Maydeu-Olivares (Eds.), *Contemporary psychometrics: A festschrift for Roderick P. McDonald* (pp. 225-273). Mahwah, NJ: Lawrence Erlbaum Associates.

McArdle, J. J. (2007). Dynamic structural equation modeling in longitudinal experimental studies. In K. van Montfort, J. Oud & A. Satorra (Eds.), *Longitudinal models in the behavioral and related sciences* (pp. 159-188). Mahwah, NJ: Lawrence Erlbaum Associates.

McArdle, J. J. (2009). Latent variable modeling of differences and changes. *Annual Review of Psychology, 60*, 577-605.

McArdle, J. J., & Bell, R. Q. (2000). An introduction to latent growth models for developmental data analysis. In T. D. Little, K. U. Schnabel & J. Baumert (Eds.), *Modeling longitudinal and multilevel data: Practical issues, applied approaches, and specific examples* (pp. 69-107). Mahwah, NJ: Lawrence Erlbaum Associates.

McArdle, J. J., Caja-Ferrer, E., Hamagami, F., & Woodcock, R. W. (2002). Comparative longitudinal structural analyses of the growth and decline of multiple intellectual abilities over the life span. *Developmental Psychology, 38,* 115-142.

McArdle, J. J., & Cattell, R. B. (1994). Structural equation models of factorial invariance in parallel proportional profiles and oblique confactor. *Multivariate Behavioral Research, 20*, 63-113.

McArdle, J. J., & Epstein, D. B. (1987). Latent growth curves within developmental structural equation models. *Child Development, 58*, 110-133.

McArdle, J. J., & Hamagami, F. (1992). Modeling incomplete longitudinal and cross-sectional data using latent growth structural models. *Experimental Aging Research, 18*, 145-166.

McArdle, J. J., & Hamagami, F. (2001). Linear dynamic analyses of incomplete longitudinal data. In L. Collins & A. Sayer (Eds.), *New methods for the analysis of change.* (pp. 137-176). Washington, DC: APA Press.

McArdle, J. J., Hamagami, F., Meredith, W., & Bradway, K. P. (2001). Modeling the dynamic hypotheses of Gf-Gc theory using longitudinal life-span data. *Learning and Individual Differences, 12*, 53-79.

McArdle, J. J., & McDonald, R. P. (1984). Some algebraic properties of the Reticular Action Model for moment structures. *The British Journal of Mathematical and Statistical Psychology, 37*, 234-251.

McArdle, J. J., & Nesselroade, J. R. (1994). Using multivariate data to structure developmental change. In S. H. Cohen & H. W. Reese (Eds.), *Life-span developmental psychology: Methodological innovations* (pp. 223-267). Hillsdale, NJ: Lawrence Erlbaum Associates.

McArdle, J. J., & Nesselroade, J. R. (2003). Growth curve analyses in contemporary psychological research. In J. Schinka & W. Velicer (Eds.), *Comprehensive handbook of psychology: Vol. 2. Research methods in psychology* (pp. 447-480). New York: Pergamon Press.

McArdle, J. J., Small, B. J., Backman, L., & Fratiglioni, L. (2005). Longitudinal models of growth and survival applied to the early detection of Alzheimer's Disease. *Journal of Geriatric Psychiatry and Neurology, 18* , 234-241.

Meredith, W., & Tisak, J. (1990). Latent curve analysis. *Psychometrika, 55*, 107-122.

Muthén, B., & Muthén, L. (2000). Integrating person-centered and variable-centered analysis: Growth mixture modeling with latent trajectory classes. *Alcoholism: Clinical and Experimental Research, 24*, 882-891.

Muthén, B., & Shedden, K. (1999). Finite mixture modeling with mixture outcomes using the EM algorithm. *Biometrics, 55*, 463-469.

Miyazaki, Y., & Raudenbush, S. W. (2000). Tests for linkage of multiple cohorts in an accelerated longitudinal design. *Psychological Methods, 5*, 24-63.

Nagin, D. (1999). Analyzing developmental trajectories: Semi-parametric, group-based approach. *Psychological Methods, 4*, 139-177.

Nesselroade, J. R., & Baltes, P. B. (Eds.) (1979). *Longitudinal research in the study of behavior and development*. New York: Academic Press.

Nesselroade, J. R., McArdle, J. J., Aggen, S. H., & Meyers, J. M. (2002). Dynamic factor analysis models for representing process in multivariate time-series. In D. S. Moskowitz & S. L. Hershberger (Eds.), *Modeling intraindividual variability with repeated measures data: Methods and applications* (pp. 235-265). Mahwah, NJ: Lawrence Erlbaum Associates.

Oud, J. H. L., & Jansen, R. A. R. G. (2000). Continuous time state space modeling of panel data by means of SEM. *Psychometrika, 65*, 199-215.

Ram, N., & Grimm, K. J. (2007). Using simple and complex growth models to articulate developmental change: Matching method to theory. *International Journal of Behavioral Development, 31*, 303-316.

Rao, C. R. (1958). Some statistical methods for the comparison of growth curves. *Biometrics, 14*, 1-17.

Rogosa, D. R. (1978). Some results for the Johnson-Neyman technique. *Dissertation Abstracts International, 38(9-A).*

Rogosa, D., & Willett, J. B. (1985). Understanding correlates of change by modeling individual differences in growth. *Psychometrika, 50*, 203-228.

Singer, J. D. (1998). Using SAS PROC MIXED to fit multilevel models, hierarchical models, and individual growth models. *Journal of Educational and Behavioral Statistics, 24*, 323-355.

Sliwinski, M., & Buschke, H. (1999). Cross-sectional and longitudinal relationships among age, cognition, and processing speed. *Psychology and Aging, 14,* 18-33.

Sullivan, E. V., Rosenbloom, M. J., Lim, K. O., & Pfefferman, A. (2000). Longitudinal changes in cognition, gait, balance in abstinent and relapsed alcoholic men: Relationships to changes in brain structure. *Neuropsychology, 14*, 178-188.

Verbeke, G., Molenberghs, G., Krickeberg, K., & Fienberg, S. (Eds.) (2000). *Linear mixed models for longitudinal data*. New York: Springer Verlag.

Wedel, M., & DeSarbo, W. S. (1995). A mixture likelihood approach for generalized linear models. *Journal of Classification, 12*, 21-55.

Willett, J. B., & Sayer, A. G. (1994). Using covariance structure analysis to detect correlates and predictors of individual change over time. *Psychological Bulletin, 116*, 363-381.

Wohlwill, J. F. (1973). *The study of behavioral development.* Oxford, England: Academic Press.

Wohlwill, J. F. (1991). The merger of developmental theory and method. In P. Van Geert & L. P. Mos (Eds.), *Annals of theoretical psychology: Vol. VII* (pp. 129-138).

Zeger, S. L., & Harlow, S. D. (1987). Mathematical models from laws of growth to tools for biologic analysis: Fifty years of growth. *Growth, 51*, 1-21.

Zill. N. (1990). Behavior problem index based on parent report. In *National Health Interview Survey. Child Health Supplement*. Washington, DC: National Center for Health Statistics.

Chapter 9
Structural Interdependence and Unobserved Heterogeneity in Event History Analysis

Daniel J. Blake, Janet M. Box-Steffensmeier, and Byungwon Woo

Abstract This chapter introduces how latent variables are handled in event history analysis, a popular method used to examine both the occurrence and the timing of events. We first emphasize why event history models are popular and what kinds of research questions the model can be used to answer. We also review the major estimation issues, briefly trace the development of event history models, and highlight the differences and similarities across various types of event history models. We then consider how latent variables are handled in event history analysis and demonstrate this with an example of latent variable analysis. In the conclusion we consider possible areas for future research.

9.1 Introduction

Event history models focus on the duration of time until an event of interest occurs. An event is commonly defined as a qualitative transition from an original state to a destination state at a specific point in time. An event history is a longitudinal record of the time until an event happens (or does not happen) for each observation. Event history models have become a popular method of empirical investigation and have been widely used in many scientific disciplines.

Daniel J. Blake
Department of Political Science, Ohio State University, Columbus, USA
e-mail: blake.165@polisci.osu.edu

Janet M. Box-Steffensmeier
Department of Political Science, Ohio State University, Columbus, USA
e-mail: steffensmeier.2@polisci.osu.edu

Byungwon Woo
Department of Political Science, Ohio State University, Columbus, USA
e-mail: woo.54@polisci.osu.edu

K. van Montfort et al., *Longitudinal Research with Latent Variables*,
DOI 10.1007/978-3-642-11760-2_9, © Springer-Verlag Berlin Heidelberg 2010

There has been significant progress in the development of estimation techniques for event history models since the 1960s that have led to a broad range of scientific disciplines such as biostatistics, mechanical engineering, labor economics, demography, criminology, and political science to utilize event history models to study a diverse range of research questions. One of the more recent and challenging directions in the development of event history techniques has been the development of estimation approaches for multiple events. Scholars have addressed the repetition of events as well as the possibility for an individual observation to experience multiple different events (competing risks). Importantly, scholars have also sought to uncover structural independencies amongst multiple events and, in this context, have endeavored to take into account unobserved heterogeneity and the effects of latent variables.

Simultaneous equation modeling is the most prominent approach to handling structural interdependencies and unobserved heterogeneity in an event history context. One sees applications of this approach in demography, sociology, labor economics, finance, political science, and transportation engineering, particularly where scholars have modeled systems of multiple event history equations believing that multiple event history processes are interdependent, i.e. the time to one event depends on the time to another related event. Studies have examined structural interdependence between the duration of marriage and fertility timing, between the duration of breast-feeding and the duration of maternal leave, between the competing risks of getting jobs from old or new employers, and between trip and stop times in shopping activities. In short, the need for simultaneous duration models is widespread and we pay particular attention to simultaneous event history models in this chapter. However, scholars have also jointly estimated event history and non-event history models and have pursued modeling strategies such as seemingly unrelated regression (SUR) in such cases to reveal the effects of latent variables on their outcomes of interest. We address these approaches as well and replicate an existing study that employs a SUR approach to study the direction and timing of U.S. legislators' positions towards ratification of the North American Free Trade Agreement (NAFTA).

The chapter is organized as follows. In the next section, we introduce the basic elements of event history models, emphasizing why event history models are popular and what kind of research questions can be answered using event history techniques. We also briefly review several important estimation issues, briefly trace the development of event history models and highlight the differences and similarities across different types of event history models. In the third section, we consider how latent variables are handled in event history analysis. In the fourth section, we provide an example of latent variable analysis in an event history context before concluding the chapter with a discussion of possible areas for further methodological developments.

9.2 Event History Models

9.2.1 Event History Models

Event history models focus on the time until an event of interest occurs. An event is a change from one state to another, such as marriage (from single to married), divorce (from married to divorce), war (from peace to military conflict), or unemployment (from employed to unemployed). An event history is then a longitudinal record of when events of interest happened, such as a time until one's marriage or a time until one's divorce. Event history models take into account not only whether or not the event of interest occurs to an observation, but also when the event occurs and allow an investigation into timing of the occurrence of the event. The dependent variable in an event history model is the time until the event occurs. Event history models are also referred to as duration, survival, failure time, and reliability models.

Event history models have many attractive features that make them a popular choice for researchers. Compared to models that allow researchers to investigate the occurrence of an event, such as binomial logit or probit models, event history models provide opportunity to exploit the rich data of the "histories" of events in addition to the occurrence of events. As the histories provide valuable additional information, event history models help researchers better understand the causes and processes of the event of interest. For instance, a labor economist might be interested in understanding the dynamics of employment. One can see that how long it takes a job seeker to find a new job is valuable information in addition to the information about the occurrence of new employment. Event history models allow the labor economist to investigate not only what makes employment more likely but also what makes an individual more likely to find employment sooner rather than later.

Modern event history techniques can handle censored observations nicely and easily incorporate time varying covariates. One may think that ordinary least squares (OLS) regression might be able to capture factors influencing duration quite nicely, as it allows researchers to study continuous dependent variables. But event history data pose several challenges for traditional OLS regression. For one, duration data are often right skewed and the OLS approach requires an arbitrary transformation of data. A more serious problem is data truncation. Data truncation happens when researchers do not know either the exact entry time of an observation (left truncation) or the end time (right censored). Right censoring is present in almost all event history data sets as there are often observations that have not experienced an event of interest by the time data is collected. It is a problem because it results in a missing value for the dependent variable (time until event). For example, one may be interested in what causes former inmates to commit another crime and return to jail. To study this, researchers may collect data for a year after inmates are released. For those inmates that commit crimes during that year, the researchers obtain values for time until inmates re-offend. A researcher using OLS needs to treat left truncated observations as if they have the equivalent entry time with other observations and to deal with right censored observation either by dropping all the observations that have not experienced the event or by capping the history by assuming the event has occurred at the conclusion of the period of data collection. Both of these arbitrary assumptions will cause biases. A second problem for cross-sectional OLS estimation of time until events occur emerges from the potential for the values of some independent variables to change as time passes.

Event history models can handle cases of left truncation and right censoring and can accommodate time varying covariates (TVCs). Due to these advantages over more common statistical methods, event history models have garnered increasing popularity among researchers from diverse disciplines. Biostatisticians have used event history models to study the effects of medical treatments on patients' recovery time after suffering a particular disease. In clinical studies, event history models are commonly called survival models, because they are often used to study the survival of patients. In engineering, event history models have been applied to investigate times until machines or some electronic components break down. Thus, event history models in engineering are often referred to reliability or failure time models. Economists have used event history models to study durations of employment and unemployment, demographers have used them to study durations of education, and time until marriage or child bearing. Meanwhile, criminologists have used event history techniques to study the time for released inmates to commit another crime and political scientists use them to investigate such diverse topics as the timing of the dissolution of coalition governments, the breakdown of cease-fire or peace agreements between countries, and candidates' decision to run for an election.

9.2.2 Key Contributions to the Development of Event History Methods

As early as the early 1900s, life tables were used by actuaries. In the late 1950s and early 1960s, more modern methods for event history analysis were actively developed by biomedical scientists and engineers. The former developed these methods to analyze survival data gathered through clinical trials while the latter needed new statistical techniques to analyze data on the breakdown of machines and electronic components (Allison 1984, 11–12). These two research traditions effectively merged in the 1970s and, as noted earlier, event history methods have since been employed in a wide range of disciplines.

Fleming and Yin (2000) provide a summary of the most important developments in event history modeling, focusing on the work done in biostatistics. Chief amongst these is the development of the Kaplan-Meier method (Kaplan & Meier 1958) "for estimating the survival function, log-rank statistic for comparing two survival distributions (Mantel 1966)", and the Cox proportional hazard model for "quantifying the effects of covariates on the survival time" (Cox 1972). Oakes' (2001) also credits these contributions with forming the foundation of modern event history techniques, noting that "Kaplan and Meier (1958) who formalized the product-limit estimator and Cox (1972) who introduced the proportional hazards model, are primarily responsible" for the present state of art of event history models. The Kaplan-Meier estimator solved the problem of estimating a distribution function with censored data via nonparametric maximum likelihood. The comparison of two survival distributions was critical given the need to provide reliable comparisons of two populations, such as whether a medical intervention led to longer life outcomes. Cox's work has been called ingenious for his semiparametric approach that allows assessment of the influence of covariates on censored outcomes. Cox leaves the baseline hazard function unspecified and discarded the times of observed events and the number of events at those times along with an assumption that censoring is independent and uninformative. This means that the partial likelihood is based on the cases that fail at each event time given the number failing and the number of cases at risk at that time. Fleming and Yin also highlight the importance of the counting process martingale theory pioneered by Aalen (1975, p. 971) as "providing a unified framework for studying the small- and large-sample properties of survival analysis statistics." This is because it allows "one to obtain simple expressions for moments of complicated statistics and asymptotic distributions for test statistics and estimators and to examine the operating characteristics of of censored data regression method" (Fleming & Lin 2000). These important statistical developments have been instrumental to the spread of event history methods across multiple fields.

Social scientists were largely unaware of earlier developments in biostatistics and engineering. A turning point for sociology comes in the late 1970s, when Tuma (Tuma 1976) introduced "explanatory variables into continuous time Markov models, an innovation that effectively bridged the gap between the sociological approach and what had already been done in biostatistics and engineering" (Allison 1984, p. 12). In economics, early applications of event history models appeared in the late 1970s and were mostly used to explain labor force dynamics. In other social science disciplines, the adoption of event history models came later. For example, event history techniques have become increasingly popular in political science since the 1990s thanks to the work of Box-Steffensmeier who introduced event history models to the field. Software packages for survival data analysis have been widely available since the early 1980s (Allison 1984) and, currently, many common software packages support estimations of event history models.

9.2.3 Basic Elements of Event History Models

There are a wide variety of event history models, but all event history models share the same structure. There are some important technical differences between continuous and discrete time models, and between parametric and nonparametric models, but the following structure and its basic elements are shared by all. Our discussion here employs continuous time notation.

Let T be a single lifetime variable. T can be thought of as the time until an event happens and can range from 0 to a theoretical end point. Let $f(t)$ denote the probability density function of T. Then $f(t)$ denotes the probability of the event of interest occurring at any given time point t where t is an element of T. The cumulative density function of T can be obtained by integrating $f(t)$ from 1 to t.

$$F(t) = Pr(T < t) = \int_0^t f(x)\,dx \tag{9.1}$$

This is the probability of the event having occurred between time 0 and t. Then the probability of survival or the probability for an observation not experiencing the event until t can be obtained by simply subtracting $F(t)$ from 1. The probability of an individual surviving to time t is referred to as the survival function.

$$S(t) = Pr(T \geq t) = 1 - F(t) \tag{9.2}$$

The hazard rate $h(t)$ is the probability of the event happening at time t given the observation has not experienced the event until time t. In terms of the equations introduced above, the hazard rate $h(t)$ is equal to $f(t)/S(t)$, the conditional probability of an event occurring given it has not happened up until time t.

$$h(t) = \lim_{\Delta x \to +0} \frac{Pr(t \leq T \leq t + \Delta t | T \geq t)}{\Delta t} = \frac{f(t)}{S(t)} \tag{9.3}$$

The hazard is an unobserved variable, much the same way that the $Pr(Y = 1)$ is an unobserved variable in binomial logit or probit models, yet the hazard controls both the occurrence and the timing of events.

9.2.4 Different Models of Event History

There are both parametric and semiparametric event history models. The main difference between parametric and semiparametric models is that parametric models make assumptions about the structure of the baseline hazard rate once covariates are included in the model. In comparison, semiparametric models do not make such assumptions. The choice between parametric and semiparametric approaches depends on how confident researchers are of the shape of the baseline hazard, which ideally is guided by theory.

There are a wide variety of parametric models and some models are nested in other, more flexible models with more parameters. For instance, the exponential model assumes that the baseline hazard is flat across time. This means that the probability of an event occurring at time t conditional on the event not having occurred is constant over time. The exact value then depends on included covariates.

The Weibull model is more flexible than the exponential model and it allows the baseline hazard rate to be monotonically increasing, monotonically decreasing, or flat over time. This is done by inserting a linear function of t into the right hand side of the equation. When the coefficient of the t term is 0, the Weibull model becomes the exponential model. Thus, the Weibull model is nested in the exponential model.

Since the Weibull model is flexible, it is a popular choice in social science applications. Yet, in some settings, the monotonicity assumption may not be appropriate.

When one suspects that the monotonicity assumption is not defendable, the log-logistic and the log-normal models can be used. These models allow hazard rates to first increase and then decrease as t passes. Neither of these models have the proportional hazards property. The generalized gamma model can be useful to adjudicate among different parametric models as several parametric models are nested within the generalized gamma model. The exponential, the Weibull, the log-normal, and the gamma models are all special forms of the generalized gamma model. When one has no a priori theoretical justification about how the baseline hazard rate varies across time, the generalized gamma model is more likely to be useful. If the fit is correct, parametric models generally have smaller standard errors than their semi-parametric counterparts. On the other hand, as parametric models require a priori assumptions about the shape of the baseline hazard, when assumptions are not correct, the estimation will be biased.

Recently, the Cox (1972, 1975) semiparametric model has become the most commonly used in social science applications (Therneau & Grambsch 2001, Singer & Willett 1993, Box-Steffensmeier & Jones 2004). The Cox semiparametric model's primary advantage is that it relaxes the assumption that the time until an event occurs follows a specific distribution. Larsen and Vaupel (1993) point out that "in the analysis of duration data, if the functional form of the hazard has the wrong shape, even the best-fitting model may not fit the data well enough to be useful" (p. 96).

A key concept for understanding the Cox model is the hazard rate. Recall, that the hazard rate can be thought of as the probability that an event will occur for a particular observation at a particular time, or the rate at which an event occurs for an observation at time t given that the observation has survived through time $t-1$. In the Cox model, the hazard rate for the ith individual is

$$h_i(t) = h_0(t)\exp(\beta'\mathbf{x}), \tag{9.4}$$

where $h_0(t)$ is the baseline hazard function, and $\beta'\mathbf{x}$ are the covariates and regression parameters. A Cox model does not report an intercept as it is absorbed into the baseline hazard function. The ratio of two hazards (or hazard ratio) can be written as,

$$\frac{h_i(t)}{h_0(t)} = \exp(\beta'(\mathbf{x}_i - \mathbf{x}_j)), \tag{9.5}$$

which demonstrates that this ratio is a fixed proportion across time. Box-Steffensmeier and Jones (2004) point out that when the proportional hazards assumption holds in the Cox model, the particular form of the baseline hazard rate, $h_0(t)$ is assumed to be unknown and is left unparameterized. More accurately the duration times are parameterized in terms of a set of covariates, but the particular distributional form of the duration times is not parameterized, hence the term "semi-parametric".

Proportional hazards is the major assumption of the Cox model, as well as many parametric models. This assumption is tested with both the global model test defined by Therneau and Grambusch (1994) and Harrell's rho for individual covariates (Box-Steffensmeier & Jones 2004, 135). If the assumption is found to be violated, the offending covariate(s) is interacted with time and the model is re-estimated. Like all models, there is a series of general diagnostics for the Cox model. These include assessments of model fit, functional form, and influence (Therneau, Grambsch, & Fleming 1990, Grambsch, Therneau, & Fleming 1995, Grambsch & Therneau 1994).

Parameters in the Cox model are estimated using a *partial likelihood* approach. The partial likelihood method is based on the assumption that the intervals between successive duration times (or failure times) contributes no information regarding the relationship between the covariates and the hazard rate (Collett 2003), which comports to the arbitrary form assumed for the baseline hazard. Because the Cox model only uses "part" of the available data ($h_0(t)$ is not estimated), the likelihood function for the Cox model is a "partial" likelihood function. In contrast, consider the more typically encountered likelihood function which gives hypothetical population value that maximizes the likelihood of the observed sample using all of the data. That is, the maximum likelihood estimate is the value that is the most likely to generate the sample that is observed.

To derive the partial likelihood function for a data set of size n with k distinct failure times, the data are first sorted by the ordered failure time, such that $t_1 < t_2 < \ldots < t_k$, where t_i denotes the failure time for the ith individual. For now, we assume that there are no "tied" events: each uncensored case experiences an event at a unique time. For censored cases, we define δ_i, see 9.7, to be 0 if the case is right-censored, and 1 if the case is uncensored, that is, the event has been experienced. Finally, the ordered event times are modeled as a function of covariates, $\mathbf{x}$.

The partial likelihood function is derived by taking the product of the conditional probability of a failure at time t_i, given the number of cases that are at risk of failing at time t_i. More formally, if we define $R(t_i)$ to denote the number of cases that are at risk of experiencing an event at time t_i, that is, the "risk set," then the probability that the jth case will fail at time T_i is given by

$$\Pr(t_j = T_i \mid R(t_i)) = \frac{e^{\beta' \mathbf{x}_i}}{\sum_{j \in R(t_i)} e^{\beta' \mathbf{x}_j}}, \tag{9.6}$$

where the summation operator in the denominator is summing over all individuals in the risk set. Taking the product of the conditional probabilities in (9.6) yields the partial likelihood function,

$$\mathscr{L}_p = \prod_{i=1}^{K} \left[\frac{e^{\beta' \mathbf{x}_i}}{\sum_{j \in R(t_i)} e^{\beta' \mathbf{x}_j}} \right]^{\delta_i}, \tag{9.7}$$

with corresponding log-likelihood function,

$$\log L_p = \sum_{i=1}^{K} \delta_i \left[\beta' \mathbf{x}_i - \log \sum_{j \in R(t_i)} e^{\beta' \mathbf{x}_j} \right]. \tag{9.8}$$

By maximizing the log-likelihood in (9.8), estimates of the β are obtained. As (Collett 2003) notes, the likelihood function in (9.7) is not a true likelihood. This is because the actual survival times of censored and uncensored cases are not directly incorporated into the likelihood. Nevertheless, Cox (1972, 1975) famously demonstrated that maximum partial likelihood estimation produces parameter estimates that have the same properties as maximum likelihood estimates - asymptotically normal, asymptotically efficient, consistent, and invariant (see also Collett 2003).

The logic underlying the partial likelihood method is seen by considering the data presented in Table 9.1 (this part of the presentation is directly adapted from (Collett 2003)). We reproduce this here because of the clarity of Collett's example. The survival times for nine cases are provided. Of these nine cases, six of them experience an event, i.e., they "fail", and three of them are right-censored. The failure times can be ordered such that $t_1 < t_2 < \ldots < t_6$. Note that the censored cases do not contribute a failure time. Each of the nine cases are at risk of experiencing an event up to the first failure time, t_1. After the first failure in the data set, the risk set decreases in size by 1; thus, the risk set up to the second failure time, t_2, includes all cases except case 7. By the fourth failure time in the data, t_4, the risk set includes only cases 1, 6, and 8; cases 2 and 9 are right-censored before the fourth failure time is observed and do not contribute any information to this part of the likelihood function. By the last failure time, only case 6 remains in the risk set. Using the notation from Collett (2003, 64), let $\psi = \exp(\beta' \mathbf{x}_i)$. Then the partial likelihood function for these data would be equivalent to

$$\begin{aligned}
\mathscr{L}_p = & \frac{\psi(7)}{\psi(1)+\psi(2)+\psi(3)+\psi(4)+\psi(5)+\psi(6)+\psi(7)+\psi(8)+\psi(9)} \times \\
& \frac{\psi(4)}{\psi(1)+\psi(2)+\psi(3)+\psi(4)+\psi(5)+\psi(6)+\psi(8)+\psi(9)} \times \\
& \frac{\psi(5)}{\psi(1)+\psi(2)+\psi(3)+\psi 5+\psi(6)+\psi(8)+\psi(9)} \times \\
& \frac{\psi(3)}{\psi(1)+\psi(3)+\psi(6)+\psi(8)} \times \\
& \frac{\psi(1)}{\psi(1)+\psi(6)} \times \\
& \frac{\psi(6)}{\psi(6)}.
\end{aligned}$$

Again we see that the partial likelihood function is based on ordered duration times and censored observations contribute information to the "risk set" but contribute

Table 9.1 Data Sorted by Ordered Failure Time

Case Number	Duration Time	Censored Case
7	7	No
4	15	No
5	21	No
2	28	Yes
9	30	Yes
3	36	No
8	45	Yes
1	46	No
6	51	No

no information regarding failure times. In terms of the likelihood function in (9.7), censored observations contribute information to the denominator, but not to the numerator.

"Ties," or coterminous event occurrences, cannot be accounted for in the partial likelihood function, as presented in (9.7). This is true for any continuous time model. However, the literature has adapted a number of approximations to take this into account. For example, numerically computing "what if" this tied event occurred first, then the computing the same "what if" for the next tied event and so on. The Efron method is a popular choice for handling ties.[1]

In sum, event history models allow scholars to more fully capture the process surrounding the occurrence (or nonoccurrence) of events. We can investigate whether covariates speed up or slow down the timing of events and gain a more complete understanding of the process with event history models.

9.3 Statistical Models for System of Equations

In general, simultaneous equation models are used when there is a system of relationships, such as a two-way flow of influence. For simplicity, consider a variable A that affects another variable B and that is also affected by variable B. In this case, we need to consider a two equation setup where there is one equation for each interdependent or endogenous variable. When estimating the parameters for simultaneous equation models, information from both (or all if there are more than two endogenous variables) equations have to be taken into account. If this is not done, biased and inconsistent estimators are the result.

A major hurdle for estimation of simultaneous equations is the identification problem. In short, the identification problem arises because the same set of data may be compatible with different models. The identification problem needs to be

[1] See Golub and Collett (2002) for further discussion of this issue of ties.

addressed before considering estimation strategies. Identification occurs through the introduction of a priori information into the analysis. There are a variety of techniques for estimating simultaneous equations and it is a vibrant and long-standing area of econometric research, as is shown in this volume and in van Montfort, Oud, and Satorra (2004).

Simultaneous Equation (SEQ) models are used when the equations for each part of the system are interdependent (also referred to as a substantive approach). Arguably, such interdependence is ubiquitous in the world around us and the questions studied by social scientists. In this case, it is more than just the disturbances that are related. For a complete system of equations, the number of equations needs to equal the number of endogenous variables. Joint estimation of all the equations in the system provides fully efficient approaches.

The Seemingly Unrelated Regression (SUR) model is an alternative approach to SEQ estimation of interdependent processes. The SUR model has a series of equations that are linked through correlated error terms (also referred to as a nuisance approach) and generalized least-squares (GLS) estimation is used to gain efficiency. The higher the correlation of the disturbances, the greater the efficiency gain in using GLS (Zelner 1962, Dwivedi & Srivastava 1978). SUR estimation is "simply the application of generalized least-squares estimation to a group of seemingly unrelated equations. The equations are related through the nonzero covariances associated with error terms across different equations at a given point in time" (Pindyck & Rubinfeld 1991, 326). Both autocorrelation and heteroscedasticity can be accommodated in the SUR model.

Both the SUR and SEQ approaches are central to the estimation of the effects of latent variables in an event history context. However, the choice of approach depends on the type of interdependence between processes the researcher assumes is present. We discuss the choice between different approaches in more depth and illustrate how these techniques have been used in the extant literature to ascertain the presence and effect of latent variables in the next section.

9.4 System of Equations, Interdependent Processes and Latent Variable Analysis

There are numerous occasions where we want to study the duration of an event within a framework of a system of equations using either SEQ or SUR approaches. First, we may wish to model multiple endogenous event history processes simultaneously. This can occur in two different ways depending on the relationship between the events of interest. In many cases, scholars will be interested in jointly modeling the duration processes for multiple events where the events of interest are not mutually exclusive. Lillard (1993) conceives of this as modeling "multiple clocks," which refers to one process depending on the duration of a related process. For example, one may be interested in modeling the time until a woman completes her education and the time until a woman becomes pregnant. The time until a woman completes her education may be affected by whether she is pregnant or not and the timing of a women's becoming pregnant may be affected by whether she completes her education or not. There are plenty of research questions that features multiple endogenous event histories: marital duration and the time to marital conceptions; duration of breast-feeding and duration of maternal leave; and duration of women's education and the time to entry into a first union.

In other contexts, events may appear as competing risks for a common duration process. Compared to multiple event history processes, competing risks models have a single duration process that can end with multiple events, whereby the occurrence of one event necessarily rules out the occurrence of another event. For example, one could model the duration of military conflict as a competing risks process: military conflict can end with an invading country's victory or a defending country's victory. As two states cannot win simultaneously on the battlefield, when an event (winning by a country), occurs, it is not longer at risk of experiencing the other event. In the simultaneous equations context, the competing risks can be related to each other. For instance, an invader's decision to continue fighting for a victory may be dependent on a defending country's decision to continue fighting and vice versa. In this case, the two hazards of the two competing risks need to be jointly estimated. Rosholm and Svarer (2001) estimate unemployment durations with two competing risks: the risk of being recalled by the previous employer and the risk of being hired by a new employer. As they theoretically expect that the hazard of getting a new job is dependent on the hazard of being recalled by the previous employer (the hazard for recall should reduce the hazard of new jobs as those who see higher probability of being recalled will be less active in pursuing new jobs), they put the hazard for recall in estimating the hazard for new jobs when constructing a system of equations and estimate structural dependency between the two hazards. The simultaneous competing risks models are useful in many situations: duration of economic sanctions where the duration can end either with target's capitulation or sender's lifting economic sanctions; duration of hospitalization where the duration can end with different events.

When jointly estimating multiple event histories, the equations in the system are all structured as some form of duration model. However, we may also wish to model duration processes jointly with other non-duration processes. In many cases, the non-duration model we wish to model attempts to estimate some important aspect of the event itself. For instance, we may wish to simultaneously model the time until a government calls an election and the result of that election, whereby timing is modeled using an event history technique and the result, measured as vote share for the incumbent government, is estimated using ordinary least squares regression (see Fukumoto 2009).

In sum, SUR and SEQ approaches may be usefully employed in event history settings when scholars are interested in estimating a system of interrelated event history models, a system of event history models with mutually exclusive outcomes, and a system of models containing both event history and non-duration equations.

When deciding which specific modeling strategy to pursue in each of these three cases, researchers need to be clear about their assumptions regarding the nature of interdependence between their multiple processes of interest. If scholars believe the processes are independent purely through their stochastic components, than a SUR approach may be appropriate as a SUR approach assumes that outcomes are interdependent because the errors of both processes share a single joint probability distribution (Hays & Kachi 2009). With respect to latent variables, since a SUR approach focuses on correlation in the errors terms, it allows scholars to identify the presence or absence of unobserved factors acting upon the dependent variables of interest. Fred Boehmke's (2006) study of the timing and position of U.S. legislators' towards the ratification of the North Atlantic Free Trade Agreement (NAFTA), which we replicate below, is an excellent example of a SUR approach. Boehmke argues that unobserved bargaining dynamics and competing pressures on legislators jointly influenced their positioning and timing on NAFTA, causing the two processes to be positively related and he finds evidence that Democratic legislators who declared their positions later in time, also tended to come out in favor of NAFTA. In another study, Fukumoto (2009) uses copula techniques to model dependence in latent variables between event history models and models of the event themselves. One advantage of copula approach is that asymmetric interdependence can be captured and modeled. Asymmetric interdependence occurs when one actor/process is more dependent on a second actor/process than the second actor/process is dependent on the first. For example, unobserved heterogeneity in the duration and event models may be such that latent variables in the duration model affect the event, but not vice versa. Both Fukumoto and Boehmke employ SUR approaches to model interdependent processes where only one of those processes is an event history process. However, in another study employing copula functions, Quiroz Flores (2008) estimates two event history models - the tenure in office of chief executives and the tenure of their foreign ministers - and finds that the tenure of individuals in both offices are closely correlated.

The main alternative to a SUR approach is to generate simultaneous equation models (SEQ) of interdependent processes of interest (Hays & Kachi 2009). The SEQ is preferable when endogeneity extends beyond stochastic components and one wishes to explicitly model the interdependence among outcomes of interest. In a system of simultaneous equations, endogenous variables appear on the right hand side of the equations which has implications for our analysis of the influence of unobserved, or unobservable, variables. This is because the variances an covariances among errors, in the reduced form of the structural equations, "need to be consistent with the structural relationship among endogenous variables" (Hays & Kachi 2009, p. 4).

The SEQ approach is by far the most popular approach in the extant literature for scholars interested in modeling multiple interdependent duration processes. Many researchers have chosen to build off the model developed in Lillard (1993) which has been quickly established as a classic article on combining simultaneous equations and duration models. Lillard (1993) presents a comprehensive model of the dynamics of marriage duration and marital fertility, i.e., the timing of marital conceptions taking into account a number of time-varying covariates, a set of exogenous covariates, and a set of endogenous covariates. He proposes that the hazard of conception if influenced by the hazard of marriage dissolution along with other variables and conversely, the hazard of marriage dissolution is influenced by the duration and outcome of marital fertility. As is common in the demography literature, he models the baseline hazard with a Gompertz distribution, then builds a system of equations, and obtains the reduced form equation.[2] As there are variables that influence one hazard but not the other, he can obtain identification of the structural dependence parameter. Lillard addresses the across-duration interdependence by allowing the baseline hazards to take flexible shapes through the use of splines.

The errors in Lillard's model represent residual, unobserved heterogeneity and their distribution is conditioned on the relationship among the endogenous variables. Joint normality of the error terms in the two separate hazard equations of fertility and marriage, is assumed[3],

$$\begin{pmatrix} \varepsilon \\ \eta \end{pmatrix} \sim \mathrm{N}\left(\begin{pmatrix} 0 \\ 0 \end{pmatrix}, \begin{pmatrix} \sigma_\varepsilon^2 & \sigma_{\varepsilon\eta} \\ \sigma_{\varepsilon\eta} & \sigma_\eta^2 \end{pmatrix}\right) \tag{9.9}$$

where ε is the error term in the hazard equation for dissolution of marriage, η is the error term in the hazard equation for conception (Lillard 1993). No assumption is made regarding the correlation or lack thereof between ε and η. However, the structural relationship between endogenous variables introduces some correlation in the residuals, which, in the reduced form of the simultaneous equation, are distributed as follows:

$$\begin{pmatrix} \varepsilon \\ \eta + \lambda\varepsilon \end{pmatrix} \sim \mathrm{N}\left(\begin{pmatrix} 0 \\ 0 \end{pmatrix}, \begin{pmatrix} \sigma_\varepsilon^2 & \sigma_{\varepsilon\eta} + \lambda\sigma_\varepsilon^2 \\ \sigma_{\varepsilon\eta} + \lambda\sigma_\varepsilon^2 & \sigma_\eta^2 + 2\lambda\sigma_{\varepsilon\eta} + \lambda^2\sigma_\varepsilon^2 \end{pmatrix}\right) \tag{9.10}$$

where λ is the coefficient for the endogenous hazard of marital disruption in the hazard model for conception. Thus, it is through the inclusion of λ that the distribution of errors is consistent with the structure of endogeneity between marriage dissolution and conception. It is here that the distinction between SUR and SEQ is perhaps most important for scholars who wish to investigate the presence of latent variables through examining the relationship between the residuals of multiple processes of interest. A SUR approach assumes no endogeneity in independent variables and therefore λ is never estimated.

Scholars who have modeled simultaneous duration processes, but are interested in the impact of latent variables have frequently followed Lillard's modeling approach and then investigated whether or not there is correlation in the heterogeneity terms, or errors. For example, Baizán, Aassve, and Billari (2004), interested in the time until the formation of cohabitation or marital relationships and the time until a couple's first child is born, find a positive correlation between the errors in the model of union formation and the model of childbirth. Other studies have looked at

---

2 Olshanksy and Carnes (1993) discusses the appropriateness of the Gompertz distribution for demography based on its U-shape.

3 Recall that the SUR approach also assumes the error terms of multiple models follow a joint distribution.

the correlation between heterogeneity components in models of duration of education and time to union formation (Billari & Philipov 2004, Coppola 2004), time to migration and to union dissolution (Boyle, Kulu, Cooke, Gayle, & Mulder 2008), and time between children and child mortality (Maitra & Pal 2007). An important attribute of the Lillard approach is that it allows scholars to both model endogenous processes and investigate the presence of correlated but unobserved heterogeneity. This makes his approach very suitable if researchers perceive their processes of interest as being "characterized by 1) mutual influence - that is events in one process trigger events in the other process - and 2) common time-constant influencing factors - which are usually not observed especially in retrospective surveys and which represent sources of potential endogeneity" (Billari & Philipov 2004, p. 92).

Many scholars adopting a SEQ approach display a theoretical interest in the endogeneity of the multiple processes, the presence of unobserved heterogeneity and the possible correlation of this heterogeneity across models. However, some scholars simply wish to statistically account for endogenous right hand side variables and the potential effects of unobserved factors. An instrumental variables approach is often pursued in such cases. For example, in their study of the duration of breast-feeding, Adair, Popkin and Guilkey (1993) argue that independent variables such as child health and the use of oral contraception influence how long mothers breast-feed their newborn infants, but are aware that the duration of breast feeding in turn shapes the health of the child and the decision to use contraception. Their methodological strategy is to generate predicted values of the potentially endogenous covariates by estimating separate OLS and logistic regression models for each endogenous variables using a battery of household factors, only some of which they include in their model of decisions regarding breast-feeding. These predicted values are then used to estimate the main discrete time logit hazard model of breast-feeding, with bootstrapped standard errors to overcome the problem of conditional standard errors. Addison and Portugal (1989) take a similar approach in their study of post-unemployment wage displacement, however, their main dependent variable of interest is not time to failure, but their potentially endogenous right hand side variable, duration of unemployment, is produced by a survival process. Thus, they generate predicted failure times for unemployment, which they then use in an OLS model of wage displacement. By enabling researchers to identify and statistically control for endogeneity, an instrumental variables approach allows them to account for potentially unobserved factors that affect both the value of endogenous right hand side variables and the dependent variable of interest.

Rosholm and Svarer (2001) also specify and estimate a simultaneous equations model for hazards. Yet, they do not examine structural dependency of two district temporal processes; they estimate the two interdependent hazards of unemployment duration. They consider unemployment ending with two exits (recall from the previous employer or a new job) and model the structurally dependent competing risks between the two processes. They find that the recall hazard affects the new job hazard negatively when taking the structural dependency into account. They suggest that the structurally dependent competing risks model is a fruitful alternative to the standard competing risks model.

There is exciting work in the field of political methodology on the topic of simultaneous duration models. In addition to the study of NAFTA positioning by Boehmke (2006) mentioned above, Boehmke, Morey, and Shannon (2006) start with Gumbel's bivariate exponential distribution to model non-random selection when the outcome of interest is duration. They apply this approach to the study of the effect of leaders' decisions to go to war on their subsequent post-crisis tenure. One of the shortcomings of the Boehmke (2006) and Boehmke et al. (2006) modeling strategy is that it does not allow researchers to identify the direction of influence between models when interdependence is asymmetric (Hays & Kachi 2009). Knowing the direction and strength of that asymmetry is often of theoretical interest.

Hays and Kachi (2009) add structure to the empirical models to estimate precise causal effects of one process on the other and vice versa. They present a simultaneous equations model for multiple interdependent duration processes using the Weibull distribution and derive its full information maximum likelihood estimator (FIML). The FIML estimator is shown to be efficient compared to a two stage least squares approach in the Monte Carlos. In their substantive application, they examine the interdependence between the duration of coalitional government formation and the duration of governmental survival. In their simultaneous equations setup, they create a system of structural equations and present the corresponding reduced form that is used to derive the likelihood function. With the estimation result of the simultaneous model, they conclude that government survival causes bargaining duration rather than the reverse, and thus that the positive covariance between the two durations is attributable to strategic bargaining. When parties expect a longer government, they bargain harder, which results in a longer negotiation duration.

Importantly, Hays and Kachi allow across-unit interdependence (in addition to across-duration dependence). A major contribution of their work is that they introduce a general approach that allows inclusion of both across-duration and across-unit dependences in one model. For example, there can be multiple interdependent duration processes, and the observational units within each duration process can be interdependent as well. They also compare the duration seemingly unrelated regression (SUR) models, such as Boehmke (2006), Boehmke et al. (2006), and Quiroz Flores (2008) and the simultaneous equations (SEQ) framework. Thus, they bring together various strands of work.

9.5 Duration and Discrete Choice: Timing and Direction of Position Taking by Legislators

We illustrate the duration and discrete choice model estimation and interpretation by replicating Boehmke (2006), who derives a seemingly unrelated discrete-choice duration estimator (SUDCD). His work is used to evaluate the presence of unobserved processes in a duration and a discrete choice context that are *not* independent. This is an excellent example of the SUR approach and a good illustration of why latent variable estimation may be important and how it can be incorporated in an event history setting. The empirical motivation for his model comes from an already existing study that employs event history techniques. This study is an article by Box-Steffensmeier, Arnold, and Zorn (1997) which explains the timing and direction of positions taken by members of the United States Congress on the ratification of the North American Free Trade Agreement (NAFTA) using two separate models: a Cox model to estimate timing and a probit model to estimate direction (for or against NAFTA). Boehmke argues that two unobservable, but strategic, processes that link together position choice and the timing of position announcement are present. First, if legislators face competing pressures regarding which position to take, they may delay selecting a position to see if their vote will have a significant effect on the final outcome before deciding which pressure to give in to. Second, legislators may delay taking a position if they are indifferent towards the outcome of the vote, but hope to induce side-payments from other representatives and actors with more at stake. Such legislators refrain from declaring a position early in case the vote should appear to be very close, thus increasing the importance of their vote and the amount of side-payments they can extract from both sides. Ultimately, these legislators will take a position based on which side offered them the best deal. These two processes are both unobservable and affect both the timing and content of a legislator's position on NAFTA.

To better evaluate these processes, Boehmke derives an estimator that follows a bivariate distribution which allows for nonzero correlation between the duration and discrete outcome equations. The Cox model cannot be used because it does not have a parametric assumption about the distribution of errors. So, two estimators are constructed using two different parameterizations of the baseline hazard that are commonly used in event history models: the Weibull and log-normal distributions[4].

9.5.1 Boehmke's Derivation of the Weibull SUDCD Estimator

The structure of the likelihood function is of the form (D_i, V_i) where D_i is the timing of a position for an individual and V_i is the position on whether to support or reject NAFTA. The duration equation is the same for individuals who reject or support NAFTA and thus the critical difference is between individuals' positions on NAFTA. The likelihood can be written using the marginal density of the duration and conditional probabilities of support for NAFTA.

[4] We address only the Weibull version in detail as it is the most common parametric model. Please note that while the Weibull's companion discrete choice model's errors follow a bivariate exponential distribution, the companion for the log-normal is a standard probit model.

$$Pr(\mathbf{D},\mathbf{V}) = \prod_{i=1}^{n} P(D_i = d_i, V_i = v_i)$$
$$= \prod_{i=1}^{n} P(D_i = d_i) \times P(V_i = 0|D_i = d_i)^{1-v_i} \times P(V_i = 1|D_i = d_i)^{v_i} \quad (9.11)$$

For the estimator, the joint density and conditional probabilities are calculated using the bivariate exponential distribution, where the cumulative and probability density functions take the form:

$$F_{exp}(x,y) = (1-\mathrm{e}^{-x})(1-\mathrm{e}^{-y})(1+\alpha\mathrm{e}^{-x-y}), \quad (9.12)$$
$$f_{exp}(x,y) = \mathrm{e}^{-x-y}[1+\alpha(2\mathrm{e}^{-x}-1)(2\mathrm{e}^{-y}-1)] \quad (9.13)$$

It is worth noting that the correlation between x and y is given by $\rho = \alpha/4$ and as α is bound between -1 and 1, the value of the correlation parameter, ρ, is restricted to $-0.25 \leq \rho \leq 0.25$.

The first part of the estimator is a standard exponential duration equation that is adapted to allow for Weibull duration dependence. The exponential distribution is given by: $d_i = \exp(\mathbf{x}_i\beta)\varepsilon_i$ where $\mathbf{x}_i$ is a vector of independent variables and ε_i has an exponential distribution. The marginal density of observing a duration d_i is given by:

$$f_w(d_i|\lambda_i) = p\lambda_{2i}^{p} d_i^{p-1} \exp[-\lambda_{2i} d_i^{p}], \quad (9.14)$$

where $\lambda_{2i} = \exp(-\mathbf{x}_i\beta)$ and p is a shape parameter that causes a Weibull distributed variable, u_i, to follow an exponential distribution.

The second part of the estimator is the discrete choice equation, developed to allow the errors to follow an exponential distribution. As the exponential distribution is not defined for negative numbers, the outcome is modeled:

$$V_i = \begin{cases} 1, \text{ if } \exp(\mathbf{w}_i\gamma)\eta_i > 1 \\ 0, \text{ otherwise,} \end{cases} \quad (9.15)$$

where η_i follows an exponential distribution. The marginal probability that $V_i = 0$ can is given by:

$$P(\exp(\mathbf{w}_i\gamma)\eta_i \leq 1) = p(\eta_i \leq \exp(\mathbf{w}_i\gamma))$$
$$= 1 - \exp(-\lambda_{1i}), \quad (9.16)$$

where $\lambda_{2i} = \exp(-\mathbf{w}_i\gamma)$. The likelihood function for SUDCD can now be written out in full, combining the two parts, as follows:

$$L(\beta,\gamma,p,\alpha|\mathbf{X},\mathbf{W},\mathbf{D},\mathbf{V}) = \prod_{i=1}^{n} p\lambda_{2i}^{p} d_i^{p-1} \exp[-\lambda_{2i} d_i^{p}](1-\pi_i^1)^{1-v_i}(\pi_i^1)^{v_i}, \quad (9.17)$$

$$\ln L(\beta,\gamma,p,\alpha|\mathbf{X},\mathbf{W},\mathbf{D},\mathbf{V}) = \sum_{i=1}^{n}(\ln(p)+p\ln(\lambda_{2i})+(p-1)\ln(d_i)-(\lambda_{2i}d_i)^p)$$
$$+[(1-v_i)\ln(1-\pi_i^1)+v_i\ln(\pi_i^1)], \quad (9.18)$$

where $\pi_i^1 = \exp(-\lambda_{1i})\{1+\alpha[2\exp(-(\lambda_{2i}d_i)^p)-1][\exp(-\lambda_{1i})-1]\}$ is the conditional probability that V_i is one.

With respect to right-censored observations, "their contribution to the overall likelihood is the joint probability of surviving until right censoring occurs and the probability of the observed discrete-choice outcome" (Boehmke 2006, p. 7), calculated as follows:

$$Pr(D_i \geq d_i^c, V_i = 1) = 1 - F_{\exp}(\lambda_{1i}) - F_{\exp}((\lambda_{2i}d_i^c)^p) + F_{\exp}(\lambda_{1i},(\lambda_{2i}d_i^c)^p), \quad (9.19)$$
$$Pr(D_i \geq d_i^c, V_i = 0) = F_{\exp}(\lambda_{1i}) - F_{\exp}(\lambda_{1i},(\lambda_{2i}d_i^c)^p), \quad (9.20)$$

where d_i^c is the censoring point.

9.5.2 Practical Application: Position Taking on NAFTA

Box-Steffensmeier, Arnold, and Zorn (1997) identify a range of covariates that potentially influenced the position members of Congress took on NAFTA, when they chose to publicly announce that position, or both. These independent variables, and the expected direction of their influence on timing and direction, are summarized in Table 9.2, adapted from Box-Steffensmeier, Arnold, and Zorn[5] (1997). The reader is directed to this article and Boehmke (2006) for further discussion of the theoretical reasoning underlying the expected effects of these *observable* factors.

To investigate the effect of *unobserved*, but related processes influencing the timing and direction of positions on NAFTA, three models are estimated. The first model treats the two equations for timing and direction separate. The second model uses the SUDCD Weibull estimator derived above where the parameter for correlation of the errors of the discrete choice and duration models is constant. The third model allows this correlation to be different for Republican and Democratic representatives. The relationship between timing and position direction is potentially stronger for Democrats as the President was a pro-NAFTA Democrat and therefore likely to apply pressure on members of Congress from his own party to vote in favor of NAFTA and/or offer inducements and side-payments to do so. This provides Democratic representatives with a greater incentive to hold out on declaring a position to see if their votes are critical to the final outcome and/or to receive greater

[5] One new independent variable, *net endorsements* is added by Boehmke (2006).

Table 9.2 Independent Variables in Models of NAFTA Position Timing and Direction

Independent Variable	Description	Expected Effect On Direction	On Timing
Union Membership	Level of unionization in a representative's district, measured as the percentage of workforce that is unionized	–	+ (if high or low)
Mexican Border	Records whether or not a representative's district shares a border with Mexico	+	+
Perot Vote	Support in the representative's district for Ross Perot (known for his strong anti-NAFTA stance) in the 1992 U.S. presidential election	–	+ (if high or low)
Household Income	Median household income in the representative's district, divided by 10,000	+	+ (if high or low)
Corporate Contributions	Share of representative's total campaign contributions accounted for by business interests	+	+
Labor Contributions	Share of representative's total campaign contributions accounted for by labor interests	–	+
NAFTA Committee	Representative is a member of a congressional committee that took up the matter of membership in NAFTA		+
Democratic Leadership	Representative is part of the Democratic congressional leadership		+/–
Republican Leadership	Representative is part of the Republican congressional leadership		+
Party Affiliation	Dummy variable coded 1 if the representative is a Democrat; 0 if Republican	+	
Ideology	Representative's ideological position on a conservative-liberal scale based on his/her congressional voting record	+	+
Net Endorsements	Difference between number of representatives that have already declared in favor of NAFTA and those that have declared against	+	

For timing, the expected effect is on the hazard rate; thus a "+" indicates an earlier declaration of a position.

side-payments from the President. Furthermore, greater pressure and side-payments from the President are likely to cause Democrats to vote in favor of NAFTA. Results for all three models are presented in Table 9.3.

Table 9.3 Separate and SUDCD Weibull Models of Timing and Direction of Positions on NAFTA

	Separate		Weibull SUDCD[6]		Weibull SUDCD	
Covariate	Estimate	s.e.	Estimate	s.e.	Estimate	s.e.
Vote						
Labor Contributions (net %)	-2.929***	(0.621)	-2.913***	(0.615)	-2.938***	(0.615)
Mexican Border	0.521	(0.600)	0.571	(0.598)	0.606	(0.597)
Union Membership	-6.089***	(1.477)	-6.143***	(1.465)	-6.046***	(1.466)
Household Income	2.749**	(1.084)	2.721**	(1.080)	2.783**	(1.081)
Democrat	-0.558***	(0.211)	-0.568***	(0.209)	-0.584***	(0.210)
Net Endorsements	0.011**	(0.005)	0.010*	(0.005)	0.010**	(0.005)
Constant	0.825***	(0.177)	0.793***	(0.177)	0.806***	(0.177)
Timing						
Corporate Contributions	0.140**	(0.068)	0.144**	(0.068)	0.150**	(0.068)
Labor Contributions	-0.116	(0.077)	-0.112	(0.077)	-0.112	(0.077)
Mexican Border	-0.220***	(0.039)	-0.220***	(0.039)	-0.217***	(0.039)
Democratic Leadership	-0.023	(0.030)	-0.023	(0.030)	-0.022	(0.030)
Republican Leaderhsip	-0.071**	(0.032)	-0.072**	(0.031)	-0.072**	(0.032)
NAFTA Committee	-0.004	(0.014)	-0.004	(0.014)	-0.003	(0.014)
Ideology	0.005	(0.017)	0.005	(0.017)	0.005	(0.017)
Union Membership	-0.352**	(0.146)	-0.337**	(0.145)	-0.328**	(0.145)
Union Mem. * Ideology	0.486**	(0.236)	0.449*	(0.234)	0.459*	(0.235)
Household Income	0.035	(0.108)	0.041	(0.107)	0.043	(0.107)
Income * Ideology	-0.021	(0.016)	-0.022	(0.015)	-0.021	(0.015)
Constant	6.059***	(0.018)	6.058***	(0.018)	6.057***	(0.018)
Correlation ($Z^{-1}(\alpha)$)						
Intercept			0.365	(0.230)	0.075	(0.338)
Democrat					0.480	(0.463)
Duration dependence ($ln(p)$)	2.086***	(0.0436)	2.089***	(0.0435)	2.090***	(0.0435)
N	434		433		433	

Estimates for duration equations have a time-to-failure interpretation.
$***p < 0.01, **p < 0.05, *p < 0.1$

We see that there is very little difference in the estimates for the observed covariates, and their statistical significance, across the models. The only notable difference is that the interaction term between union membership and ideology is statistically significant when the duration model is estimated separately from the discrete-choice model at the 0.05 level but is only significant at the 0.1 level in the combined SUDCD models.

Of most interest here is the estimation of the correlation between the stochastic elements of the discrete choice model of NAFTA positioning and the duration model of position timing. The constant correlation is parameterized in the log-likelihood function by $Z^{-1}(\alpha)$. Fisher's Z transformation is employed, as the correlation must be between -1 and 1, such that:

$$\alpha = Z(\alpha^*) = (\exp(2\alpha^*) - 1)/(\exp(2\alpha^*) + 1) \tag{9.21}$$

and the correlation parameter $\rho = \alpha/4$. The SUDCD estimator reports $Z^{-1}(\alpha) = \alpha^*$. Thus to calculate $\hat{\rho}$, when not differentiating between representatives according to party affiliation, one would substitute the reported value for the intercept (0.365) into equation (9.21) for α^* to generate a value for α, which when divided by four gives a correlation coefficient of 0.087.

The parameterization of the correlation changes when we wish to calculate different different values of $\hat{\rho}$ for Democratic representatives and Republican representatives. When this is done, the parameterization of the correlation is:

$$\alpha^* = Z^{-1}(\alpha) = \alpha_0^* + \alpha_1^* \times \text{Democrat}_i \tag{9.22}$$

with α_0^* equal to the reported intercept (0.075) and α_1^* equal to the value reported for Democrat (0.480). Substituting these values into equation (9.22), we find that $\hat{\rho}$ for Republican members of Congress equals 0.019 and for Democrats it equals 0.126. The overall estimated parameter is 0.56 (0.48 + 0.075) with a χ^2 value of 2.88 and a p-value of 0.09.

Substantively, the results indicate several things. First, that the correlation between the errors in the two models is positive and statistically significant at the 0.1 level, indicating that the unobservable influences on NAFTA vote choice and the timing of that choice are positively related. Those unobservable factors such as side-payments and competing political pressures that caused legislators to hold out longer before declaring their position, also caused them to vote in favor of NAFTA. Furthermore, these factors had a much greater impact on the Democratic members of Congress than on Republicans as is evident by the comparatively small value of $\hat{\rho}$ for Republican representatives. This accords with Boehmke's expectations. Democrats, more than Republicans, faced competing pressures from their party, and the President, to approve NAFTA while their constituents lobbied them to reject the agreement. As the final vote on NAFTA was close, Democrats that held out until as late as possible to see if their vote would be crucial were forced to cast their vote in favor of NAFTA to make sure the measure passed. If the vote did not appear to be close, they could have voted against NAFTA and thus neither angered their support base nor the President. Furthermore, the positive relationship between a late declaration of position and taking a pro-NAFTA stance supports the argument that an unobservable process whereby indifferent Democratic members held out on taking a position in order to extract side-payments from NAFTA supporters, and the President, before agreeing to cast their votes in favor of NAFTA.

9.6 Promising Future Directions

This review has focused on the modeling of systems of equations using one or more duration equations as a technique for uncovering the effects of latent variables on our outcomes of interest, which in the event history context is typically the time until an event occurs. In the extant literature, these models are generally based on SUR or SEQ approaches which jointly estimate equations that model interdependent processes (e.g. time to event). However, Hays and Kachi (2009) have gone one step further to consider interdependence of durations between actors (or units), which means that the time to a particular event for one actor depends on the time to the same event for other actors. They provide several persuasive examples which suggest that accounting for unobservable inter-unit interdependence is important. For instance "the time it takes for states to enter wars [often] depends on the time it takes other states to make these decisions" (2), and the time for states to decide on a policy issue, such as allowing casino gambling, may depend on when other states have adopted similar policies. Similarly, the decision to lower gas prices may depend on when competitors do the same. While Hays and Kachi (2009) argue that their estimation approach applies to both interdependence between units and interdependence between times to events, the former has previously been modeled with a spatial duration model with correlated errors (Darmofal 2009) and with a spatial lag model (Honoré & de Paula, in press), both of which may be of interest to readers.

Note that the literature on estimating duration models in systems of equations is dominated by parametric approaches for the duration equations. In contrast, most of the single equation duration literature in the social and behavioral sciences is dominated by the use of the Cox model. The advantage of relaxing the distributional assumption about the time until an event occurs suggests that incorporation of the Cox model into a simultaneous setup would be promising.

Combinations of different types of duration models and simultaneous equations is also promising, particularly, the use of the competing risks duration model, which allow for more than one type of event. Problems such as the duration of education and the time to form a union where a union can be cohabitation or marriage would be an example of two durations where one of them (union formation) requires the use of a competing risks duration model. Similarly, the competing risks of the duration of cohabitation where cohabitation can end with either marriage or break-up and the discrete choice model for fertility may prove to be a useful model.

In short, the area of simultaneous equations and duration models is a flourishing area of research with wide applicability to key questions in the social sciences. Recent modeling developments have provided new questions as well as new answers to old questions.

Appendix: Computer Code

Example: Duration and Discrete Choice in the NAFTA Study

The results for the NAFTA study examining interdependence between duration and discrete choice models were generated using STATA. Below is code, adapted from Boehmke's own do-files, that allows for replication of his analysis for the models presented here.

Installing the estimation program SUDCD:

```
net from http://myweb.uiowa.edu/fboehmke/stata/sudcd
net install sudcd
```

Defining the likelihood functions for the discrete exponential estimator:

```
program define expdisc
version 7
args lnf theta1
quietly replace `lnf' = ln(exp(-exp(-`theta1')))
  if $ML_y1==1
quietly replace `lnf' = ln(1-exp(-exp(-`theta1')))
  if $ML_y1==0
end
```

Separate estimation of the discrete choice model of NAFTA support using the discrete exponential estimator:

```
ml model lf expdisc (vote = contdiff mexbordr pscenter
  hhcenter partyid numdiff)
ml search
ml maximize

predict ystar_exp if e(sample), xb
generat yhat_exp = exp(-exp(-ystar_exp))
recode yhat_exp 0/0.5=0 0.5/1=1
tab yhat_exp vote, matcell(crosstab)
```

Separate estimation of the duration model of NAFTA position timing:

```
stset timing, failure(position)

streg corptpct labtpct mexbordr dleader rleader
  ncomact ideol pscenter
inter1 hhcenter inter2, d(weibull) time
```

Estimating the combined SUDCD models:
(Right-censoring is hard coded into the likelihood functions, which requires explicitly declaring the _rtcens variable using a dummy variable, rtcsensr, which is set to one if the observation is right censored.[7])

```
gen _rtcens = rtcensr
```

Joint estimation of the duration and discrete choice equations without estimating different coefficients of correlation for subsets of the sample:

```
sudcd timing corptpct labtpct mexbordr dleader rleader
  ncomact ideol pscenter inter1
hhcenter inter2, discrete(vote= contdiff mexbordr pscenter
  hhcenter partyid numdiff)
dist(weibull) time rtcensor(rtcensr)
```

Joint estimation of the duration and discrete choice equations, estimating different coefficients of correlation for Democrats and Republicans. The variable name for party affiliation is *partyid*:

```
sudcd timing corptpct labtpct mexbordr dleader rleader
  ncomact ideol pscenter inter1
hhcenter inter2, discrete(vote= contdiff mexbordr
  pscenter hhcenter partyid numdiff)
dist(weibull) time rtcensor(rtcensr) rho(partyid)

display "Correlation for Republicans (rho): "
((exp(2*([Z_alpha]_b[_cons]))-1)/
  (exp(2*([Z_alpha]_b[_cons]))+1))/4

display "Correlation for Democrats (rho): "
((exp(2*([Z_alpha]_b[_cons] + [Z_alpha]_b[partyid]))-1)
  /(exp(2*([Z_alpha]_b[_cons] + [Z_alpha]_b[partyid]))+1))/4

test [Z_alpha]partyid + [Z_alpha]_cons = 0
```

Acknowledgements We thank Aya Kachi and Kentaro Fukumoto for correspondence and discussion.

[7] To run without right-censoring, just set this variable equal to zero.

References

Aalen, O. O. (1975). *Statistical inference for a family of counting processes.* PhD thesis, University of California, Berkeley.

Adair, L. S., Popkin, B. M., & Guilkey D. K. (1993). The duration of breast-feeding: How is it affected by biological, sociodemographic, health sector, and food industry factors? *Demography, 30,* 63-80.

Addison, J. T., & Portugal, P. (1989). Job displacement, relative wage changes, and duration of unemployment. *Journal of Labor Economics, 7,* 281–302.

Allison, P. D. (1984). *Event history analysis: Regression for longitudinal event data.* Thousands Oaks: Sage.

Baizán, P., Aassave, A., & Billari, F. C. (2004). The interrelations between cohabitation, marriage and first birth in Germany and Sweden. *Population and Environment, 25,* 531-561.

Billari, F. C., & Philipov D. (2004). Women's education and entry into a first union: A simultaneous-hazard comparative analysis of Central and Eastern Europe. *Vienna Yearbook of Population Research,* 91-110.

Boehmke, F. J. (2006). The influence of unobservable factors in position timing and content in the NAFTA vote. *Political Analysis, 14,* 421-428.

Boehmke, F. J., Morey, D. S., & Shannon, M. (2006). Selection bias and continuous-time duration models: Consequences and a proposed solution. *Journal of Political Science, 50,* 192-207.

Box-Steffensmeier, J. M., & Jones, B. S. (2004). *Event history modeling: A guide for social scientists.* New York: Cambridge University Press.

Box-Steffensmeier, J. M., Arnold, L. W., & Zorn, C. J. W. (1997). The strategic timing of position taking in Congress: A Study of the North American Free Trade Agreement. *American Political Science Review, 91,* 324-338.

Boyle, P. J., Kulu, H., Cooke, T., Gayle, V., & Mulder, C. H. (2008). Moving and union dissolution. *Demography, 45,* 209-222.

Collett, D. (2003). *Modelling survival data in medical research.* London: Chapman & Hall.

Coppola, L. (2004). Education and union formation as simultaneous processes in Italy and Spain. *European Journal of Population, 20,* 219-250.

Cox, D. R. (1972). Regression models and life tables. *Journal of the Royal Statistical Society, B, 34,* 187-220.

Cox, D. R. (1975). Partial likelihood. *Biometrika, 62,* 269-276.

Darmofal, D. (2009). Bayesian spatial survival models for political event processes. *American Journal of Political Science, 53,* 241-257.

Dwivedi, T. D., & Srivastava, V. K. (1978). Optimality of least squares in the seemingly unrelated regression equation model. *Journal of Econometrics, 7,* 391-395

Fleming, T. R., & Lin, D. Y. (2000). Survival analysis in clinical trials: Past developments and future directions. *Biometrics, 56,* 971-983.

Fukumoto, K. (2009). *What happens depends on when it happens: Continuous or ordered event history analysis.* Working paper: Faculty of Law, Gakushuin University, Tokyo.

Golub, J., & Collett, D. (2002). Institutional reform and decision making in the European Union. In M. Hosli & A. van Deemen (Eds.), *Institutional Challenges in the European Union.* London: Routledge.

Grambsch, P. M., & Therneau, T. M. (1994). Proportional hazards tests and diagnostics on weighted residuals. *Biometrika, 81,* 515-526.

Hays, J. C., & Kachi, A. (2009). *Interdependent Duration Models in Political Science.* Paper presented at the Annual Meeting of the American Political Science Association, Toronto, Sept. 3-6, 2009.

Honoré, B., & de Paula, A. (in press). Interdependent durations. *Review of Economic Studies.*

Kaplan, E. L., & Meier, P. (1958). Nonparametric estimation for incomplete observations. *Journal of the American Statistical Association, 53,* 457-481.

Larsen, U., & Vaupel, J. W. (1993). Hutterite fecundability by age and parity: Strategies for frailty modeling of event histories. *Demography, 30,* 81-102.

Lillard, L. A. (1993). Simultaneous equations for hazards. *Journal of Econometrics, 56,* 189-217.

Maitra, P., & Sarmistha P. (2007). *Birth spacing, fertility selection and child survival: Analysis using a correlated hazard model.* Institute for the Study of Labor (IZA), Discussion Paper No. 2878.

Mantel, N. (1966). Evaluation of survival data and two new rank order statistics arising in its consideration. *Cancer Chemotherapy Report, 50,* 163-170.

Oakes, D. (2001). Biometrika centenary: Survival analysis. *Biometrika, 88,* 1.

Olshanksy, S. J., & Carnes, B. A. (1997). Ever since Gompertz. *Demography, 34,* 1-15.

Pindyck, R. S., & Rubinfeld, D. L. (1991). *Econometric models and economic forecasts.* New York: McGraw-Hill.

Quiroz Flores, A. (2008). *Copula functions and bivariate distributions for survival analysis: An application to government survival.* NYU Department of Politics. Working paper.

Rosholm, M., & Svarer, M. (2001). Structurally dependent competing risks. *Economics Letters, 73,* 169-173.

Singer, J. D., & Willett, J. B. (1993). It's about time: Using time survival analysis to study duration and the timing of events. *Journal of Educational Statistics, 18,* 155-195.

Therneau, T. M., Grambsch, P. M., & Fleming, T. R. (1990). Martingale based residuals for survival models. *Biometrika, 77,* 147-160.

Therneau, T. M., & Grambsch, P. M. (2001). *Modeling survival data: Extending the Cox model.* New York: Springer-Verlag.

Tuma, N. B. (1976). Rewards, resources, and the rate of mobility: A nonstationary multivariate stochastic model. *American Sociological Review, 41,* 338-360.

van Montfort, K., Oud, J., & Satorra, A. (Eds.) (2004). *Recent developments on structural equation models: Theory and Applications.* Mahwah, NJ: Lawrence Erlbaum Associates.

Zellner, A. (1962). An efficient method of estimating seemingly unrelated regressions and tests for aggregation bias. *Journal of the American Statistical Association, 57,* 348-368.